ICME-13 Monographs

Series editor

Gabriele Kaiser, Faculty of Education, Didactics of Mathematics, Universität Hamburg, Hamburg, Germany

Each volume in the series presents state-of-the art research on a particular topic in mathematics education and reflects the international debate as broadly as possible, while also incorporating insights into lesser-known areas of the discussion. Each volume is based on the discussions and presentations during the ICME-13 Congress and includes the best papers from one of the ICME-13 Topical Study Groups or Survey Teams.

More information about this series at http://www.springer.com/series/15585

Alexander Soifer
Editor

Competitions for Young Mathematicians

Perspectives from Five Continents

With the Foreword by Gabriele Kaiser

Springer

Editor
Alexander Soifer
Osborne Center for Science & Engineering
University of Colorado, Colorado Springs
Colorado Springs, CO
USA

ISSN 2520-8322 ISSN 2520-8330 (electronic)
ICME-13 Monographs
ISBN 978-3-319-56584-2 ISBN 978-3-319-56585-9 (eBook)
DOI 10.1007/978-3-319-56585-9

Library of Congress Control Number: 2017936346

Printed on acid-free paper

This Springer imprint is published by Springer Nature
The registered company is Springer International Publishing AG
The registered company address is: Gewerbestrasse 11, 6330 Cham, Switzerland

This book is dedicated to all those people around the world
Who are passing baton to next generations of mathematicians

Foreword

Mathematical competitions are a chance for mathematically talented young scholars to experience mathematics as a research-oriented discipline. These competitions offer the chance to get insight into the beauty of mathematical structures at a high level, which many of these young mathematicians usually will not experience at home. Furthermore, these competitions allow to meet other talented young mathematicians, exchange their ideas with them and experience that they are not singular and isolated youngsters, but part of an important community.

Despite this high importance of mathematical competitions, either as mathematical Olympiad or as mathematical tournament of towns or other kinds of mathematical competitions, there exists hardly any scientific research about mathematical competitions. This is surprising, because these mathematical competitions have a long tradition and a high influence on the promotion of young talented mathematicians.

At the occasion of the 13th International Congress on Mathematical Education (ICME-13) a Topic Study Group on Mathematics Competitions took place, at which famous researchers working in this field met and exchanged about the state-of-the-art in this field. This intensive work together with papers from related groups forms the basis of this book.

The book provides an excellent overview about the current discussion, topical themes and experiences with mathematical competitions. It starts with reflections on goals of mathematics education, problems coming from geometry or combinatorics being used in mathematical competitions. The next parts reflect on the role of competitions in the classroom, this theme is hardly researched so far. Then two examples of mathematical competitions are analyzed. The last two parts focus on the present state of mathematical competitions and its future and a bridge between competitions and 'real' mathematics.

To summarize, this book is more than overdue and reflects from an academic perspective on the potential of mathematical competitions for mathematics education in general.

I wish to congratulate the editor—Alexander Soifer—and the contributors to this timely and excellent book.

Hamburg, Germany

Gabriele Kaiser
Convenor of the 13th International Congress
on Mathematical Education, University of Hamburg

Preface

The role and usefulness of competitions in mathematics instruction has been debated for decades. If memory holds, I attended a deep and entertaining debate on this topic between a distinguished mathematician Peter John Hilton and a renowned math educator Gilah C. Leder at ICME-6, held in 1988 in Budapest. As this volume demonstrates, competitions problems can be used to enrich classroom instruction, to offer our students an exciting pastime, to raise interest in mathematics, and to enable students to commence their mathematical research. If not for Moscow State University Olympiads and a mathematical circle conducted by Nikolai Konstantinov (one of the authors in this volume!), I would have become a classical pianist and composer and not a mathematician. (By no means am I suggesting here that mathematics is better than music—they both belong to the Pantheon of the Arts.)

I am duty bound to add one warning. If a student does consistently well in mathematical Olympiads, s(he) clearly has a talent, and with a good measure of interest and hard work will go far. However, no discouraging conclusion could be made about a student, who has not sparkled in the Olympiads. Young people develop at diverse speeds. Moreover, mathematics competitions inevitably have an element of sports, the necessity to perform under pressure and within a limited time. High speed of thinking is attractive, but it is not an essential property for a future successful researcher.

This book includes plenary talks and some of the best presentations made in the Topic Study Group 30: Mathematics Competitions of the International Congress on Mathematical Education (ICME-13) in Hamburg, and some of the best presentations from related groups, dedicated to work with gifted students and mathematical enrichment. Each of the chapters, on request of this editor, includes not only original ideas of pedagogy and state-of-the-art methods of mathematical instruction, but also original problems and their beautiful solutions. I believe that this volume will be a valuable addition to the mathematics literature for secondary teachers and university professors around the world, and their gifted students of all levels, from secondary to graduate students, seeking problems to start their research careers.

The authors of this book comprise a group that impresses me enormously. It includes seven laureates of the Paul Erdős Award and one of the David Hilbert Award presented by the World Federation of National Mathematics Competitions (WFNMC); three past or present Presidents of WFNMC; five past or present WFNMC's Vice Presidents; three WFNMC's Secretaries; laureates of numerous other awards, leaders of and contributors to ICMI studies; authors of many books and countless articles, organizers of the International Mathematical Olympiad (IMO). In fact, in 1994 and 2016, K. P. Shum was the Organizer of two IMO's held in Hong Kong; while in 2013 Maria Falk de Losada served as the President of the International Jury at the Colombian IMO. The authors include many leaders and deputy leaders of national teams IMO teams, coordinators of IMO, organizers of numerous national and international competitions, conferences and congresses, etc.

Each of the 14 chapters addresses many issues and contributes to a multitude of directions, which makes a partition of the material into parts nearly impossible. I attempted to identify the main direction of each chapter and thus help the reader by partitioning the book into seven parts. As you can see, Francisco Bellot-Rosado (Spain) and Kar-Ping Shum (P.R. China) present problems of geometry; Kiril Bankov (Bulgaria), and Luis F. Cáceres-Duque, Jose H. Nieto-Said, and Rafael Sánchez-Lamoneda (Puerto Rico) share combinatorial problems. Role of competitions for a classroom is described by Robert Geretschläger (Austria); Ingrid Semanišinová, Matúš Harminc, and Martina Jesenská (Slovakia); and Iliana Tsvetkova (Bulgaria). Nikolai Konstantinov and Sergei Dorichenko (Russia), describe their famous International Mathematical Tournament of Towns; V.M. Sholapurkar (India) presents a relatively recent competition for college students. Romas Kasuba (Lithuania) shares his lifetime experiences with competitions; while Peter Taylor (Australia) classifies problems of mathematics competitions. Maria Falk De Losada (Colombia) collects valuable observations of the influence of mathematics competitions on their contestants, destined to become world's leading mathematical researchers. Alexander Soifer (USA) opens the book with his view of goals and means of mathematics instruction and closes the book with examples of bridges between problems of mathematical Olympiads and research problems of 'real' mathematics.

It was a delight to organize and run the Topic Study Group jointly with Maria Falk de Losada, thank you, Maria! My gratitude goes to my referees, encompassing four continents, who helped the authors to improve their chapters in a significant way. I thank all the officials and volunteers of ICME-13 in Hamburg, who allowed us all a pleasure of sharing knowledge and experiences during this Olympics-like forum of nearly 4,000 professionals from 109 countries. My special thanks go to the Convenor and the Chair of the International Program Committee of the ICME-13 Prof. Dr. Gabriele Kaiser for creating the Congress and arranging this splendid opportunity for my group of 18 authors from five continents to unite in a truly Olympic spirit and produce this volume, and to Springer for making it possible for us to preserve the wonderful memories of the Hamburg Congress in the form of this book.

On behalf of all the authors of this book, I wish you, our reader, to get much pleasure of mathematical kind from this book and many other books written by these 18 authors.

Colorado Springs, USA
January 2017

Alexander Soifer

Contents

Part I
Goals of Mathematics Instruction

Chapter 1
Goals of Mathematics Instruction: Seven Thoughts and Seven Illustrations of Means

Alexander Soifer

Abstract The goal of this chapter is to present what the author sees as the state-of-the-art approach to mathematics instruction, and the state-of-the-art use of mathematical Olympiads in bringing instruction closer to 'real' mathematics and identifying young talents. One of the principle goals of mathematics instruction ought to be showing in a classroom what mathematics is and what mathematicians do. This cannot be achieved by teaching but rather by creating an environment in which students learn mathematics by doing it. As in 'real' mathematics, this ought to be done by solving problems that require not just plugging numbers into memorized formulas and one-step deductive reasoning, but also by experimenting, constructing examples, and utilizing synthesis in a single problem of ideas from various branches of mathematics, built on high moral foundations. The author's eight recent Springer books present fragments of 'live' mathematics, and illustrations of these ideas. The chapter also describes the role of mathematical olympiads in instruction and includes some problems used at the Colorado Mathematical Olympiad over the past 34 years.

This essay is an expanded version of the Plenary Talk in the *Topic Study Group 30: Mathematics Competitions* at the 13th International Congress on Mathematical Education, Hamburg, Germany, July 2016. Prof. Dr. Gabriele Kaiser was the Convener of this very successful Congress. The early version appeared in the journal of the World Federation of National Mathematics Competitions 29(1), 2016, 7–30.

A. Soifer (✉)
University of Colorado, 1420 Austin Bluffs Parkway, Osborne Center for Science, Office A423, Colorado Springs, CO 80918, USA
e-mail: asoifer@uccs.edu
URL: http://www.uccs.edu/~asoifer/

A. Soifer (ed.), *Competitions for Young Mathematicians*,
ICME-13 Monographs, DOI 10.1007/978-3-319-56585-9_1

Keywords Colorado mathematical olympiad · Problem solving · Gifted students · Goals of instruction · Goals of life

1.1 Part I: Seven Thoughts on Mathematics Instruction

Give a man a fish, and you will feed him for a day.
Teach a man how to fish, and you will feed him for a lifetime.

– 老子 (Lǎozǐ, VI century BC)

1. The Purpose of Life Implies the Purpose of Instruction

Before we address the purpose of mathematics instruction, it makes sense to ask ourselves, what is the purpose of life itself? It seems to me that the purpose of life is to discover and express ourselves, and in so doing contribute to high culture of our planet. The ultimate purpose of instruction is therefore to aid our students in their quest for self-discovery and self-expression.

2. A Typical Instruction: Dishing out a Collection of Facts a la "Give a Man a Fish"

Instruction is often reduced to memorization of a certain collection of facts: dates in history, theorems in mathematics, etc. While memorization and knowledge are of value, they seem to be overestimated in instruction. I agree with the great Chinese Sage Lǎozǐ: giving a man a fish will not solve man's problem of survival.

3. Lǎozǐ and a Skill Approach to Life: "Teach a Man How to Fish"

Lǎozǐ proposes to teach a man fishing as a method of solving the problem of survival. This does go further than giving a man a fish. However, is it good enough in today's world?

4. Beyond Lǎozǐ: Enable a Man to Learn How to Solve Problems

Not every education is as good an investment as another. We ought to go beyond Lǎozǐ and his universally celebrated lines. Is teaching skills good enough? Not quite, dear Sage, not in today's rapidly changing world. What if there is no more fish? What if the pond has dried out while your man has only one skill, fishing?

A problem solver will not die if the fish disappears in a pond—he'll learn to hunt, grow crop, solve whatever problems life puts in his way. And so, we will go a long way by putting emphasis not on training skills but on creating environment for developing problem solving abilities and attitudes. This is the state-of-the-art. The proverb for today's world ought to be:

Give a man skills, and you will feed him in the short run.

Let a man learn solving problems, and you will feed him for a lifetime.

5. Mathematics and Life

Every day we confront and solve a myriad of problem. Life *is* about solving problems. And mistakes in solving life's problems could be quite costly: a bridge could collapse, electrical grid could get overloaded, traffic could get to a halt, etc. This is where mathematics comes in handy. Mathematics allows us to learn how to think creatively, how to solve problems. And once our student masters problem solving in mathematics, s(he) will be better prepared to confront problems in any human endeavor.

6. Are the Two Popular Approaches to Mathematics Instruction Good Enough?

Today's discussions of mathematical instruction seem to be reduced to two competing approaches, "Embrace the Technology" versus "Back to the Basics."

"Back to the Basics" is not the best solution, for it emphasizes mind numbing drill, and treats students as robots, who need to be pre-programmed with a set of skills. In the newer "Embrace the Technology" approach, I support taking a teacher off the lectern and letting students work on their own. This approach too more often than not treats students like robots, and pre-programs them with skills of today. However, technology nowadays changes rapidly, as do the societal demands for particular skills.

Providing public education is not only an ethical thing to do—it is a profitable investment. Are there many jobs today for computer-illiterate persons? And yet just one generation ago, computers were a monopoly of researchers, and one generation before that did not exist at all. And so, we will go a long way by putting emphasis not on training skills but on creating atmosphere for developing problem solving abilities and attitudes.

Observe, one *cannot teach* mathematics, or anything else for that matter. State-of-the-art in mathematics instruction is about creating an atmosphere where students can learn mathematics by doing it, with a gentle guidance of a teacher.

7. The True Goal of Mathematics Instruction is to Demonstrate What Mathematics Is and What Mathematicians Do

Standardized three-letter tests, such as SAT, ACT, GRE, KGB, CIA (well, the latter two triples are from a different opera:-) can only inform us how well a student does on these tests. Is this the goal of instruction? We ought to abandon standardized multiple choice testing of skills. There are more important things to assess. Over the past 34 years, *The Colorado Mathematical Olympiad* has been offering middle and high school students 5 original problems of increasing difficulty and 4 hours to think, to invent, and to solve. We "test" predominantly not knowledge, not skills, but creativity and originality of thought (Soifer 2011–2; Soifer 2017).

Is the goal "teaching to the test," as the past USA President George W. Bush believed? Not really. We all agree that problem solving is the means of instruction. However, what is *problem solving*? A typical secondary school problem asks to "find the hypotenuse of a right triangle, whose legs are 3 and 4, by using Pythagoras Theorem." No, not any more, you would reply. Nowadays, at the Age of Technology, a typical secondary school problem asks to "find the hypotenuse of a right triangle, whose legs are 3.1 and 4.2, by using Pythagoras Theorem and your smartphone." Would you call it a progress?

More generally, a secondary school problem has the structure $\mathbf{A} \Rightarrow \mathbf{B}$, i.e., given **A** prove **B** by using theorem **C**. In real life, no one gives a research mathematician a **B**; it is discovered by intuition and is based on experimentation. And of course, no one knows a **C** since nobody solved the problem: a research mathematician is a pioneer, moving along an untraveled path!

And so, we ought to bring our secondary and college mathematics, which often looks so superficial, as close as possible to the 'real' mathematics. We ought to let our students experiment in our classroom-laboratory. We ought to let them develop intuition and use it to come up with a conjecture **B**. And we ought to let our students find those tools **C** that do the job of deductive proving the conjecture **B**. In my opinion, the true goal of mathematics instruction is to demonstrate in the classroom *what mathematics is*, and *what mathematicians do*.

8. What Can Mathematical Olympiads Bring to Mathematics Instruction?

Let us first of all define the term. A *mathematical olympiad* is a competition where contestants are required to write essay-type complete solutions of the problems. Number of problems offered to Olympians is relatively small, usually between 4 to 6, and the time allowed is relatively long, usually from 4 to 9 hours. This does not completely eliminate time as a factor affecting performance, but substantially reduces it, especially compared to multiple choice or answer-only competitions with their speed-guessing as the main virtue. I often see best Olympians continuing to think about difficult problems after the Olympiad ends. In fact, I know some of them, who have been thinking about a Colorado Mathematical Olympiad problem and its research generalizations for many years. This process and the Olympiad influence may last a lifetime. While I see value in quick-type mathematical competitions and its sporty attraction for television broadcasting, I personally do not think they faithfully represent what mathematics is and what mathematicians do.

Olympiads allow us to introduce secondary students to topics, ideas, and methods of 'real' mathematics in the context and terminology of secondary mathematics, in the form that is digestible by them. Problems of mathematical Olympiads—as not much else—demonstrate beauty and elegance of mathematics. At the age of 14, I switched from writing and performing piano music to mathematics due exclusively to *The Moscow Mathematical Olympiad*. In March 1989 in Colorado Springs, Paul Erdős told me that "the Olympiads create a new enthusiasm toward mathematics, and in this sense are very valuable."

At *The Colorado Mathematical Olympiad*, we have been often asked a natural question: how does one create a mathematical Olympiad? This and other related questions are clarified by the University of Colorado, which produced the film "*Thirtieth Colorado Mathematical Olympiad—30 Years of Excellence*" that can be found on the Olympiad's homepage http://olympiad.uccs.edu/.

9. The Moral Foundation Is Critical

There is an opinion shared by many of my colleagues that all that matters is mathematics, *Mathematik über Alles*, if you will, above all moral concerns. In my opinion, there is no good science or good art unless it is built on the foundation of high ethical principles. Luitzen Egbertus Jan Brouwer, a great Dutch mathematician and philosopher, wrote in his 1929 letter: "It is my opinion that the tiniest moral matter is more important than all of science, and that one can only maintain the moral quality of the world by standing up to any immoral project."

We have seen in history time and again how evil the usage of science could be if it is not built on high moral foundation. Atrocities of Nazi Germany alone provide countless examples of science, technology and even art used for ill deeds. My book (Soifer 2015) is dedicated to moral dilemmas of a scholar in the Third Reich and in the world of today. Lessons of history ought to enter our classrooms and give moral guidance to our students today. I value education, however, I must admit that

Fine education does not guarantee high culture,

And high culture does not guarantee humanity.

In order for creative work to be good, it must also serve the good. It ought to be humane. It has to be grounded in morality, empathy, compassion, and kindness. The Great Russian poet Alexander Pushkin (1799–1837) beautifully wrote about it. Let me translate his lines for you:[1]

And people will be pleased with me for years to come,

For I awakened kindness with my lyre,

For in my cruel age I Freedom praised and sang

And urged I mercy for the fallen people.

And so we ought to pass to our students the baton of mercy and humanity, so that our students by their creative work contribute to the high culture of our small endangered planet.

1.2 Part II: Seven Illustration of Means

Alright, but what kind of problems should we offer our students? What approaches should we present in our classrooms? Permit me to illustrate seven essential components of the state-of-the-art classroom.

1. Experiment in Mathematics

First of all, we ought to set up a *mathematical laboratory*, where students conduct mathematical experiments, develop inductive reasoning and an insight needed to create conjectures. Some illustrations of it can be found in (Soifer 2010–1). For example, a short experiment allows us to conjecture a formula for the sum of cubes of consecutive integers:

$$
\begin{aligned}
1^3 &= 1^2\\
1^3+2^3 &= 3^2\\
1^3+2^3+3^3 &= 6^2\\
1^3+2^3+3^3+4^3 &= 10^2
\end{aligned}
$$

We observe that the sums of consecutive cubes are perfect squares. But squares of what numbers? If you are not able to develop a conjecture yet, continue to experiment: $1^3 + 2^3 + 3^3 + 4^3 + 5^3 = 15^2$. You will soon

---

[1]И долго буду тем любезен я народу,
Что чувства добрые я лирой пробуждал,
Что в мой жестокий век восславил я Свободу
И милость к падшим призывал.

notice that $1^3 + 2^3 + 3^3 + 4^3 + 5^3 = (1 + 2 + 3 + 4 + 5)^2$. This kind of equality holds for all the values in our experiment, and the conjecture is ready:

$$1^3 + 2^3 + \cdots + n^3 = (1 + 2 + \cdots + n)^2.$$

We can now prove, for example, by mathematical induction, that both the left side and the right side of the conjectured equality is equal to $\left(\frac{n(n+1)}{2}\right)^2$. ■

2. Construction of Examples in Mathematics

Construction of counterexamples is almost non-existent in secondary education and even university, whereas counterexamples play a major role in mathematics, amounting to circa 50% of its results. In fact, the Great Russian mathematician Israel M. Gelfand once said, "Theories come and go; examples live forever."

You would agree that practically the entire school mathematics consists of analytical proofs. In order to bring instruction closer to the 'real' mathematics we ought to include in education construction of examples and counterexamples. Let me share one example, where a construction solves the problem (Soifer 2011–2).

Positive² (18th Colorado Mathematical Olympiad, Soifer 2001). Is there a way to fill a 2001 × 2001 square table T with pluses and minuses, one sign per cell of T, such that no series of interchanging all signs in any 1000 × 1000 or 1001 × 1001 square of the table can fill T with all pluses?

Solution. Having created this problem and its solution for the 2001 Colorado Mathematical Olympiad, I felt that another solution was possible using an invariant, but failed to find it. Two days after the Olympiad, on April 22, 2001, the past double-winner of the Olympiad Matthew Kahle, now a Professor at Ohio State University, found the solution that eluded me. It is concise and beautiful.

Define (see Fig. 1.1) Φ = {the set of all cells of T, except those in the middle row}. Observe that no matter where a 1000 × 1000 square S is placed in the table T, it intersects Φ in an even number of cells, because there are 1000 equal columns in S. Observe also that no matter where a 1001 × 1001 square S' is placed in T, it also intersects Φ in an even

Fig. 1.1 .

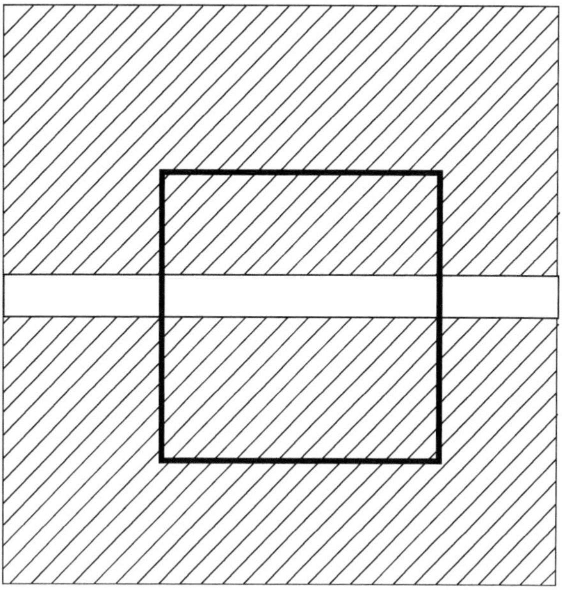

Fig. 1.2 .

number of unit squares, because there are 1000 equal rows in *S'* (one row is always missing, since the middle row is omitted in *S*.)

Now we can easily create the required assignment of signs in *T* that cannot be converted into all pluses. Let Φ have any assignment with an *odd* number of + signs, and the missing in Φ middle row be assigned signs in any way. No series of operations can change the parity of the number of pluses in Φ, and thus no series of allowed operations can create all pluses in Φ. ■

3. Utilizing Analogy

A sense of analogy could be a powerful tool. Here is one example from (Soifer 2009–2).

Problem 2 Prove that a map formed in the plane by finitely many circles can be 2-colored (Fig. 1.2).

Fig. 1.3 .

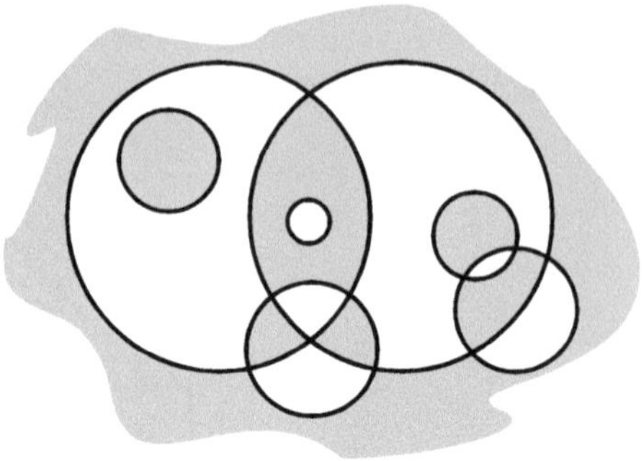

Fig. 1.4 .

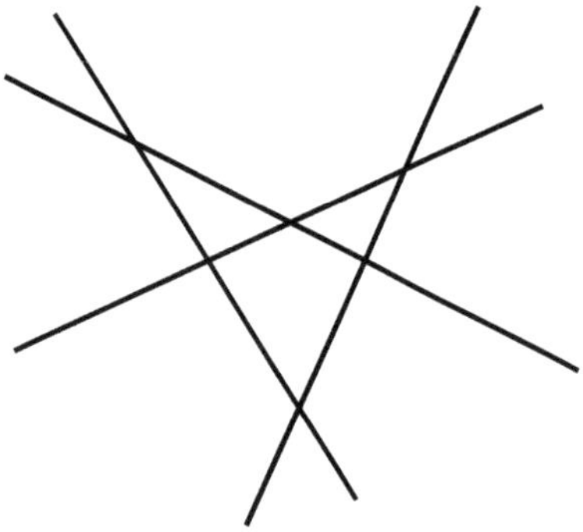

Proof We partition regions of the map into two classes (Fig. 1.3): those contained in an even number of circles (color them gray), and those contained in an odd number of circles (leave them white). Clearly, neighboring regions got different colors because when we travel across their boundary line, the parity changes. ■

I am sure you realize that the shape of a circle is of no consequence. We can replace circles in problem 2 by simple closed curves. However, can we replace simple closed curves by straight lines?

Problem 3 Prove that a map formed in the plane by finitely many straight lines is 2-colorable (Fig. 1.4).

An inductive proof is well known, but, as is usually the case with proofs by mathematical induction, it does not provide an insight. Decades ago I found a 'one-line' proof that takes advantage of similarity between simple closed curves and straight lines.

Proof Attach to each line a vector perpendicular to it (Fig. 1.5). Call the half-plane *inside* if contains the vector, and *outside* otherwise. Repeat the proof of problem 2 word-by-word to complete the proof (Fig. 1.6). ■

Fig. 1.5 .

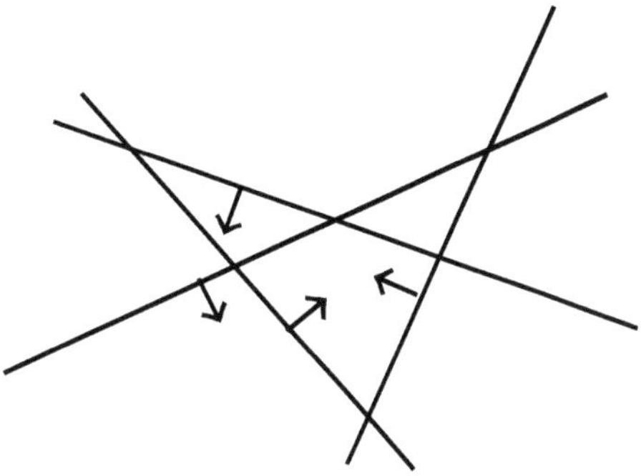

Fig. 1.6 .

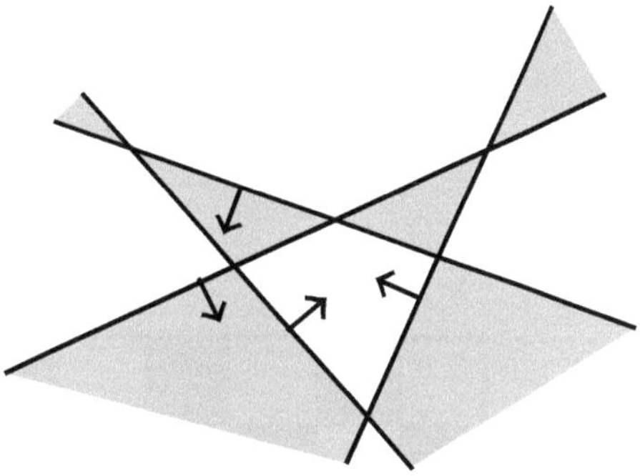

4. Method and Anti-method

Tiling with Dominoes. (Method). Can a chessboard with two diagonally opposite squares missing, be tiled by dominoes (Fig. 1.7)?

Solution. Color the board in a chessboard fashion (Fig. 1.8). No matter where a domino is placed on the board, vertically or horizontally, it would cover one black and one white square. Thus, it is necessary for tileability to have equal numbers of black and white squares in the board—but they are not equal in our truncated board. Therefore, the required tiling does not exist. ■

It is impressive and unforgettable for a student to see for the first time how coloring can solve a mathematical problem. However, I noticed that once a student learns a coloring idea, s(he) always resorts to it when a chessboard and dominoes are present in the problem. This is why I created the following 'Anti-Method' Problem and used it in the Colorado Mathematical Olympiad (Soifer 2011–2).

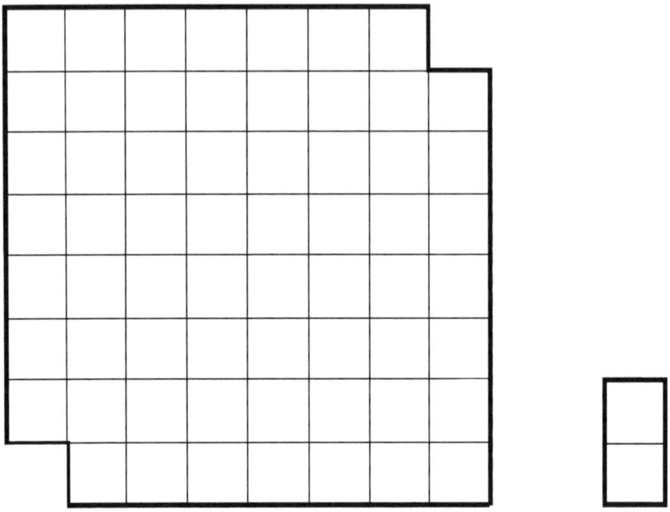

Fig. 1.7 .

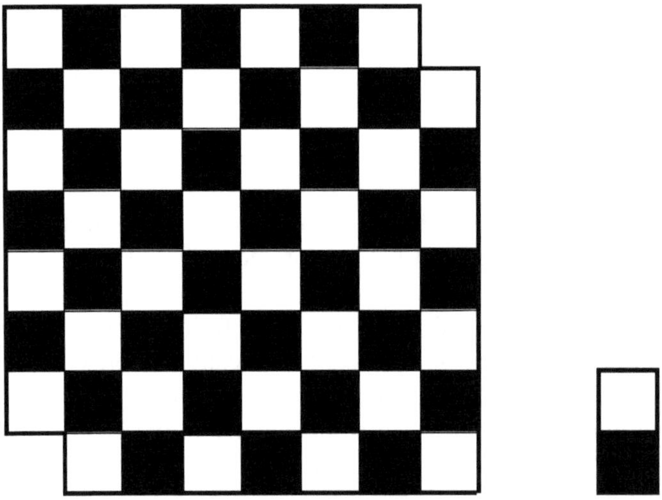

Fig. 1.8 .

The Tiling Game (**Anti-method**, 6th Colorado Mathematical Olympiad, Soifer 1989). Mark and Julia are playing the following tiling game on a 1988×1989 chessboard. They in turn are putting 1×1 square tiles on the board. After each of them made exactly 100 moves (and thus they

Fig. 1.9 .

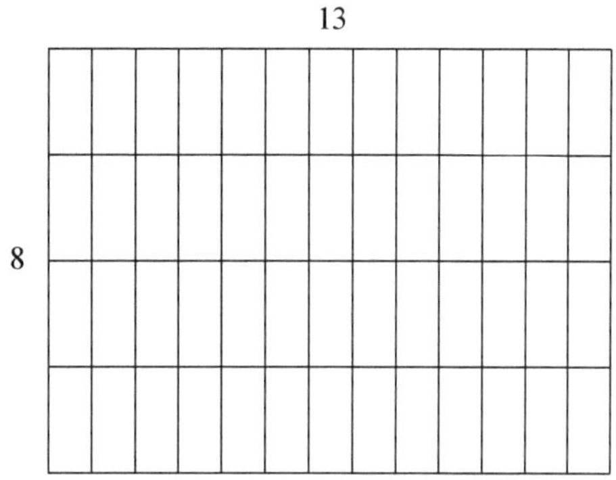

Tiling template T for a 8 x 13 chessboard

Fig. 1.10 Winning strategy

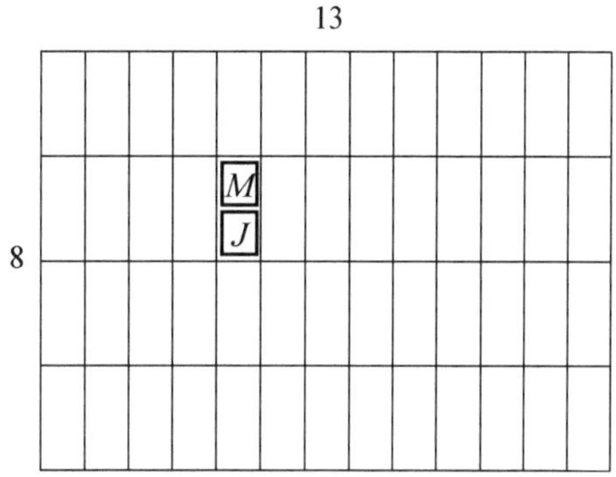

covered 200 squares of the board) a winner is determined as follows: Julia wins if the tiling of the board can be completed with dominoes. Otherwise Mark wins. (Dominoes are 1×2 rectangles, which cover exactly two squares of the board.) Can you find a strategy for one of the players allowing him to win regardless of what the moves of the other player may be? You cannot? Let me help you: Mark goes first!

Solution. Julia (i.e., the second player) has a strategy that allows her to win regardless of what Mark's moves may be. All she needs is a bit of home preparation: Julia creates a tiling template showing one particular way, call it T, of tiling the whole 1988×1989 chessboard with dominoes. Figure 1.9 shows one such tiling template T for an 8×13 chessboard.

The strategy for Julia is now clear. As soon as Mark puts a 1×1 tile M on the board, Julia puts her template T on the board to determine which domino of the template T contains Mark's tile M. She then puts her 1×1 tile J to cover the second square of the same domino (Fig. 1.10). When each player makes 100 moves, 100 dominoes will be covered, and the template T will show how to complete the tiling of the board. ■

5. Synthesis and Combinatorial Geometry

Secondary school mathematics consists predominantly of problems with single-idea solutions, found by analysis. We ought to introduce a sense of *mathematical reality* in the classroom by presenting *synthesis*, by offering problems that require for their solution ideas from a number of mathematical disciplines: geometry, algebra, number theory, trigonometry, linear algebra, etc.

And here comes *Combinatorial Geometry* to the rescue. It offers an abundance of problems that sound like a 'regular' secondary school geometry, but require for their solutions synthesis of ideas from geometry, algebra number theory, trigonometry, ideas of analysis, etc. See for example (Soifer 2009–1; 2009–3; 2011–1). Moreover, combinatorial geometry offers us open-ended problems. It offers problems that any geometry student can understand, and yet no one has yet solved! Let us stop this discrimination of our students based on their young age, and allow them to touch and smell, and work on 'real' mathematics and its unsolved problems. They may find a partial advance into solutions; they may settle some open problems completely. And they will then know the answer to what ought to become the fundamental questions of mathematical instruction: What is mathematics? What do mathematicians do?

In fact, I would opine that every discipline is about problem solving. And so the main goal of every discipline ought to be to enable students to learn *how to think within the discipline*, *how to solve problems of the discipline*, and finally *what that discipline is about*, and *what the professionals within the discipline do*. And mathematics to all sciences does what gymnastic does to all sports: Mathematics is gymnastics of the mind. Doing mathematics develops a universal approach to problem solving and intuition that go a long way in preparing our students for solving problems they will face in their lives.

6. Open Ended and Open Problems

As a junior at the university, I approached my supervisor Professor Leonid Yakovlevich Kulikov with an open problem I liked—he was my supervisor ever since my freshman year. He replied, "Learn first, the time will come later to enter research." He meant well, but politically speaking, this was a discrimination based on my young age. Seeing my disappointment, Kulikov continued, "It does not look like I can stop you from doing research. All right, whatever results you obtain on this open problem, I will count as your course paper." Soon I received my first research results, and my life in mathematics began.

We ought to allow our students to learn what mathematicians do by offering them not just unrelated to each other exercises but rather series of problems leading to a deeper and deeper understanding. And we ought to let students 'touch' unsolved problems of mathematics, give them a taste of the unknown, a taste of adventure and discovery. Combinatorial geometry serves these goals well by providing us with easy-to-understand, hard-to-solve—or even unsolved—problems. I will formulate here two examples. You can find their solutions in my Springer books listed in references.

Points in a Triangle (Soifer 2009–3). Out of any n points in or on the boundary of a triangle of area 1, there are 3 points that form a triangle of area at least ¼.

(a) Prove this statement for $n = 9$.

(b) Prove this statement for $n = 7$.

(c) Prove this statement for $n = 5$.

(d) Show that the statement is not true for $n = 4$, thus making $n = 5$ best possible.

Chromatic Number of the Plane (Soifer 2009–1). No matter how the plane is colored in n colors, there are two points of the same color distance 1 apart.

(a) Prove this statement for $n = 2$.

(b) Prove this statement for $n = 3$.

(c) Disprove this statement for $n = 7$.

(d) The answer for $n = 4, 5$, and 6 is unknown to man—this is a forefront of mathematics!

7. Beauty of 'Real' Mathematics Can Be Transplanted to Olympiads for Secondary Schools

New Olympiad problems occur to us in mysterious ways. This problem came to me one summer morning of 2003 as I was reading a never published

1980s manuscript of a Ramsey Theory monograph, while sitting by a mountain lake in Bavarian Alps. It all started with my finding a hole in a lemma, which prompted a construction of a counterexample (part b of the present problem). Problem (a) is a corrected particular case of that lemma, translated, of course, into a language of a nice 'real' story of a chess tournament. I found three distinct striking solutions of (a) and an even more special solution of (b). As a result, this problem became for me the most beautiful Olympiad problem I have ever created. What is more, the journey that led me from Ramset Theory to problems of mathematical Olympiads, continued to a mathematical problem of Finite Projective Planes! I will show here only one of my three solutions—read other solutions in (Soifer 2004) or (Soifer 2017).

Chess 7 × 7 (21st Colorado Mathematical Olympiad, April 16, 2004, A. Soifer).

(a) Each member of two 7-member chess teams is to play once against each member of the opposing team. Prove that as soon as 22 games have been played, we can choose 4 players and seat them at a round table so that each pair of neighbors has already played.

(b) Prove that 22 is the best possible; i.e., after 21 games the result of (a) cannot be guaranteed.

Solution. This solution harnesses the power of combinatorics. In the selection and editing process, Col. Dr. Bob Ewell suggested to use a 7 × 7 table to record the games played. We number the players in both teams. For each player of the first team we allocate a row of the table, and for each player of the second team a column. We place a checker on the table in the

Fig. 1.11 .

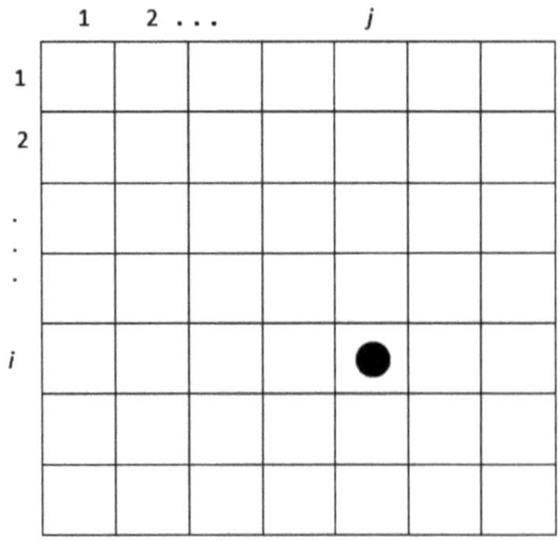

Fig. 1.12 .

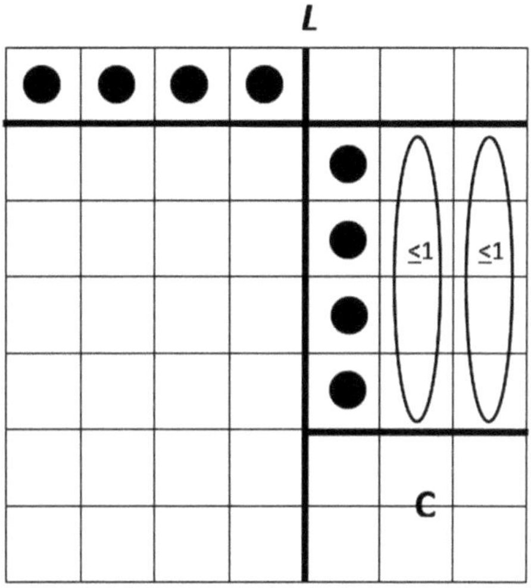

location (i, j) if the player i of the first team played the player j of the second team (Fig. 1.11).

If the required four players are found, this would manifest itself in the table as a rectangle formed by four checkers, a *checkered rectangle*! (Sides of the checkered rectangle are required to be parallel to the lines of the grid.) The problem thus translates into the new language as follows:

A 7 × 7 table with 22 checkers must contain a checkered rectangle.

Assume that a table has 22 checkers but does not contain a checkered rectangle. Since 22 checkers are contained in 7 rows, by the Pigeonhole Principle, there is a row with at least 4 checkers in it. Observe that *interchanging rows or columns does not affect the property of the table to have or have not a checkered rectangle*. By interchanging rows, we make the row with at least 4 checkers first. By interchanging columns, we make all checkers of the first row to appear consecutively from the left side of the board. We consider two cases.

(1) Top column contains exactly 4 checkers (Fig. 1.12).

Draw a bold vertical line L after the first 4 columns. To the left from L, top row contains 4 checkers, and all other rows contain at most 1 checker each, for otherwise we would have a checkered rectangle (that includes the top row). Therefore, to the left from L we have at most $4 + 6 = 10$

Fig. 1.13 .

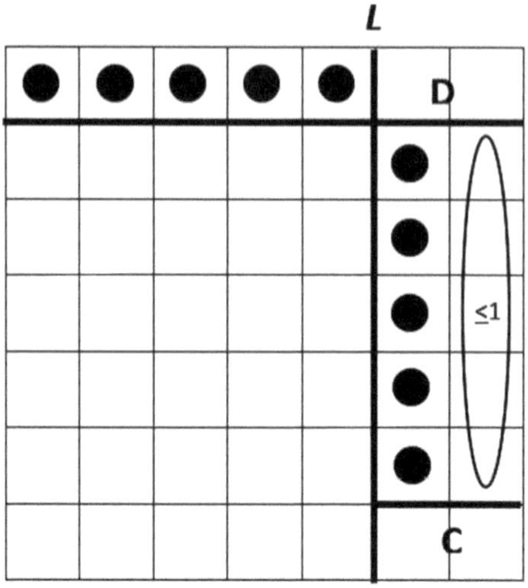

checkers. This leaves at least 12 checkers to the right of L, thus at least one of the three columns to the right of L contains at least 4 checkers; by interchanging columns and rows we put them in the position shown in Fig. 1.12. Then each of the two right columns contains at most 1 checker total in the rows 2 through 5, for otherwise we would have a checkered rectangle. We thus have at most $4 + 1 + 1 = 6$ checkers to the right of L in rows 2 through 5 combined. Therefore, in the lower right 2×3 part C of the table we have at least $22 - 10 - 6 = 6$ checkers—thus C is completely filled with checkers and we get a checkered rectangle in C in contradiction with our assumption.

(2) Top column contains at least 5 checkers (Fig. 1.13).

Draw a bold vertical line L after the first 5 columns. To the left from L, top row contains 5 checkers, and all other rows contain at most 1 checker each, for otherwise we would have a checkered rectangle (that includes the top row). Therefore, to the left from L we have at most $5 + 6 = 11$ checkers. This leaves at least 11 checkers to the right of L, thus at least one of the two columns to the right of L contains at least 6 checkers; by interchanging columns and rows we put 5 of these 6 checkers in the position shown in Fig. 1.13. Then the last column contains at most 1 checker total in the rows 2 through 6, for otherwise we would have a checkered rectangle. We thus have at most $5 + 1 = 6$ checkers to the right of L in rows 2 through 6 combined.

Therefore, the upper right 1×2 part D of the table plus the lower right 1×2 part C of the table have together at least $22 - 11 - 6 = 5$ checkers—

Fig. 1.14 .

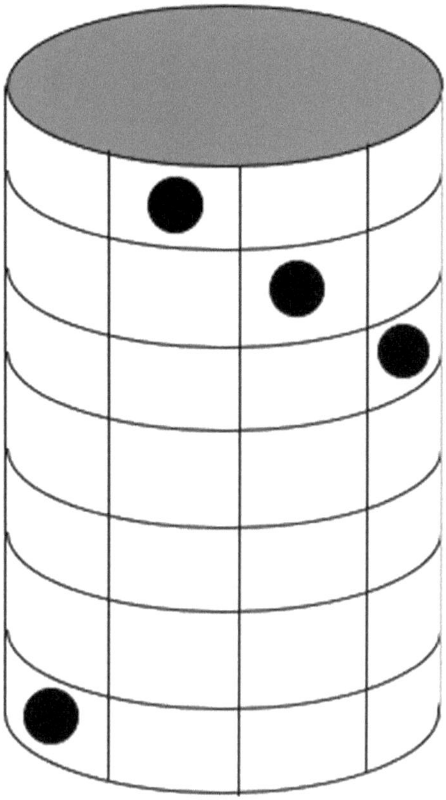

but they only have 4 cells, and we thus get a contradiction. ■

Solution of Part (b). Glue a cylinder (!) out of the board 7×7, and put 21 checkers on all squares of the 1st, 2nd, and 4th diagonals (Fig. 1.14 shows the cylinder with one such checkered diagonal; Fig. 1.15 shows, in a plane representation, the cylinder with all three checkered diagonals).

Assume that 4 of the placed checkers form a rectangle on our 7×7 board. Since these four checkers lie on 3 diagonals, by the Pigeonhole Principle, two checkers lie on the same (checkers-covered) diagonal D of the cylinder. But this means that *on the cylinder our 4 checkers form a square*! Two other (opposite) checkers a and b thus must be symmetric to each other with respect to D, which implies that the diagonals of the cylinder that contain a and b must be symmetric with respect to D—but no two checker-covered diagonals in our checker placement are symmetric with respect to D. (To see that, observe Fig. 1.16 which shows the top rim of the

Fig. 1.15 .

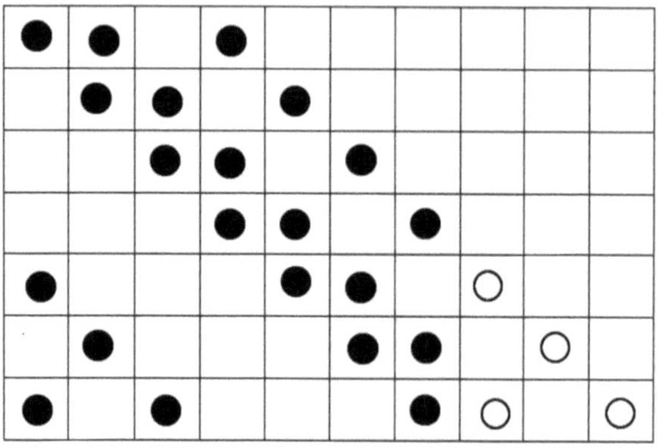

Fig. 1.16 .

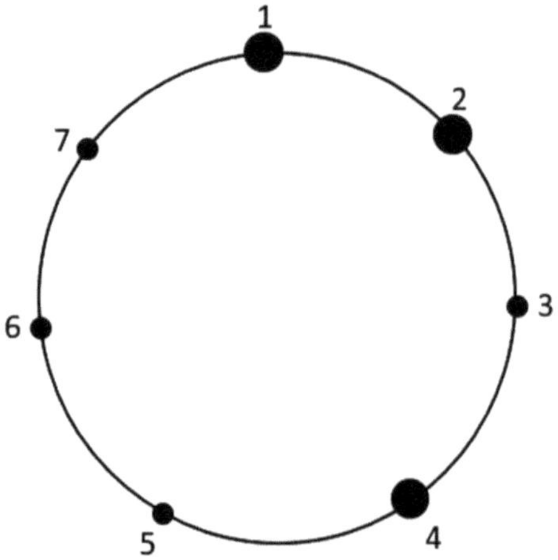

cylinder with bold dots for checkered diagonals: distances between the checkered diagonals measured in unit squares clockwise, are 1, 2, and 4.) This contradiction implies that there are no checkered rectangles in our placement. Done! ■

Observe: Obviously, any solution of part (b) can be presented in a form of 21 checkers placed on a 7 × 7 board (see, for example, the left 7 × 7 part with 21 black checkers in Fig. 1.15). It is not at all obvious that the solution is *unique*, i.e., by a series of interchanges of rows and columns, *any* solution of this problem can be brought to match precisely the one I presented in Fig. 1.15! Of course, such interchanges mean merely renumbering of players of the same teams.

The uniqueness of the solution of problem (b) is precisely another way of stating the uniqueness of what is known in mathematics as the *Projective*

Fig. 1.17 .

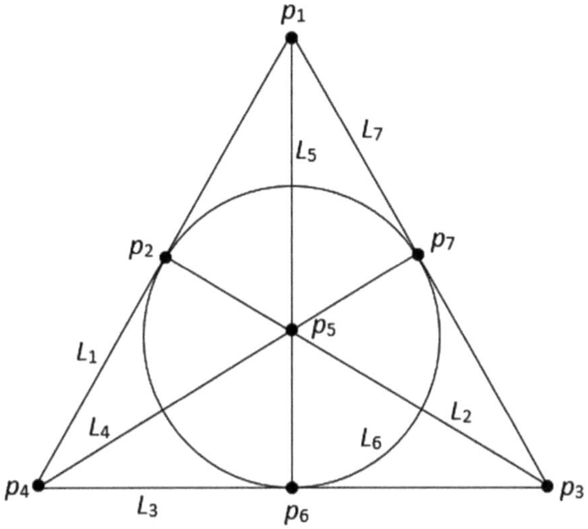

Plane of Order 2, the so called "Fano Plane," denoted by *PG*(2, 2). It was named after Gino Fano (1871–1952), the Italian geometer who pioneered the study of finite projective geometries.

The Fano Plane is an abstract construction, with symmetry (duality) between points and lines: it consists of 7 points and 7 lines. You can think of rows and columns of our 7 × 7 table as lines and points respectively, with 3 points on every line and 3 lines through every point. See in Fig. 1.17 a traditional depiction of the Fano Plane where a circle depicts one of the lines.

Observe that if on our 7 × 7 board we replace each of the 21 checkers by 1 and the rest of the squares fill with zeroes, we get the so-called *Incidence Matrix* of the Fano Plane.

This problem reminds me *Mary-Go-Round*: it originates in a 'real' mathematics, Ramsey Theory, generates excitement of Olympiad kind, and ends in another branch of 'real' mathematics, Finite Projective Planes!

Acknowledgements I thank Col. Dr. Robert Ewell for converting my hand-drawn sketches into computer-aided illustrations.

References

Soifer, A. (2004). A journey from Ramsey theory to mathematical olympiad to finite projective planes. *Mathematics Competitions, 17*(2), 8–16.

Soifer, A. (2009–1). *The mathematical coloring book: Mathematics of coloring and the colorful life of its creators*. New York: Springer.
Soifer, A. (2009–2). *Mathematics as problem solving*. New York: Springer.
Soifer, A. (2009–3). *How does one cut a triangle?* New York: Springer.
Soifer, A. (2010–1). *Geometric etudes in combinatorial mathematics* (2nd ed.). New York: Springer.
Soifer, A. (2011–1). *Geometric etudes in combinatorial mathematics*. New York: Springer.
Soifer, A. (2011–2). *The Colorado mathematical olympiad and further explorations: From the mountains of colorado to the peaks of mathematics*. New York: Springer.
Soifer, A. (2015). *The scholar and the state: In search of Van der Waerden*. Basel: Birkhäuser.
Soifer, A. (2017). *The Colorado mathematical olympiad, the third decade and further explorations: From the mountains of colorado to the peaks of mathematics*. New York: Springer.

Part II
Geometry for Competitions

Chapter 2
From a Mathematical Situation to a Problem

Francisco Bellot-Rosado

Abstract The approach to problems creation starting from a mathematical situation is developed, with several examples of such situations and problems arising from this, with solutions (if the problem is not open).

Keywords Geometrical situation · Mathematical problem

2.1 Introduction

The teaching of mathematics on the basis of problem solving is a periodically repeated subject in ICMEs, as TG or WG. Within this general frame, we will consider in this chapter an approach to problems creation that we will call "From a mathematical situation to a problem".

In *Mathematical Competitions*, the journal of the WFNMC, the question of the creation of problems has been studied many times; in particular, between 1986 and 1999, more than 20 papers on this subject were been published. The paper by Engel (1987) *The creation of mathematical Olympiad Problems*, starts with the following sentence:

Dedication: To Eduardo Wagner (SBM, Brazil), from whom I learned to go from a mathematical situation to a problem or a theorem, and how to solve them.

F. Bellot-Rosado (✉)
Royal Spanish Mathematical Society and WFNMC, Dos de Mayo street, number 16, 8th floor, Apt. A, 47004 Valladolid, Spain
e-mail: franciscobellot@gmail.com

A. Soifer (ed.), *Competitions for Young Mathematicians*,
ICME-13 Monographs, DOI 10.1007/978-3-319-56585-9_2

> It is far more difficult to create problem than to solve it. There are very few routine methods of problem creation. As far as I know no Polya among problem creators who wrote a book with the title "How to create it".

When analyzing some examples of workshops about Learning based in problems, we notice that, although the term "situation-problem" may be used, the teacher actually presents a closed statement to the students. That is, the teacher is helping the students to find a way to gather the details of the solution of a problem from which the full statement is, sooner or later, given. It's clear that during the discussion, students can discover some alternative statements which can became new problems, and this, no doubt, improves the enrichment of the mathematical-didactical discussion which must follow. In this sense, the treatment of the question given in the book "*Pour un enseignement problematisé des Mathématiques au Lycée*" (2 vols.), APMEP, in French, no date of publication, a collective work of the group "Problématiques Lycée", is interesting.

To begin, we can take a look at an example included in the workshop *Aprendizaje basado en problemas* (Learning based on problem-solving), by Prof. Rolando Sáenz, from Ecuador. This example was presented in 2006 in Salinas (Ecuador), during the Iberoamerican Symposium (with emphasis in problem solving), a didactical activity prior to the Iberoamerican Mathematical Olympiad.

Example 1.1 $ABCD$ is a square. We take points M, N, O and P, respectively in AB, BC, CD and DA, in a such manner that $AM = BN = CO = DP$. Determine the point M such that the quadrilateral $MNOP$ have maximal area.

Maybe if the last sentence was changed to something like this: *Consider the quadrilateral MNOP*, some other statements, equally interesting, would emerge during the discussion. We invite the readers to try it by themselves.

Many times, the reading of a paper about problem creation will provide some very interesting problems, but there are rarely many explanations on how they were created, that is, what was the process which gave birth to the problem.

We can now take a look at some characteristics which a good problem should have. Gardiner (1992, p. 59) wrote this:

(a) *The ingredients (of the problem) should be simple and familiar, but the problem should not be of any standard type.*
(b) *No method of solution should be immediately obvious, but a careful survey of the given information should suggest one or two promising points of attack.*

(c) *And exploratory phase should then reveal how (or whether) these approaches can be exploited.*
(d) *The final solution when it emerges should, in retrospect, have an unexpected elegance or conceptual simplicity.*

Example 1.2 The positive numbers x, y and z satisfy

$$x^2 + xy + \frac{y^2}{3} = 25$$
$$\frac{y^2}{3} + z^2 = 9$$
$$z^2 + xz + x^2 = 16$$

Find the value of xy + 2yz + 3zx.

Note: The sources of the problems will be included in the solutions section

The readers are invited to think about this statement and to try by themselves the "promising points of attack" in the words of Gardiner.

As last part of this introduction, here is a quote of Branko Grünbaum in his introduction to the book of Soifer (1990) *How Does One Cut a Triangle?*.

> Many people find mathematics attractive because it presents to the mind the same challenge that other activities, such as sports, present to the body. In mathematics, and specially in geometry, there are abundance of topics that are accessible without much previous knowledge. They present the exploring mind with opportunities to rise to that challenge, and to experience the joy of discovery.

2.2 What Is a Mathematical Situation?

Searching in libraries, it is possible to find—at least—two types of books which can be related to the topic of the chapter.

(1) Books where mathematical situations with problems are presented (with or without solutions).
(2) Books where mathematical problems are discussed in detail, showing what should be the way by which the solution must be presented to the audience (much more detailed than the typical way in which the solution seems to appears like riding a parachute, falling down from the sky).

One of the earliest examples of books from type 1 is *Geometry for Advanced Pupils*, by Maxwell (1949).

Dr. Maxwell presents here 47 configurations from which is possible to deduce results, many problems and geometric properties of interest. He also includes examples from the Oxford and Cambridge Examinations Papers.

An interesting paper, published in *Quantum*, January/February 2001 by the late Prof. I. Sharygin, is *Where do problems come from?* (Sharygin 2001) (*The art of problem composition*). Sharygin explains in this paper some of his own procedures for composing problems (Olympiad type): by reformulation, problems built on other problems, considering special cases of a theorem; varying the problem statement; by generalization of a problem (or some result). And he says: *However, the main source of new problems is inquisitiveness, the desire to reveal the essence of a problem, the ability to look at a well-known fact from an unusual point of view. This is when the most interesting geometric problems appear, ones that can be called discoveries.*

Sharygin ends his paper with this assertion: *You don't have to be a budding mathematical genius to make geometric discoveries—some problems show that any student can do it. And this includes you!*

Another book of type 1 is *Geometry in figures*, by Akopyan (2011) (no Editorial name, but the place is Lexington, KY). This is a collection of theorems and problems of Euclidean geometry formulated in figures, without text. This is a good illustration about what a geometrical situation is. Recently (2015), the Union of Bulgarian Mathematicians published the book by Dimitrov, Lichev and Chovanov *555 problems of Geometry* (in Bulgarian) with the solutions to the problems of the book by Akopyan.

To end with the examples of publications of type 1 it is worth mentioning the book by Monk (2009). This is a very popular book among the participating countries in the IMO since it was published. The five categories of problems of the book are E (easy), 18 problems; M (moderate difficulty), 20 problems; H (hard), 18 problems; C (Computational), 24 problems; and T (Trigonometry), 18 problems.

For the books of Type 2 the situation seems to be better. There are many publications about this subject (see the References section for more titles) and some of them are really excellent. Here are a few examples:

Burns (2000).

Gardiner (1997).

Savchev and Andreescu (2003).

Nevertheless, it seems there are not many titles in libraries and bookstores which describe what a mathematical situation is. Paraphrasing Prof. Eduardo Wagner, Brazilian expert in problem solving: *As important as teaching Mathematics is to create new problems, interesting and challenging. Problems are new questions, of different aspect to the usual one and which should stimulate the development of the reasoning. To create one problem a big effort, enough time to try many attempts, and good luck are required. With continued work and much reading, the ability to create problems is developed and the ideas can emerge in our mind more easily. This work is not different to other sciences or artistic work. To acquire any ability, everybody needs specific training.*

The "Office of creating problems", promoted by the OEI (Iberoamerican States Organization) in the years 1994 to 1997, is an introduction to the art of creating problems. With its own methodology, the participants have the opportunity of experimenting with real problem creation situations, and they then developed their own methods.

A Mathematical situation is not yet a problem. It consists of a set of mathematical objects, linked by some certain relations. With this basis, the participants (in the Office) must investigate the properties of the proposed situation, adding if necessary other elements, and to create one of more problems. In this way, with the reasoning focused in a particular situation, the activity was followed with the biggest interest by the participants and some new problems of different degrees of difficulty were created. End of the quote, taken from Wagner (1997).

Prof. Wagner was the coordinator of the "Office" in the years 1994, 1995 and 1996. The Mexican Prof. Alejandro Bravo was the coordinator in 1997.

The next section of the chapter provides examples of mathematical situations, which are deliberately left open, in order that readers can experiment by themselves creating new problems (this would be truly excellent!). In the subsequent sections, the problems arising from these situations are presented and the section of detailed solutions will follow.

2.3 Several Examples of Mathematical Situations

Situation 3.1 In the acute triangle *ABC*, let *AM* be the median (*M* belongs to the side *BC*), and let *AD* be the internal bisector of angle *A*. (*D* belongs to the side *BC*). From *B* the perpendicular to *AD* is drawn, meeting *AD* at *J*, to *AM* at *L* and to *AC* at *K*.

Situation 3.2 The most important carpet seller of Orient is very worried. His device to measure the carpets has been stolen and so he can't measure the new carpet recently received, for one of his best clients. The carpet is rectangular, but the dimensions are unknown. If he display the carpet in the floor of two of the rooms of his house, one after the other, in a convenient way, the four corners of the carpet are located on each one of the 4 walls of each room.

Situation 3.3 The quadrilateral *ABCD* has an inscribed circle, being *K*, *L*. *M* and *N* the tangency points with the sides *AB, BC, CD* and *DA*, respectively. The lines *DA* and *CB* intersect at *S*, and the lines *BA* and *CD* intersect at *P*.

Situation 3.4 Let *M* and *N* be points of the side *BC* of the triangle *ABC*, such that $BM = CN$ (point *M* is located between *B* and *N*). Let *P* and *Q* be points located respectively on *AN* and *AM* such that $\angle PMC = \angle MAB$ and $\angle QNB = \angle NAC$.

Situation 3.5 Consider the sum $\sum_{i=1}^{n} x_i y_i$, where the values of the $2n$ variables $x_1, \ldots, x_n; y_1, \ldots, y_n$ are only 0 either 1. Let *I*(*n*) be the number of $2n$-tuples $x_1, \ldots, x_n; y_1, \ldots, y_n$ such that the sum is an odd number, and *P*(*n*) the number of $2n$-tuples $x_1, \ldots, x_n; y_1, \ldots, y_n$ such that the sum is an even number. Consider $\frac{P(n)}{I(n)}$.

Situation 3.6 In the triangle *ABC*, *G* is the point of intersection of the medians and *K* the point of intersection of the symmedians. The lines *AG* and *AK* intersect again the circumcircle of *ABC* at *M* and *N*, respectively. Let $P = BC \cap GN$, $R = BC \cap KM$ and $S = GR \cap KP$.

Situation 3.7 The acute triangle *ABC* is inscribed in a circle. The point *P* is inside the triangle. Lines *AP*, *BP* and *CP* intersect again the circumcircle of *ABC* at *X*, *Y* and *Z*, respectively. Consider the triangle *XYZ*.

Situation 3.8 Consider the sequence of real numbers $\{x_n\}$ with x_0 arbitrary and $x_{n+1} = 2(x_n)^2 - 1$.

Situation 3.9 Lines r and s lie mutually orthogonal and do not are in the same plane. Let AB be its common perpendicular ($A \in r$, $B \in s$). Consider the sphere of diameter AB. The points $M \in r$ and $N \in s$ are variable, with the condition that MN is tangent to the sphere. Let T be the point of tangency.

Situation 3.10 Let ABC be a triangle inscribed in a circle, and I is the incenter of the triangle. Lines BI and CI intersect again the circumcircle at M and N, respectively. Line MN intersect AB at P and AC at Q, respectively.

Situation 3.11 With center in the incenter I of the triangle ABC, a circle is drawn, intersecting in two points each side of the triangle: to BC at D and P (being D the most near to B), to CA at E and Q (being E the most near to C), and to AB at F and R (being F the most near to A). Let S be the point of intersection of the diagonals of the quadrilateral $EQFR$, and T the point of intersection of the diagonals of the quadrilateral $FRDP$. Finally, let U be the intersection of the diagonals of the quadrilateral $DPEQ$.

Situation 3.12 $ABCD$ is a convex quadrilateral, and $M = AC \cap BD$. The internal bisector of $\angle ACD$ intersects BA at K. Suppose $MA \cdot MC + MA \cdot CD = MD \cdot MB$.

2.4 Some Problems Arising from the Mathematical Situations of Sect. 2.3

Problem 4.1 In the acute triangle ABC, let AM be the median (M belongs to the side BC), and let AD be the internal bisector of angle A (D belongs to the side BC). From B the perpendicular to AD is drawn, meeting AD at J, to AM at L and to AC at K. Show that AB and DM are parallel.

Problem 4.2 The most important carpet seller of Orient is very worried. His device to measure the carpets has been stolen and so he can't measure the new carpet recently received, for one of his best clients. The carpet is rectangular, but the dimensions are unknown. If he display the carpet in the floor of two of the rooms of his house, one after the other, in a convenient way, the four corners of the carpet are located on each one of the 4 walls of each room. If the sides of the first room are 55 and 50, and those of the second room are 55 and 38, find the dimensions of the carpet.

Problem 4.3 The quadrilateral $ABCD$ has an inscribed circle, being K, L. M and N the tangency points with the sides AB, BC, CD and DA, respectively. The lines DA and CB intersect at S, and the lines BA and CD intersect at P. If S, K and M are collinear, prove that P, N and L are also collinear.

Problem 4.4 Let M and N be points of the side BC of the triangle ABC, such that $BM = CN$ (point M is located between B and N). Let P and Q be points located respectively on AN and AM such that $\angle PMC = \angle MAB$ and $\angle QNB = \angle NAC$ Would it be always true that $\angle QBC = \angle PCB$?

Problem 4.5 Consider the sum $\sum_{i=1}^{n} x_i y_i$, where the values of the $2n$ variables $x_1, \ldots, x_n; y_1, \ldots, y_n$ are only 0 either 1. Let $I(n)$ be the number of $2n$-tuples $x_1, \ldots, x_n; y_1, \ldots, y_n$ such that the sum is an odd number, and $P(n)$ the number of $2n$-tuples $x_1, \ldots, x_n; y_1, \ldots, y_n$ such that the sum is an even number. Show that $\frac{P(n)}{I(n)} = \frac{2^n+1}{2^n-1}$.

Problem 4.6 In the triangle ABC, G is the point of intersection of the medians and K the point of intersection of the symmedians. The lines AG and AK intersect again the circumcircle of ABC at M and N, respectively. Let $P = BC \cap GN$, $R = BC \cap KM$ and $S = GR \cap KP$. Show that $AGSK$ is a parallelogram.

Problem 4.7 The acute triangle ABC is inscribed in a circle. The point P is inside the triangle. Lines AP, BP and CP intersect again the circumcircle of ABC at X, Y and Z, respectively. Determine the position of the point P for that XYZ be equilateral.

Problem 4.8 Consider the sequence of real numbers $\{x_n\}$ with x_0 arbitrary and $x_{n+1} = 2(x_n)^2 - 1$. Show that, if $|x_0| \leq 1$, then $|x_n| \leq 1$. Find a closed expression for x_n.

Problem 4.9.1 Lines r and s are mutually orthogonal and do not are in the same plane. Let AB be its common perpendicular ($A \in r$, $B \in s$). Consider the sphere of diameter AB. The points $M \in r$ and $N \in s$ are variable, with the condition that MN is tangent to the sphere. Let T be the point of tangency. Show that $TM.TN$ is constant.

Problem 4.9.2 Lines r and s are mutually orthogonal and do not are in the same plane. Let AB be its common perpendicular ($A \in r$, $B \in s$). Consider the sphere of diameter AB. The points $M \in r$ and $N \in s$ are variable, with the condition that MN is tangent to the sphere. Let T be the point of tangency. Determine the geometrical locus of the point T.

Problem 4.10 Let ABC be a triangle inscribed in a circle, and I is the incenter of the triangle. Lines BI and CI intersect again the circumcircle at M and N, respectively. Line MN intersect AB at P and AC at Q, respectively. Show that IA is perpendicular to MN.

Problem 4.11 With center in the incenter I of the triangle ABC, a circle is drawn, intersecting in two points each side of the triangle: to BC at D and

P (being D the most near to B), to CA at E and Q (being E the most near to C), and to AB at F and R (being F the most near to A). Let S be the point of intersection of the diagonals of the quadrilateral $EQFR$, and T the point of intersection of the diagonals of the quadrilateral $FRDP$. Finally, let U be the intersection of the diagonals of the quadrilateral $DPEQ$. Show that the circumcircles of the triangles FTR, DPU and EQS have one common point.

Problem 4.12 $ABCD$ is a convex quadrilateral and $M = AC \cap BD$. The internal bisector of $\angle ACD$ intersects BA at K. Suppose $MA \cdot MC + MA \cdot CD = MD \cdot MB$. Show that $\angle BKC = \angle CDB$.

2.5 Hints, Solutions and Comments to Some of the Problems and Examples

2.5.1 *Comment and Hint to Example 1.2*

The right hand side of the three equations are numbers of a Pythagorean triad. The left hand side of the equations represents the expressions of the cosine law for some convenient angles. So, the advice is to locate one point M inside a rectangle triangle with convenient sides in a such way the three equations be fulfilled, and from this, evaluate more easily xy + 2yz + 3zx.

Source of the problem: Zhang Jung-da et al., *Mathematics Competitions*, vol.10, number 2, 1997, pp. 52–63.

2.5.2 *Solution to Problem 4.1*

In the acute triangle ABC, let AM be the median (M belongs to the side BC), and let AD be the internal bisector of angle A. (D belongs to the side BC). From B the perpendicular to AD is drawn, meeting AD at J, to AM at L and to AC at K. Show that AB and DM are parallel (Fig. 2.1).

Solution (by F. Bellot)

There is no loss of generality if we suppose that angle B is bigger than angle C. First, being AD the internal bisector of angle A, $\angle BAD = \frac{A}{2}$. And as BJ is perpendicular to AD, $\angle ABJ = 90° - \frac{A}{2}$. The same argument in the triangle AJK gives us $\angle AKJ = 90° - \frac{A}{2}$. Then, triangle ABK is isosceles and $AK = AB = c$. From this, we get $KC = b - c$.

To prove that AB and DM are parallel, it is enough to prove that $\frac{AM}{ME} = \frac{BD}{DE}$, and the theorem of Thales will finish the problem.

Fig. 2.1 Figure for Problem 4.1

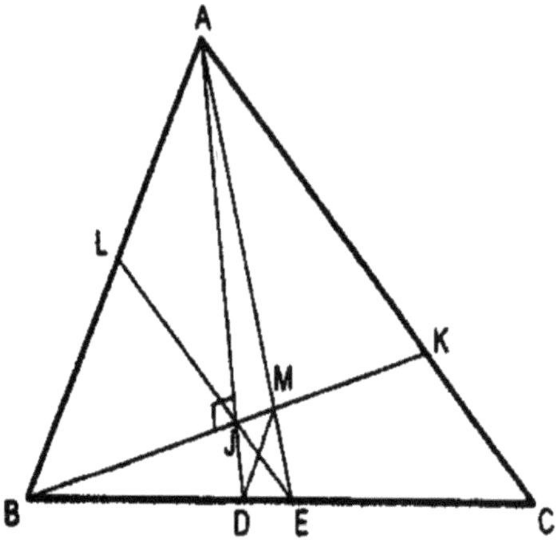

First we will compute $\frac{BD}{DE}$. From the angular bisector theorem, we have $BD = \frac{ac}{c+b}$. As E is the midpoint of BC, we have $DE = BE - BD = \frac{a(b-c)}{2(c+b)}$, therefore $\frac{BD}{DE} = \frac{2c}{b-c}$.

To compute $\frac{AM}{ME}$, we can use the Menelaus theorem in triangle AEC with the transversal KMB:

$\frac{AM}{ME} \cdot \frac{EB}{BC} \cdot \frac{CK}{KA} = 1 \Leftrightarrow \frac{AM}{ME} = \frac{2c}{b-c}$, and we are done. ∎

Source of the problem: Course on Euclidean Geometry I, University of Costa Rica, 2012.

2.5.3 *Solution of Problem 4.2*

The most important carpet seller of Orient is very worried. His device to measure the carpets has been stolen and so he can't measure the new carpet recently received, for one of his best clients. The carpet is rectangular, but the dimensions are unknown. If he display the carpet in the floor of two of the rooms of his house, one after the other, in a convenient way, the four corners of the carpet are located on each one of the 4 walls of each room. If the sides of the first room are 55 and 50, and those of the second room are 55 and 38, find the dimensions of the carpet.

Solution by María Ascensión López Chamorro, Valladolid (Spain)

We will solve the problem in a more general context, and then will apply it to the case of the carpet with the given numerical measures.

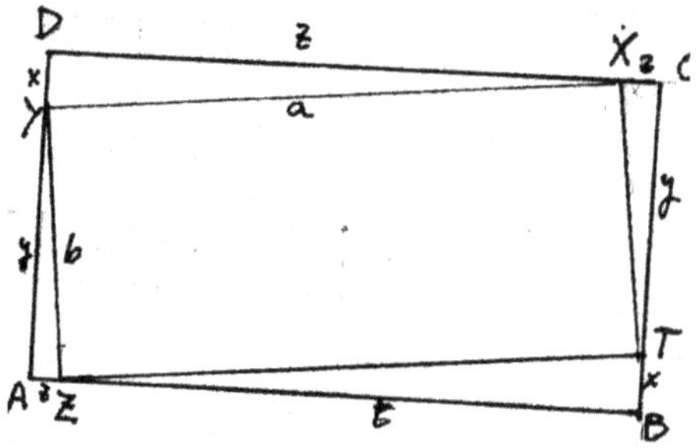

Fig. 2.2 First figure for Problem 4.2

Let *ABCD* be a rectangle, and let *XYZT* be another rectangle, inscribed in the first, with *Z* on the side *AB*, *T* on *BC*, *X* on *CD* and finally *Y* on *DA* (Fig. 2.2).

Suppose $AB = CD = l_1$; $AD = BC = l_2$; $XY = ZT = a$; $YZ = TX = b$. (In terms of the problem, a, b are the dimensions of the carpet; l_1, l_2 those of the room).

Triangles *XDY* and *ZBT* are congruent, also *YAZ* and *TXC*. This means

$$XC = AZ = z\,; XD = ZB = t; DY = BT = x; AY = TC = y$$

But moreover triangles *XDY* and *YAZ* are similar, and then $\frac{t}{y} = \frac{x}{z} = \frac{a}{b}$.

This proportion can be written as $bt = ay$; $bx = az$; and moreover the equalities $z + y = l_1$ and $x + y = l_2$ holds.

From this we obtain the two relations $\frac{bx}{a} + \frac{ay}{b} = l_1$; $x + y = l_2$ and solving them in the unknowns x and y gives us $\left(\frac{a}{b} - \frac{b}{a}\right)y = l_1 - \frac{b}{a}l_2$; $\left(\frac{b}{a} - \frac{a}{b}\right)x = l_1 - \frac{a}{b}l_2$.

The final expressions for x, y, z, t are the following:

$$x = \frac{a(al_2 - bl_1)}{a^2 - b^2};\ y = \frac{b(al_1 - bl_2)}{a^2 - b^2};\ z = \frac{b(al_2 - bl_1)}{a^2 - b^2};\ t = \frac{a(al_1 - bl_2)}{a^2 - b^2}.$$

But by Pythagora's Theorem, $x^2 + t^2 = a^2$; $y^2 + z^2 = b^2$. Both equalities given the same equation: $\left(\frac{al_2 - bl_1}{a^2 - b^2}\right)^2 + \left(\frac{al_1 - bl_2}{a^2 - b^2}\right)^2 = 1$; and developing, ordering and simplifying this can be written as

$$\left(a^2 + b^2\right)\left(l_1^2 + l_2^2\right) - 4l_1 l_2 ab = \left(a^2 - b^2\right)^2 (*)$$

If the rectangle *XYZT* also can be inscribed in another rectangle with dimensions m_1 and m_2, the same reasoning allows us to writing a second equation

$$\left(a^2+b^2\right)\left(m_1^2+m_2^2\right)-4m_1m_2ab=\left(a^2-b^2\right)^2(**)$$

Substracting (*) and (**) we get

$$\left(a^2+b^2\right)\left(l_1^2+l_2^2-m_1^2-m_2^2\right)-4ab(l_1l_2-m_1m_2)=0.$$

In order to simplify the notation we define

$$k=l_2^1+l_2^2-m_1^2\;-m_2^2;h=\,l_1l_2-m_1m_2;\quad \text{and}\,u=(b/a).$$

With this we have the quadratic equation in u

$$\left(1+u^2\right)k-4uh=0\Leftrightarrow ku^2-4uh+k=0$$

$$u=\frac{2h\pm\sqrt{4h^2-k^2}}{k}.$$

Now we make the computations with the data of the problem (crossing the fingers!):

$$l_1=55;\; l_2=50; m_1=55;\; m_2=38.$$

We get in sequence:

$$\begin{aligned} k&=88\cdot 12;\; h=55\cdot 12\\ 4k^2-h^2&=(2h+k)(2h\,-\,k)=12^2\cdot 11^2\cdot 6^2\\ u&=2\,\text{or}\,(1/2) \end{aligned}$$

and from this,

$$x=20;\; t=15\Rightarrow a=25,\; b=50$$

and for the second rectangle we get

$$x_1=7; t_1=24\Rightarrow a=25,\; b=50,$$

and so the same carpet can be placed in both rooms (Fig. 2.3). ∎

Fig. 2.3 Second figure for Problem 4.2

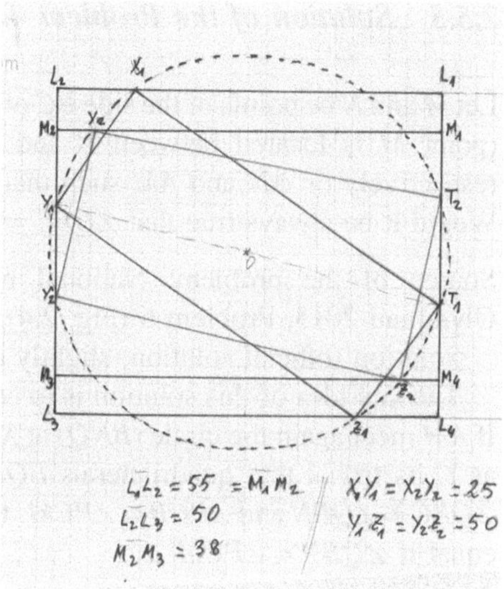

Source of the problem: Course on Euclidean Geometry I, University of Costa Rica, 2012.

2.5.4 *Solution of the Problem 4.3*

The quadrilateral *ABCD* has an inscribed circle, being *K, L. M* and *N* the tangency points with the sides *AB, BC, CD* and *DA*, respectively. The lines *DA* and *CB* intersect at *S*, and the lines *BA* and *CD* intersect at *P*. If *S, K* and *M* are collinear, prove that *P, N* and *L* are also collinear.

Source of the problem: Belarusian Math Olympiad 1996 (TST). In the booklet of this Olympiad no authorship attribution of the problem is given. In the booklet the solution of the student M. Vronski, given during the test (a long but nice metrical solution) is published. Some time after the 2002 Melbourne Conference of the WFNMC, where I presented this problem, I received the following solution:

Solution (by Andy Liu)

Let *O* be the centre of the circle and *r* its radius. Then *OS* and *LN* are perpendicular and let them meet at *R*. Also, *OP* and *SKM* are perpendicular and let them meet at *Q*. Since triangles *OLR* and *OSL* are similar, we have $OS.OR = r^2$. Similarly, $OP.OQ = r^2$. Hence *PQRS* is cyclic. Now, $\angle PRS = \angle PQS = 90° = \angle NRS$. It follows that *L, N* and *P* are collinear. ∎

2.5.5 Solution of the Problem 4.4

Let M and N be points of the side BC of the triangle ABC, such that $BM = CN$ (point M is located between B and N). Let P and Q be points located respectively on AN and AM such that $\angle PMC = \angle MAB$ and $\angle QNB = \angle NAC$ Would it be always true that $\angle QBC = \angle PCB$?

Source of the problem: National round of the Spanish Mathematical Olympiad 2015, Problem 6 (Fig. 2.4).

Solution (official solution, slightly edited by F. Bellot)

The key idea of this solution is to consider the circles (BNQ) and (PMC). If AM meet again the circle (BNQ) at X, and AN meet again the circle (PMC) at Y, its trivial that quadrilaterals $BQNX$ and $MPCY$ are cyclic. But being $\angle QBC = \angle QBN$ and $\angle PCB = \angle PCM$, then the angles of the problem will be equal if $\angle QBN = \angle PCM$

$$\text{But } \angle QBN = \angle QXN = \angle MXN \text{ and } \angle PCM = \angle PYM = \angle NYM$$

Then, the problem will be solved in affirmative sense if we prove the equality

$$\angle MXN = \angle NYM$$

and this means than the four points M, N, Y, X belong to the same circle.

So, we will try to prove that

$$AM \cdot AX = AN \cdot AY \Leftrightarrow \frac{AM}{AN} = \frac{AY}{AX} \tag{2.5.5.1}$$

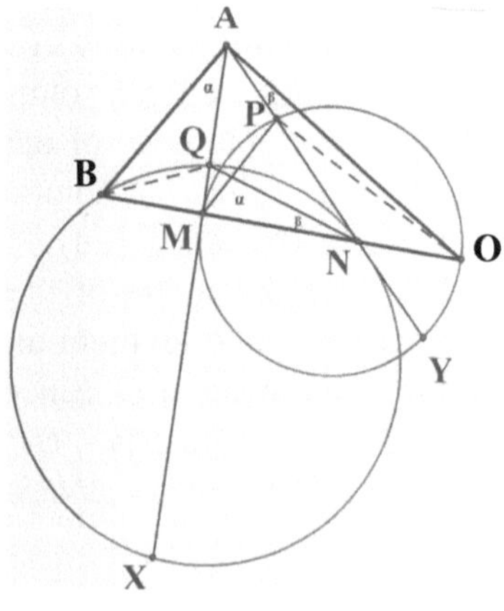

Fig. 2.4 Figure for Problem 4.4

Our argument is the following:

Triangles *ABM* and *CAN* have the same area, because their basis are equal by hypothesis and their altitudes from A are the same. So we have

$$AM \cdot AB \cdot \sin\alpha = AN \cdot AC \cdot \sin\beta \tag{2.5.5.2}$$

where $\alpha = MAB; \beta = NAC$.

For another hand, two of the angles of the triangle *ABX* are α, and $\angle BXQ = \angle QNB = \beta$ in circle (BNQ)

Similarly, two angles of triangle *ACY* are β and α. Therefore triangles *ABX* and *ACY* are similar, and we can write down the proportionality between the homologous sides as

$$\frac{AY}{AX} = \frac{CY}{AB}. \tag{2.3}$$

Finally, using the sinus law in triangle *ACY*, we get

$$\frac{AC}{\sin\alpha} = \frac{CY}{\sin\beta} \Leftrightarrow \frac{\sin\beta}{\sin\alpha} = \frac{CY}{AC}$$

and (2.5.5.2) can be written as

$$\frac{AM}{AN} = \frac{AC \cdot \sin\beta}{AB \cdot \sin\alpha} = \frac{AC}{AB} \cdot \frac{CY}{AC} = \frac{CY}{AB} = \text{by(3)} = \frac{CY}{AC}$$

and we are done. ∎

2.5.6 Solution to Problem 4.5

Consider the sum $\sum_{i=1}^{n} x_i y_i$, where the values of the $2n$ variables $x_1, \ldots, x_n; y_1, \ldots, y_n$ are only 0 either 1. Let $I(n)$ be the number of $2n$-tuples $x_1, \ldots, x_n; y_1, \ldots, y_n$ such that the sum is an odd number, and $P(n)$ the number of $2n$-tuples $x_1, \ldots, x_n; y_1, \ldots, y_n$ such that the sum is an even number. Show that $\frac{P(n)}{I(n)} = \frac{2^n+1}{2^n-1}$.

Source of the problem: This problem, created jointly by the Mexican mathematicians Gerardo Raggi and Humberto Cárdenas, was awarded with the Second Prize in the First Iberoamerican Contest of Creation of Problems, organized by the O.E.I. (Organization of Iberoamerican States for the Education, the Science and the Culture). Before this award were announced, the problem was included in the shortlist presented to the International Jury of the XII Iberoamerican Mathematical Olympiad, held at Guadalajara, Jalisco, Mexico, September 1997, and proposed to the students as problem number 4.

Official solution

First observe that for each natural number n, the recursive formula $P(n + 1) = 3P(n) + 1$ holds. This is so, because in any $2n$-uple in which the value is even, there are three possibilities of to choose the couple (x_{n+1}, y_{n+1}) to obtain one $2(n + 1)$-uple such that the value still be even; and starting with one $2n$-uple such that the value is odd, there are only one way to choose the couple (x_{n+1}, y_{n+1})—*both values equal to* 1—to complete to get an even value.

Analogously we have $I(n + 1) = 3I(n) + P(n)$.

We will use these recursive formulas and the induction over n to get the result.

The proposition is true if $n = 1$, because $P(1) = 3$ and $I(1) = 1$.

Suppose the result true for some $n \geq 1$ and we will prove it for $n + 1$. We have

$$\frac{P(n+1)}{I(n+1)} = \frac{3P(n)+I(n)}{3I(n)+P(n)} = \frac{3\left(\frac{2^n+1}{2^n-1}\right)+1}{3+\left(\frac{2^n+1}{2^n-1}\right)} = \frac{3\cdot 2^n+3+2^n-1}{3\cdot 2^n-3+2^n+1} = \frac{4\cdot 2^n+2}{4\cdot 2^n-2} = \frac{2^{n+1}+1}{2^{n+1}-1}.$$

∎

2.5.7 *Solution to Problem 4.6*

In the triangle ABC, G is the point of intersection of the medians and K the point of intersection of the symmedians. The lines AG and AK intersect again the circumcircle of ABC at M and N, respectively. Let $P = BC \cap GN$, $R = BC \cap KM$ and $S = GR \cap KP$. Show that $AGSK$ is a parallelogram.

Source of the problem: Problem proposed by Spain to the International Jury of the 12th Iberoamerican Math. Olympiad. The Problem selection committee changed the statement to the problem, changing barycenter and Lemoine's point by circumcenter and orthocenter, making it more easy. This is the originally proposed problem.

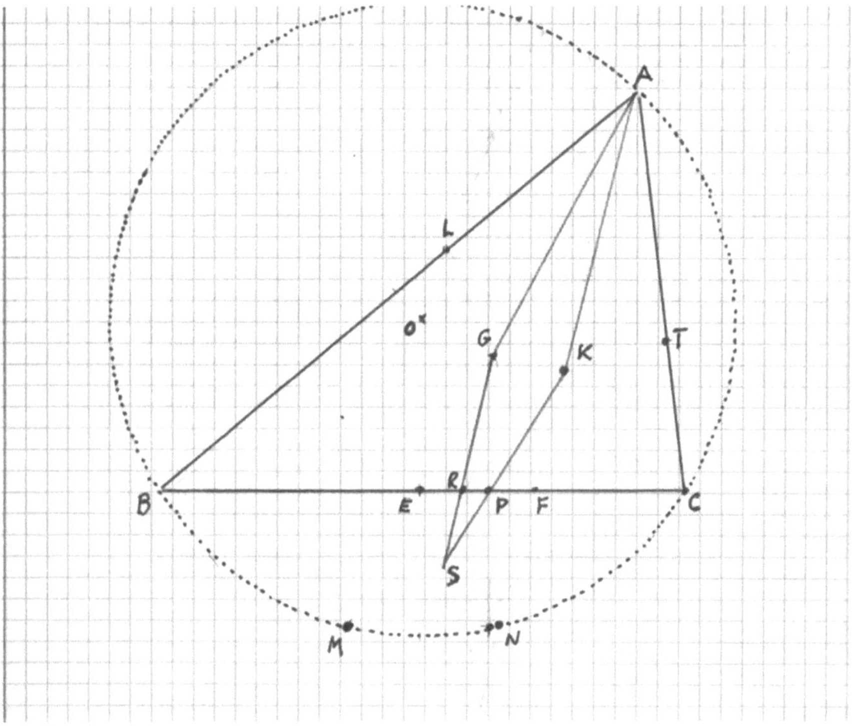

Fig. 2.5 Figure for Problem 4.6

Solution by F. Bellot (Fig. 2.5).

Let $E = BC \cap AG$, $F = BC \cap AK$, $L = AB \cap CK$ and $T = AC \cap BK$.

Taking account that cevians AE and AF are isogonal, the arcs BM and NC in the circle (ABC) are equal. From this we have that angles AEC and AMN are supplementary, due to the equalities

$$\angle AEC = \frac{1}{2}(arcAC + arcBM);\ \angle AMN = \frac{1}{2}arcAN = \frac{1}{2}(arcAC + arcNC)$$

This means that EF is parallel to MN, and as a consequence,

$$\frac{EM}{AM} = \frac{FN}{AN}. \tag{2.5.7.1}$$

For another hand, the power of point E with respect to the circle circumscribed to ABC can be written in two different ways:

$$AE \cdot EM = \frac{BC^2}{4},$$

Hence

$$EM = \frac{BC^2}{4 \cdot AE}. \tag{2.5.7.2}$$

From (2.5.7.2) we get

$$AM = \frac{4 \cdot AE^2 + BC^2}{4 \cdot AE},$$

whence, taking account that

$$AE^2 = \frac{2(AB^2 + AC^2) - BC^2}{4},$$

we get

$$AM = \frac{AB^2 + AC^2}{2 \cdot AE}. \tag{2.5.7.3}$$

From (2.5.7.2) and (2.5.7.3) we obtain

$$\frac{EM}{AM} = \frac{BC^2}{2(AB^2 + AC^2)}, \tag{2.5.7.4}$$

and by (2.5.7.2), we can write down

$$\frac{FN}{AN} = \frac{BC^2}{2(AB^2 + AC^2)}. \tag{2.5.7.5}$$

As the cevians AF, CL and BT are concurrent at K, the Van Aubel theorem allow us to write

$$\frac{AK}{KF} = \frac{AL}{LB} + \frac{AT}{TC};$$

and by the Theorem of the Symmedian,

$$\frac{AL}{LB} = \frac{AC^2}{AB^2}; \frac{AT}{TC} = \frac{AB^2}{AC^2}.$$

So we get

$$\frac{AK}{KF} = \frac{AB^2 + BC^2}{BC^2}. \qquad (2.5.7.6)$$

For another hand, the Menelaus theorem applied to the triangle *AEF* with the transversal *KM* gives us

$$\frac{ER}{RF} = \frac{AK}{KF} \cdot \frac{EM}{AM}.$$

From this, with (2.5.7.4) and (2.5.7.6), we obtain $\frac{EF}{RF} = \frac{1}{2}$, and as $\frac{EG}{GA} = \frac{1}{2}$ we have

$$\frac{EF}{RF} = \frac{EG}{GA} \qquad (2.5.7.7)$$

and therefore *GR* is parallel to *AF*, whence *GS* is parallel to *AK* (2.5.7.8).
Again the Menelaus theorem at *AEF* with *GN* gives us

$$\frac{EP}{PF} = \frac{AN}{FN} \cdot \frac{GE}{AG},$$

which with (2.5.7.5) gives us

$$\frac{EP}{PF} = \frac{AK}{KF},$$

and this means *KP* is parallel to *AE*, or that is the same, *KS* parallel to *AG* (2.5.7.9).
So (2.5.7.8) and (2.5.7.9) proves that *AGSK* is a parallelogram. ∎

2.5.8 Comments and Solution to Problem 4.7

The acute triangle *ABC* is inscribed in a circle. The point *P* is inside the triangle. Lines *AP*, *BP* and *CP* intersect again the circumcircle of *ABC* at *X*,

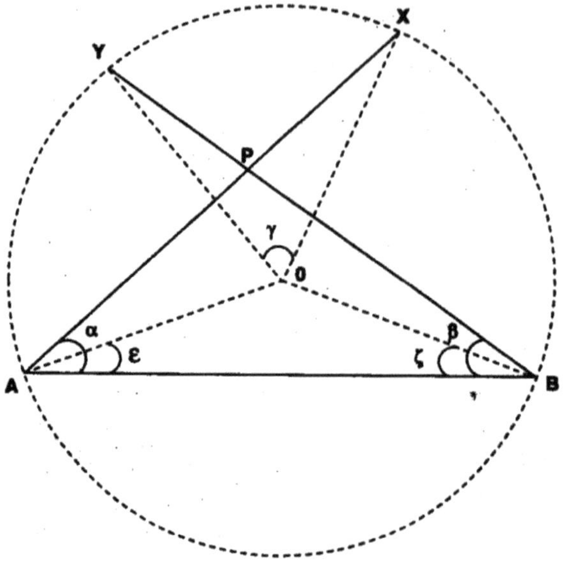

Fig. 2.6 Figure for Problem 4.7

Y and Z, respectively. Determine the position of the point P for that XYZ be equilateral.

Comments

This problem, created in the Symposium held immediately before the IXth Iberoamerican Math. Olympiad 1994, was included in the exam as problem 4. The solution below—slightly edited—was obtained by a Portuguese student, Joao Menano, during the contest (Fig. 2.6).

Solution

Consider, for instance, the side AB and the diameter of the circle (ABC) which is parallel to this side. Point C, then, must belong to the opposed semicircle to that in which A and B are located, because triangle ABC is acute. The same observation is valid for any other couple of vertices. Then we can forget the point C. We will find all the points P such that $\gamma = 120^o$. This condition is obviously necessary and sufficient for that X and Y be two vertices of the equilateral triangle XYZ (the center of the equilateral triangle must to be the center of the circle).

We have $\angle YOB = 2 \cdot \angle YAB$ and $\angle XOA = 2 \cdot \angle XBA$. In order to get $\angle XOY = 120°$. We need that

$$\angle XOY + \angle YOB + \angle AOB + \angle AOX = 360°$$

$$\angle AOB = 180° - \angle OAB - \angle OBA.$$

Observing the Fig. 2.4, this means

$$2\alpha+2\beta=60^\circ+\varepsilon+\delta \Rightarrow \alpha+\beta=\frac{60^\circ+\varepsilon+\delta}{2}.$$

The set of points P which verify this last equation are the points of an arc of circle through A and B with this measure. By means of this construction we get an arc of circle to which P belongs. Repeating this construction using other vertices, say B, C, we will get another arc of circle. The intersection of both arcs gives the position of searched point P. ∎

2.5.9 Solution to the Problems 4.8

Consider the sequence of real numbers $\{x_n\}$ with x_0 arbitrary and $x_{n+1}=2(x_n)^2-1$. Show that, if $|x_0|\le 1$, then $|x_n|\le 1$. Find a closed formula for x_n.

Solution by F. Bellot

If $|x_0|\le 1$, we can write $x_0=\cos\theta$, for some $\theta\in[0,\pi)$.

Then we get $x_1=2(\cos^2\theta)-1=\cos 2\theta$, and $|x_1|\le 1$. Continuing in this approach, we obtain $x_2=\cos(2^2\theta)$, and by induction we can prove that $x_n=\cos(2^n\vartheta)$, and we are done the two proposed problems. ∎

2.5.10 Solution to Problems 4.9.1 and 4.9.2

Lines r and s are mutually orthogonal and do not are in the same plane. Let AB be its common perpendicular ($A\in r$, $B\in s$). Consider the sphere of diameter AB. The points $M\in r$ and $N\in s$ are variable, with the condition that MN is tangent to the sphere. Let T be the point of tangency. Show that $TM.TN$ is constant. Determine the geometrical locus of the point T.

Both problems were also created during the Symposium on Creating problems, previously to the 10th Iberoamerican Mathematical Olympiad, Chili 1995. The problem was chosen by the International Jury and proposed to the students as problem 3 (Fig. 2.7).

First we will prove that $TM.TN$ is constant (this part was not included in the text of the problem 3 of the Iberoamerican Olympiad 1995).The picture can be simplified a bit (Fig. 2.8):

Fig. 2.7 Figure for Problem 4.9.1

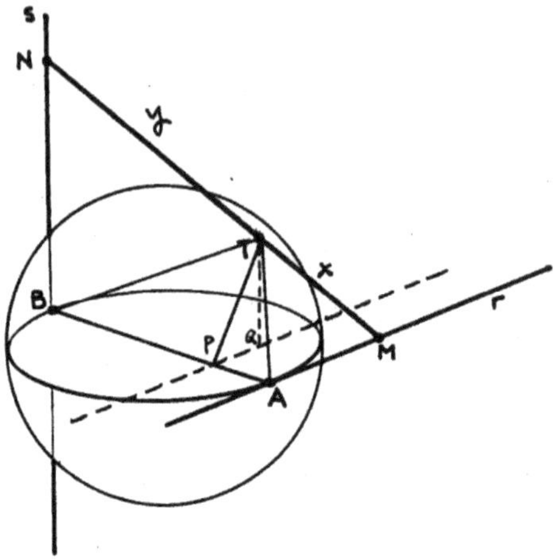

Fig. 2.8 Figure for Problem 4.9.2

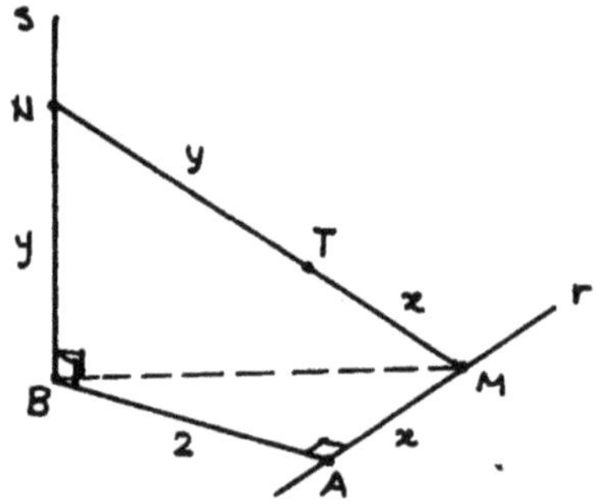

The argument is by Eduardo Wagner. If we take $AB = 2$, $MA = MT = x$, $NB = NT = y$, then we get

$$NM^2 = NB^2 + BM^2 \Leftrightarrow (x+y)^2 = y^2 + 4 + x^2 \Leftrightarrow xy = 2$$

∎

Going back to the Fig. 2.7, we will give an analytical solution of the problem. (Solution by F. Bellot during the Symposium).

Suppose $AB = 2$. We will choose the midpoint O of AB as origin of a Cartesian system of coordinates in the space, the line AB will be the x axis; the line through O parallel to the line s as "y" axis; and the perpendicular to the plan xy through O (upwards) as "z" axis. OB is the positive "x" axis.

The equation of the sphere is $x^2 + y^2 + z^2 = 1$; the equations of the line r are $(x = -1, y = 0)$; the equations of the line s are $(x = 1, z = 0)$ and the coordinates of points M and N are $M(-1, 0, m), N(1, n, 0)$.

The equations of the line MN are $\frac{x+1}{-2} = \frac{y}{-n} = \frac{z-m}{m} = \mathrm{t}$.

The condition of tangency of the line MN with the sphere is $4 = m^2n^2$, that is $mn = \pm 2$.

If $mn = 2$, the coordinates of the tangency point T are $\left(\frac{m^2-2}{m^2+2}, \frac{2m}{m^2+2}, \frac{2m}{m^2+2}\right)$ and as the second and third coordinates of T are the same, this means that T belong to the plane of equation $y = z$, and so this plane contain the line AB and make an angle of 45° with the plan xy.

If $mn = -2$, the plane to which T belongs is $y = -z$, which is orthogonal to the first one. Both planes pass through the center of the sphere, and intersect it following two maximal circles through A and B, forming angles of 45° with the plan xy. ∎

2.5.11 Solution to Problem 4.10

Let ABC be a triangle inscribed in a circle, and I is the incenter of the triangle. Lines BI and CI intersect again the circumcircle at M and N, respectively. Line MN intersect AB at P and AC at Q, respectively. Show that IA is perpendicular to MN.

Source of the problem: Problem created during the Third Iberoamerican Workshop about the creation of problems, held in San José, Costa Rica, Sept. 1996, just before the 11th Iberoamerican Math Olympiad (Fig. 2.9).

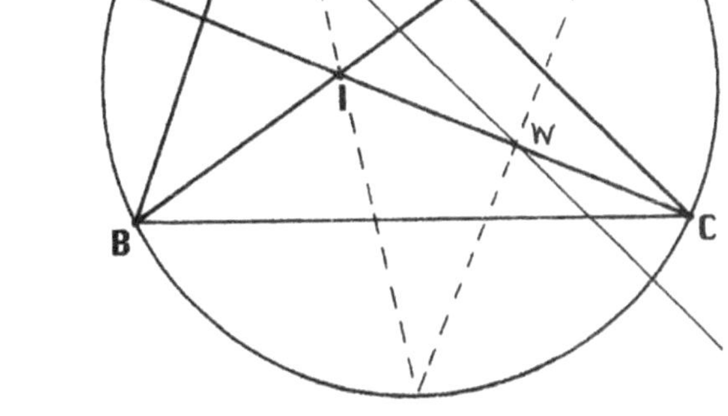

Fig. 2.9 Figure for Problem 4.10

Let L be the midpoint of the arc BC which do not contains A. The perpendicular line from N on LA intersects AL at T. The perpendicular line from N on ML intersects ML at W. Note that I is the orthocenter of the triangle LMN. The line WT is parallel to AC, and therefore is the line of the statement of the problem.

Now, if from N draw the perpendicular to AM, intersecting AM at X, the Simson line of N with respect to the triangle AML is the line which pass through T and W, that is, $X = AM \cap TW$. ∎

2.5.12 Solution to Problem 4.11

With center in the incenter I of the triangle ABC, a circle is drawn, intersecting in two points each side of the triangle: to BC at D and P (being D the most near to B), to CA at E and Q (being E the most near to C), and to AB at F and R (being F the most near to A). Let S be the point of intersection of the diagonals of the quadrilateral $EQFR$, and T the point of intersection of the diagonals of the quadrilateral $FRDP$. Finally, let U be the intersection of the diagonals of the quadrilateral $DPEQ$. Show that the circumcircles of the triangles FTR, DPU and EQS have one common point.

Source of the problem: The problem was created during the 4th Workshop of Creation of problems, held in Guadalajara, Jalisco, Mexico in September of 1997, just before the 12th Iberoamerican Mathematical Olympiad. The workshop was conducted by Prof. Alejandro Bravo. The problem was chosen by the International Jury and proposed to the students as problem number 3.

Solution by Alejandro Bravo.

As S belongs to the bisector of angle A of triangle ABC, the angles QIS and SIF are equal. But angle $QIF = 2(angle\ SIQ)$ is a central angle in the circle, and QES is inscribed and subtend the same arc FQ; therefore angle QES = angle SIQ and the four points Q, S, I and E are concyclic (Fig. 2.10).

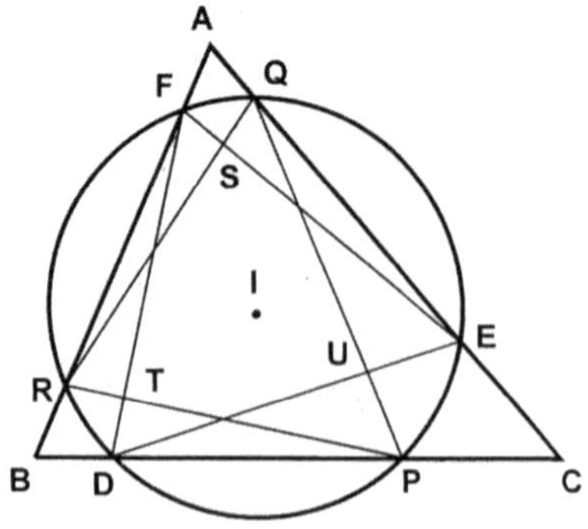

Fig. 2.10 Figure for Problem 4.11

The same argument proves that U belong to this same circle. Repeating the reasoning, the circles circumscribed to the triangles DPU, EQS and FRT pass through the incenter I of the triangle ABC. ∎

2.5.13 Solution of the Problem 4.12

$ABCD$ is a convex quadrilateral and $M = AC \cap BD$. The internal bisector of ACD intersect BA at K. Suppose $MA \cdot MC + MA \cdot CD = MD \cdot MB$. Show that $\angle BKC = \angle CDB$.

Source of the problem: Course of Euclidean Geometry 1, University of Costa Rica.

Solution by F. Bellot

First we will draw a figure in such a way that it meet the conditions of the statement of the problem (Fig. 2.11):

Drawing first the dotted circle, choose on it arbitrary points B, C and D. Choosing then the angle KCD, with K on the circle, joining K with B we will get the straight line where the point A must to be. Then, with the protractor the angle KCA equal to the angle KCD is drawn (because CK is the bisector of ACD) and so the position of the point A is determinate.

The thesis of the problem is equivalent to say that the points B, C, D and K are in the circle (and this justify the drawing) and furthermore gives an interpretation of the strange condition

$$MA \cdot MC + MA \cdot CD = MD \cdot MB \tag{2.5.13.1}$$

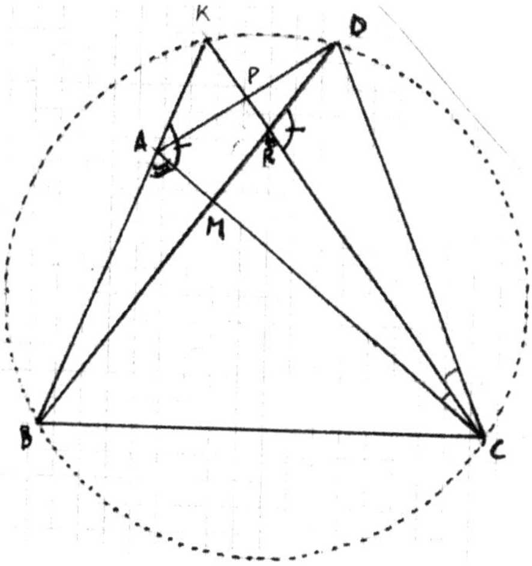

Fig. 2.11 Figure for Problem 4.12

given in the statement of the problem.

First at all, as R is the foot of the internal bisector CK of triangle MDC, we have, by the internal bisector theorem,

$$\frac{RM}{RD} = \frac{MC}{CD} \Rightarrow CD = \frac{MC \cdot RD}{MR}.$$

The value of CD is substituted in (2.5.13.1): $MA \cdot MC + MA \cdot \frac{MC \cdot RD}{MR} = MD \cdot MB.$

The left hand side can be written in the form $MA \cdot MC \cdot \left(1 + \frac{RD}{MR}\right) = MD \cdot MB$, i.e.

$$MA \cdot MC \cdot \frac{MR + RD}{MR} = MD \cdot MB \Leftrightarrow MA \cdot MC \cdot \frac{MD}{MR} = MD \cdot MB$$

which reduces to $\frac{MA \cdot MC}{MR} = MB \Leftrightarrow MA \cdot MC = MR \cdot MB$. This last equality warranty that the points B, C, A and R are in the same circle (not drawn in the picture above), and therefore the angles BAC and BRC are equal.

Consider now the triangles KAC and DRC. Both have equal the angle C (because CK is the bisector of angle ACD), and for another hand $\angle KAC = \angle DRC$, because they are supplementary of the equal angles $\angle BAC = \angle DRC$. Therefore the third angles in both triangles should to be equal, that is $\angle BRC = \angle BDC$, and we are done. ∎

References

Akopyan, A. (2011). *Geometry in figures*. Lexington, KY.

Boudine, J-P., LoJacomo, F., & Cuculière, R. (1998). *Olympiades Internationales de Mathématiques (Énoncés et solutions détaillées. Années 1988 à 1997)*. Ed. du Choix (In French).

Burns, J. C. (2000). *Seeking solutions*. Australian Mathematics Trust.

Cuculescu, I. (1984). *Olimpiadele Internationale de Matematica ale elevilor*. Ed. Tehnica, Bucarest (In Romanian).

Engel, A. (1987). The creation of mathematical olympiad problems. In *Newsletter of the WFNMC*, no. 5.

Gardiner, A. (1992). Creating elementary problems to stimulate thinking. In *Mathematics competitions* (Vol. 5, no. 1).

Gardiner, A. (1997). *The mathematical olympiad handbook* (*An introduction to problem solving*). Oxford University Press.

Ivanov, O. A. (2009). *Making mathematics come to life (A guide for teachers and students)*. AMS.

Maxwell, E. A. (1949). *Geometry for advanced pupils*. Oxford University Press.

Monk, D. (2009). *New Problems in Euclidean Geometry*. UKMT, Problems, Number 1.

Savchev, S., & Andreescu, T. (2003). *Mathematical miniatures*. The Anneli Lax New Mathematical Library (Vol. 43), Mathematical Association of America.

Sharygin, I. (2001). Where do problems come from? In *Quantum,* January–February 2001 (pp. 18–28).

Wagner, E. (1997). *Oficina de criaçao de problemas*, pg. 9–11. In *Siproma, Number 0, O.E.I. (Organización de Estados Iberoamericanos para la Educación, la Ciencia y la Cultura* [In Portuguese].

Chapter 3
Techniques for Solving Problems of Plane Geometry

K.P. Shum

Abstract In this paper, we present some problems in plane geometry, which can be solved by using analytic geometry and quadratic equations. Some of these problems have been taught to the high school students who participated the preliminary HKIMO committee selection contest.

Keywords Collinear points · Concyclic points · Ceva Theorem · Menelaus Theorem

AMS Mathematics Subject Classification 05AXX · 05B50 · 51M04 · 52A40

3.1 Introduction

In many national and regional mathematical competitions of high school mathematics, there are always problems in plane geometry in the examination. Usually, these problems are related with the collinear points, concyclic points, the mid points, Angle bisectors, the centroid, the orthocenter, the circumcenter, the in-center and some of the inequalities. Because many students face difficulty in proceeding with proof to unfamiliar problems in the contest, we (therefore) advise the students first to write down the coordinates of the points in the given diagram.

The students are encouraged to use analytic geometry to tackle the problem in case they fail to provide a proof for the given problem. We also

---

K.P. Shum (✉)
Institute of Mathematics, Yunnan University, Kunming 650091, People's Republic of China
e-mail: kpshum@ynu.edu.cn

A. Soifer (ed.), *Competitions for Young Mathematicians*, ICME-13 Monographs, DOI 10.1007/978-3-319-56585-9_3

observe that there are a number of geometry problems of IMO type, which can be solved by using the relationship of roots and coefficients and the discriminant of a quadratic equation. There are also plane geometry problems related to collinear points and concyclic points. In this paper, we will present some interesting problems to demonstrate the applications of some well known plane geometry theorems and techniques. We also propose some exercises to the readers.

3.2 Plane Geometry Problems (Moise 1990; Encyclopedia of the Solutions of Mathematics Problem 1983; Some Geometry Problems in Mathematical Olympiad Competitions 2015; Encyclopedia of Solved Problems 2016), Which Can be Solved by Analytic Geometry

In this section, we first present the following problems.

Example 2.1 In the following diagram, AD is the angle bisector of $\angle A$ of the $\triangle ABC$, AM is the median of $\triangle ABC$. BF is the perpendicular line passing through the point B to meet the X-axis at F. Assume that the angle bisector of $\angle$BAF meets BF at E. Join ED and prove that $ED // AB$ (Diagram 3.1).

Proof Let $|AB| = 2c$, $|AC| = 2b$ and $\angle CAB = 2\theta$. Let the coordinates of M be (X_M, Y_M).

Then, we have

Diagram 3.1 .

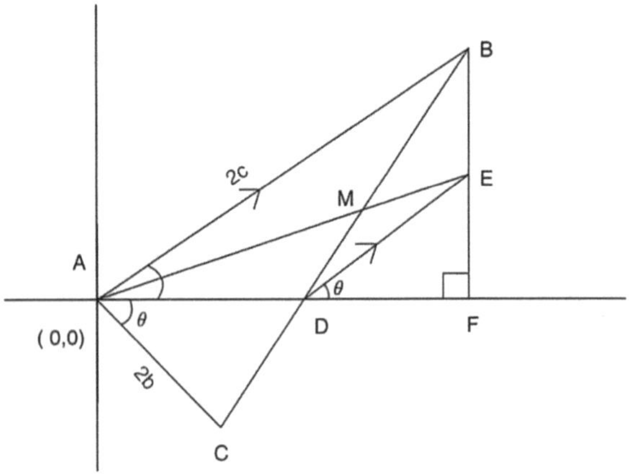

$$B\,(2c\,cos\theta,\ 2c\,sin\theta),$$
$$C\,(2b\,cos\theta,\ -2b\,sin\theta),$$
$$X_M = (b+c)cos\theta,$$
$$Y_M = (c-b)sin\theta$$

Let the coordinates of E be $(X_E,\ Y_E)$. Then $X_E = 2c\,cos\theta$.

Because the points A, M, E are collinear, we have $\dfrac{(c-b)sin\theta}{(c+b)cos\theta} = \dfrac{Y_E}{2c\,\cos\theta}$, $Y_E = \dfrac{2c(c-b)}{c+b}sin\theta$. Now, let the coordinates of D be $(x_D, 0)$. Notice that the points B, D, C are collinear, and therefore the slopes of BC and BD are the same. We have $\dfrac{2(c+b)sin\theta}{2(c-b)cos\theta} = \dfrac{2c\ sin\theta}{2c\cdot cos\theta - X_D}$, and hence $X_D = 2c$, $cos\theta - \dfrac{2c(c-b)}{c+b}cos\theta = \dfrac{4bc}{b+c}cos\theta$.

The slope of the line AB is obviously $K_{DE} = tan\theta$.

Compute the slope of DE, we have $K_{DE} = \dfrac{\dfrac{2c(c-b)}{c+b}sin\theta}{2c\ cos\theta - \dfrac{4bc\ cos\theta}{c+b}} = \dfrac{c-b}{c+b-2b}\tan\theta = tan\theta$.

Because $K_{DE} = K_{AB} = tan\theta$, therefore $DE\,//\,AB$.

Example 2.2 To Prove that two lines AB and CD are perpendicular. We usually use the product of their slope is equal to -1 that is, $K_{AB}\cdot K_{CD} = -1$

In the following trapezoid (Diagram 3.2),

Given $AB + CD = BC,\ DP = PA$. Prove that $PB \perp PC$

Proof We simply let $B = (0,0)$, $A = (a,0)$, $C = (b,c)$, $D = (b+d,c)$

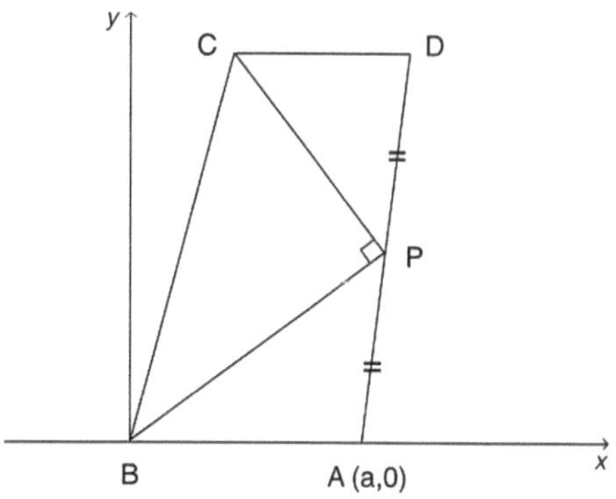

Diagram 3.2 See Encyclopedia of the Solutions of Mathematics Problem (1983), problem 62

Because it is given that $AB + CD = BC$, we have $a + d = \sqrt{b^2 + c^2}$

Thus, we get $(a+d)^2 - b^2 = c^2$. This means that the coordinates of the point P are $\left(\frac{a+b+d}{2}, \frac{c}{2}\right)$

The planes of BP and CP are

$$m_{CP} = \frac{\frac{c}{2}}{b - \frac{1}{2}(b+a+d)} = \frac{c}{b+a+d},$$

$$m_{BP} = \frac{\frac{c}{2}}{\frac{\frac{1}{2}}{(a+b+d)}} = \frac{c}{b+a+d}.$$

Hence, we have $m_{BP} \cdot m_{CP} = \frac{c^2}{b^2 - (a+d)^2} = \frac{c^2}{-c^2} = -1$. This shows that $PB \perp PC$.

The proof is completed.

The following example shows that we sometimes need to assign the coordinates of the points in the diagram with some suitable numbers.

Example 2.3 Given a square $ABCD$. Draw a line through the corner point D parallel to the diagonal CA of the square. On the bisector of the second quadrant of the X-Y plane, find the point E so that $CE = CA$. Join the points C and E to meet AD at F. Prove that $|AE| = |AF|$.

Diagram 3.3 .

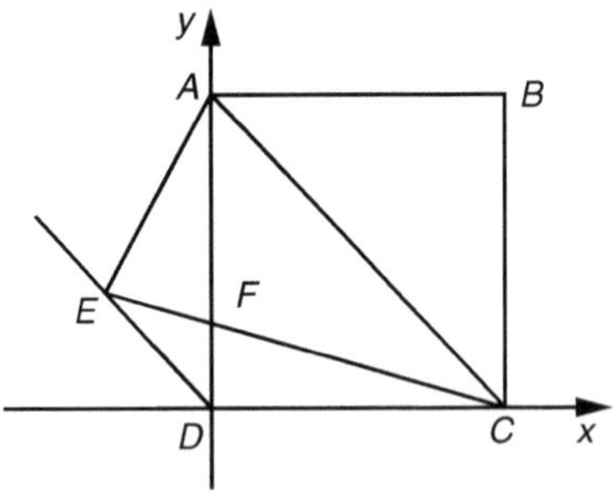

Consider the above diagram (Diagram 3.3),

Proof In Diagram 3.3, the coordinates of $A(0,a)$, $B(a,0)$, $C(a,0)$, $D(0,0)$.

Because $CA\,//\,DE$, the point E lies in the bisector of the second quadrant of the X–Y plane, we can write $E=(x,\,-x)$, $(x<0)$.

According to the give condition, we have $|CE|=|AC|=\sqrt{2a}$, hence we deduce that $(x-a)^2+x^2=2a^2$ and so $x=\frac{a}{2}\left(1-\sqrt{3}\right)$. Therefore, the coordinates of E is $\left[\frac{a}{2}\left(1-\sqrt{3}\right),\frac{a}{2}\left(\sqrt{3}-1\right)\right]$. Let the coordinates of F be $(0,y)$. Because the points E, F, C are collinear, we have

$$\begin{vmatrix} \frac{a}{2}\left(1-\sqrt{3}\right) & \frac{a}{2}\left(\sqrt{3}-1\right) & 1 \\ 0 & y & 1 \\ a & 0 & 1 \end{vmatrix}=0.$$

Therefore, $\frac{a}{2}y\left(\sqrt{3}+1\right)\cdot\frac{a^2}{2}\left(\sqrt{3}-1\right)=0$, and hence we find $y=(2-\sqrt{3})a$.

Thus,

$$\begin{aligned}|AE|&=\sqrt{\frac{a^2}{4}\left(\sqrt{3}-1\right)^2+\frac{3a^2}{4}\left(\sqrt{3}-1\right)^2}\\&=\left(\sqrt{3}-1\right)a\end{aligned}$$

Therefore, $|AE|=|AF|$.

In closing this section, we propose the following exercise.

Exercise 2.4 (See Encyclopedia of the Solutions of Mathematics Problem (1983), problem 61) Let ABC be a triangle. Construct a square $ABDE$ on AB and another square $ACFG$ on AC. Suppose that K and L are the centers of these two squares. Also, let M be the midpoint of the side BC in $\triangle ABC$. Prove that $KM\perp LM$.

Hint

(I) First draw Diagram 3.4. Put the point A on the y-axis and the line BC on the x-axis. Thus, we have

$$A=(0,a),\ B=(b,0),\ C=(c,0),\ M=\left(\frac{b+c}{2},0\right).$$

Let the center L of the square $ACFG$ be (x,y).

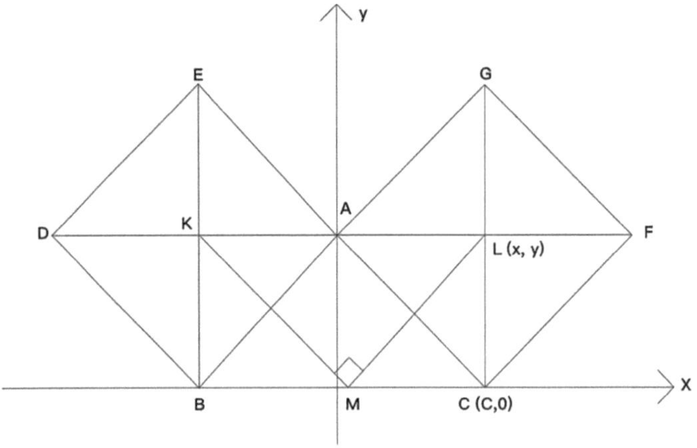

Diagram 3.4 .

(II) Observe that $AL \perp CL$ and $|AL| = |CL|$. Then we use the fact that the product of the slope of AL and the slope CL is equal to -1. Thus, we have

$$\frac{y-a}{x} \cdot \frac{y}{x-c} = -1.$$

Hence, we deduce that $x^2 + (y-a)^2 = (x-c)^2 + y^2$.
Solving the above equation, we get the following set of equations

$$\begin{cases} x = \frac{1}{2}(c-a) \\ y = \frac{1}{2}(a-c) \end{cases}; \begin{cases} x = \frac{1}{2}(c+a) \\ y = \frac{1}{2}(c+a) \end{cases}$$

(III) Now, according to the position of L in the diagram, we can let the coordinate of the point L be $\left(\frac{c+a}{2}, \frac{c+a}{2}\right)$.

Similarly, the coordinate of the point K is $\left(\frac{b-a}{2}, \frac{a-b}{2}\right)$.

(IV) Calculate the length of ML, we have

$$|ML| = \sqrt{\left(\frac{c+a}{2} - \frac{c+b}{2}\right)^2 + \left(\frac{c+a}{2}\right)^2}$$
$$= \frac{1}{2}\sqrt{(a-b)^2 + (a+c)^2}$$

This shows that $|ML| = |MK|$

(V) Calculate

$$m_{mL} \cdot m_{mx} = \frac{\frac{c+a}{2}}{\frac{c+a}{2} - \frac{b+c}{2}} \cdot \frac{\frac{a-b}{2}}{\frac{b-a}{2} - \frac{b+c}{2}} = -1$$

Hence $KM \perp LM$ and the proof is completed.

3.3 Lattice Points and Collinear Points (see Liu 1979)

A point A is called a lattice point if the coordinates of A are integers. A triangle ABC is called a lattice triangle if the points A, B, C are lattice points.

The following theorem is easy to prove.

Theorem 3.1 *A lattice triangle is not an equilateral triangle.*

Proof Suppose on the contrary that the lattice triangle ABC is an equilateral triangle. We first move the equilateral triangle to a new position so that one of its vertices A is the point of origin of the XY coordinate plane, namely $A = (0, 0)$. Now, let $B = (m_1, n_1)$, $C = (m_2, n_2)$.

Because $\triangle ABC$ *is an* equilateral triangle, all its angles are 60°. Hence,

$$tan\angle BOC = tan60° = \sqrt{3} = \frac{\frac{n_2}{m_2} - \frac{n_1}{m_1}}{1 + \frac{n_1}{m_1} - \frac{n_2}{m_2}} = \frac{m_1n_2 - m_2n_1}{m_1m_2 + n_1n_2}.$$

Because m_1, m_2, n_1, n_2 are integers, $\frac{m_1n_2 - m_2n_1}{m_1m_2 + n_1n_2}$ is clearly a rational *number.* This is clearly a contradiction! Hence, we have proved that it is impossible for a lattice triangle to be an equilateral triangle.

By using the argument in Theorem 3.1, we obtain the following corollary.

Corollary 3.2 *Let S be the area of a lattice polyhedral. Then 2S must be an integer.*

Proof We first observe that the area of any lattice polyhedral can be regarded as the area of some sum of several lattice triangles.

Diagram 3.5 See Encyclopedia of the Solutions of Mathematics Problem (1983), problem 345

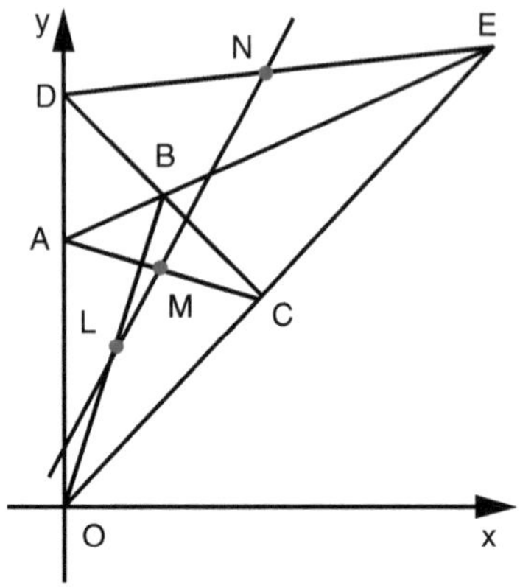

We now consider three points $A=(x_1, y_1)$, $B=(x_2, y_2)$, $C=(x_3, y_3)$. Then the following criterion gives the condition for the three points *A*, *B*, *C* to be collinear.

Criterion 3.3 *Three points A, B, C are collinear if and only if*

$$\begin{vmatrix} x_1 & y_1 & 1 \\ x_2 & y_2 & 1 \\ x_3 & y_3 & 1 \end{vmatrix} = 0$$

This shows that the three points *L*, *M*, *N* are collinear. The following example is an application of the above theorem.

Example 3.4 In Diagram 3.5, *ABCO* is a quadrilaterial with *O* be the point of origin in the XY-plane. Let *L* and *M* be the mid points of the diagonals *OB* and *AC*, respectively. Extend the side *OC* to meet *AB* at *E* and *OA* to meet *CB* at *D*. Let *N* be the midpoint of *DE*. Prove that the points *L*, *M*, *N* are collinear (see Diagram 3.5).

Proof Let
$A=(0, 2a)$, $B=(2u, 2v)$, $C=(2c, 2kc)$, $D=(0, 2d)$, $E=(2e, 2ke)$.
Then the mid points of *OB*, *AC* and *DE* are

$$L=(u, v),\ M=(c, u+kc),\ N=(c, d+ke).$$

Because *C*, *B*, *D* are collinear, we have

$$\begin{vmatrix} 2c & 2kc & 1 \\ 2u & 2v & 1 \\ a & 2a & 1 \end{vmatrix} = 0,$$

that is, $cv + du - cd - kcu = 0$.

On the other hand, we notice that the points A, B, E are collinear, so we have

$$\begin{vmatrix} 0 & 2a & 1 \\ 2u & 2v & 1 \\ 2e & 2ke & 1 \end{vmatrix} = 0$$

therefore, $ae + keu - ev - au = 0$.

Now, we consider

$$\begin{vmatrix} u & v & 1 \\ c & a+kc & 1 \\ e & d+ke & 1 \end{vmatrix}$$
$$= au + kcu + cd + cke + cv -$$
$$= -(cv + du - cd - kcu) - (ae + keu - ev - au) = 0$$

This shows that the three points L, M, N are collinear.

To determine whether any three points in the plane are collinear, the well-known Theorem of Menelaus is an important tool.

Theorem 3.5 (Menelaus' Theorem) *Let $A^{'}$, $B^{'}$, $C^{'}$ be three distinct points in the three extension lines of the three sides of $\triangle ABC$. Then the points $A^{'}$, $B^{'}$, $C^{'}$ are collinear if and only of* $\frac{BA^{'}}{A^{'}C} \cdot \frac{CB^{'}}{B^{'}A} \cdot \frac{AC^{'}}{C^{'}B} = 1$

Because the Menelaus' Theorem is a well-known theorem, we hence omit the proof (see Benitez 2007; *Grunbaum and Shephard* 1955; *Klamkin and Liu* 1992).

The application of Menelaus' Theorem can be shown in the following example. This example is one of the IMO examination questions in geometry (The 35th IMO problem in Hong Kong, 1994).

Example 3.6 Let $\triangle ABC$ be an isosceles triangle with $AB = AC$ and M is the mid point of BC. Suppose that O is a point in the extension of line AM. Draw $OB \perp AB$. Let Q be an arbitrary point in BC which is a distinct point from the points B and C. Also, let E be a point in AB and F a point in AC such that E, Q, F are collinear (Diagram 3.6).

Prove that OQ is perpendicular to EF if and only if $QE = QF$.

Proof We first draw the following Diagram 3.6.

In Diagram 3.6, we observe that the four points O, E, B, Q are concyclic points because $\angle EBO = \angle EQO = 90°$.

Thus, $\angle OEQ = \angle OCQ = \angle OFQ = \angle OEQ$ and hence $OE = OF$.

Diagram 3.6 .

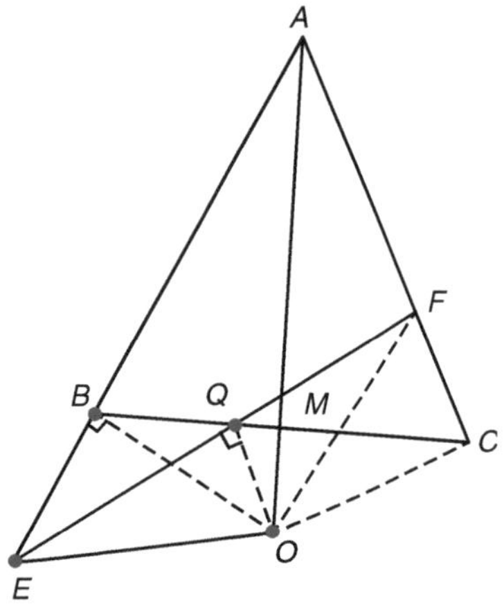

Conversely, we suppose that $EQ = QF$ and recall that $AB = AC$. In $\triangle AEF$ with the cutting line BQC. We apply the Theorem of Menelaus, and we immediately obtain $1 = \frac{AB}{BE} \cdot \frac{EQ}{QF} \cdot \frac{EC}{AC} = \frac{FC}{BE}$, that is, $BE = CF$.

Thus the right angled $\triangle OBE$ is congruent to the right angled $\triangle OCF$, consequently $OE = OF$ and therefore $OQ \perp EF$.

The proof is completed.

It is well known that Menelaus Theorem is a powerful tool in solving problems involving collinear points. We present below another application of this theorem in the following example.

Example 3.7 (IMO Preliminary contest problem of Hong Kong 2011) In the following diagram the circle DEF is an inscribed circle inside the $\triangle ABC$ (Diagram 3.7).

Join lines BE and CF to meet the inscribed circle at P and Q respectively. Prove that the points R, P, Q are collinear.

Proof Because F and E are the tangent points of the inscribed circle of $\triangle ABC$, we have $AE = AF$. Because EFR is also a cutting line of $\triangle ABC$. By applying the theorem of Menelaus, we have immediately $\frac{AF}{FB} \cdot \frac{BR}{RC} \cdot \frac{CE}{EA} = 1$, that is, $\frac{BR}{RC} = \frac{EA}{CE} \cdot \frac{FB}{AF} = \frac{FB}{CE}$.

Since BE and CF meet at S, and given that $\triangle EFC \sim \triangle QEC$, $\triangle FEB \sim \triangle PFB$, $\triangle SEQ \sim \triangle SFP$, we have $\frac{CQ}{EQ} = \frac{CE}{EF} \cdot \frac{FP}{PB} = \frac{FE}{FB}$, $\frac{SP}{SQ} = \frac{FP}{EQ}$.

Consider $\triangle SBC$ and the points R, P, Q on its sides CB, SB and SC, we have

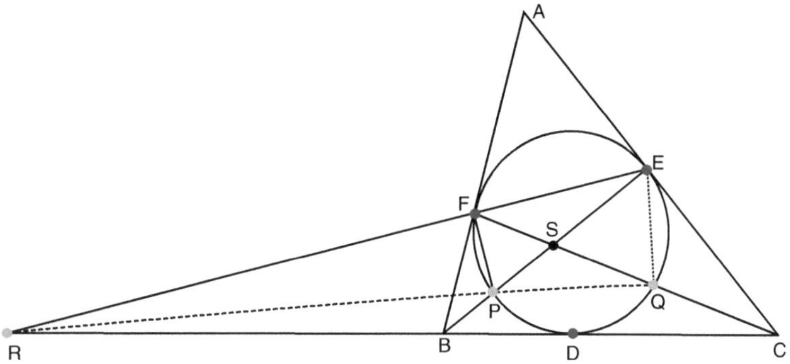

Diagram 3.7 .

$$\frac{SP}{PB}\cdot\frac{BR}{RC}\cdot\frac{CQ}{QS}=\frac{SP}{SQ}\cdot\frac{CQ}{PB}\cdot\frac{BR}{RC}=\frac{FP}{EQ}\cdot\frac{CQ}{PB}\cdot\frac{FB}{CE}=\frac{FP}{PB}\cdot\frac{CQ}{QE}\cdot\frac{FB}{CE}$$
$$=\frac{FE}{FB}\cdot\frac{CE}{EF}\cdot\frac{FB}{CE}=1.$$

Then, by the converse part of Menelaus' Theorem, we know that P, Q, R are collinear points.

Now, we mention the Simpson line theorem, which is frequently used to solve IMO problems in plane geometry. The following theorem is the well-known Simpson line theorem.

Theorem 3.8 *Let P be a point in the circumcircle of a* $\triangle ABC$ *which is distinct from points A, B, C. Draw the perpendicular lines from the point P meeting the lines AB, BC and CA in N, L and M respectively. Then, the points L, M, N are collinear.*

Proof There are many methods to prove this theorem, we provide two proofs (see Diagram 3.8).

(I) Draw Diagram 3.9. Join the points L, M, N, P, B, P, A and P, C. Then we see that $\angle PMN=\angle PAN=\angle PAB=\angle PCB=\angle PCL$ because P, N, A, M are concyclic on the circle. Also, we notice that P, M, C, L four appoints are concyclic on the circle; we have $\angle DML=\angle PCL$, that is L, N, M are collinear.

(II) In Diagram 3.9, we let $\angle PBC=\alpha$, $\angle PCB=\beta$, $\angle PCM=\gamma$. Then, we have $\angle PAM=\alpha$, $\angle PAN=\beta$, $\angle PBN=\gamma$. Clearly, $BL=PB\cdot cos\alpha$; $LC=PC\,cos\beta$; $CM=PC\,cos\gamma$, $MA=PA\cdot cos\alpha$, $AN=PA\cdot cos\beta$, $NB=PB\cdot cos\gamma$

Diagram 3.8 .

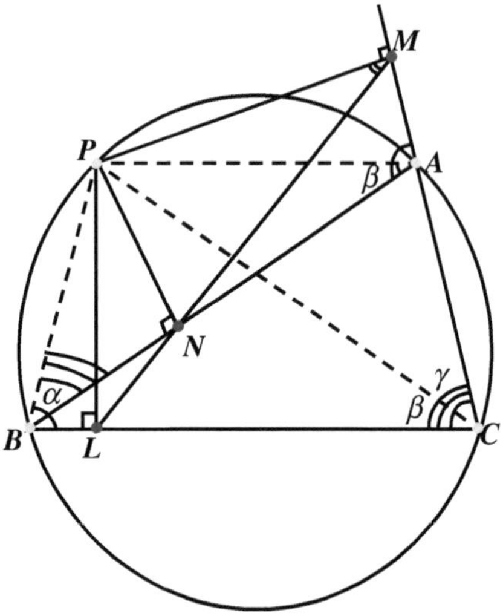

Diagram 3.9 .

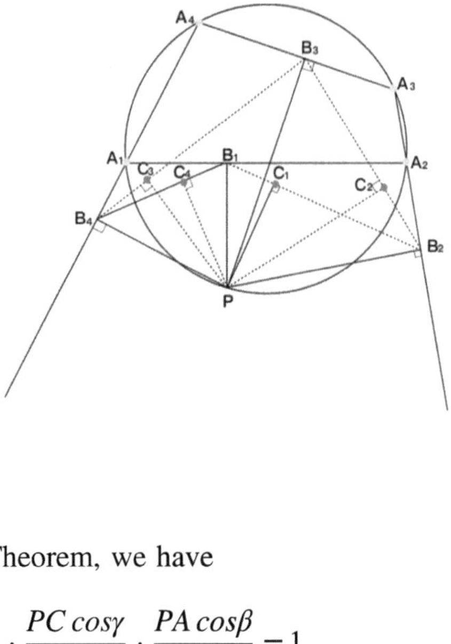

Now, in $\triangle ABC$, by use Menelaus Theorem, we have

$$\frac{BL}{LC}\cdot\frac{CM}{MA}\cdot\frac{AN}{NB}=\frac{PB\,cos\theta}{PC\,cos\alpha}\cdot\frac{PC\,cos\gamma}{PA\,cos\alpha}\cdot\frac{PA\,cos\beta}{PB\,cos\gamma}=1$$

Thus, by the converse part of Menelaus Theorem, we know that the perpendicular foot points L, N, M are collinear.

Remark 3.9 The converse of Simpson line theorem also holds.

Proof The proof is trivial. Just let P be a point not in $\triangle ABC$. From the point P draw perpendicular lines meeting AB, BC and CA at the points N, M and L, respectively. Then we see that P, B, L, N and P, N, A, M are concyclic points. Finally, we notice that $\angle PBC = \angle PBL = \angle PNM$. Then, we conclude that the four points P, B, C, A are concyclic. Therefore, the point P must be on the circumcircle of $\triangle ABC$. The proof is completed.

As an application of Simpson line theorem, we give the following example.

Example 3.9 Let P be a point on the circumcircle of quadrilateral $A_1A_2A_3A_4$. From the point P draw the perpendicular to the lines A_1A_2, A_2A_3, A_3A_4, A_4A_1 at the points B_1, B_2, B_3, B_4, respectively. Also, let the projections of the point P on the lines B_1B_2, B_2B_3, B_4B_4 and B_4B_1 be C_1, C_2, C_3, C_4, respectively. Prove that the four points C_1, C_2, C_3, C_4 are collinear.

Proof We first draw Diagram 3.9.

Through the point P we draw the perpendicular lines to meet the lines A_1A_2, A_3A_2, A_3A_4, A_4A_1, at B_1, B_2, B_3, B_4

Draw the line through the point P perpendicular to the line B_2B_3, we get C_2.
Draw the line through the point P perpendicular to the line B_1B_2, we get C_1.
Draw the line through the point P perpendicular to the line B_4B_1, we get C_4.
Draw the line through the point P perpendicular to the line B_4B_3, we get C_3.

In $\Delta A_1A_2A_3$, the Simpson line through the point is the line B_3QB_4 where Q is the foot point of the perpendicular line through $P \perp A_1A_3$. Similarly, the point P is a point on the Simpson line B_3QB_4 in $\Delta A_1A_2A_3$.

Since $\angle A_1B_4P = \angle A_1BP_1$, we see that the point P is on the circumcircle of ΔQB_1B_4.

Hence, by Simpson line theorem, we see that the points C_1, C_3, C_4 are collinear. Similarly, the three points C_1, C_2, C_4 are collinear. Thus, the four points C_1, C_2, C_3, C_4 are collinear.

The following exercises are applications of Simpson line theorem.

Exercise 3.10 From the vertex A of $\triangle ABC$, draw the perpendicular lines to meet the internal and external angle bisectors at the points F, G, E, D. Prove that these four points are collinear.

Hint: We first draw the diagram. Then, we extend the lines BE and CD to meet at the point K. Suppose that CG and BE meet at I, with $\angle CKI = 90° - \angle CIK = 90° - \left(\frac{1}{2}\angle B + \frac{1}{2}\angle C\right) = \frac{1}{2}\angle A$. The four points A, I, C, K are concyclic. Apply the Simpson line theorem to the $\triangle ICK$ and the point

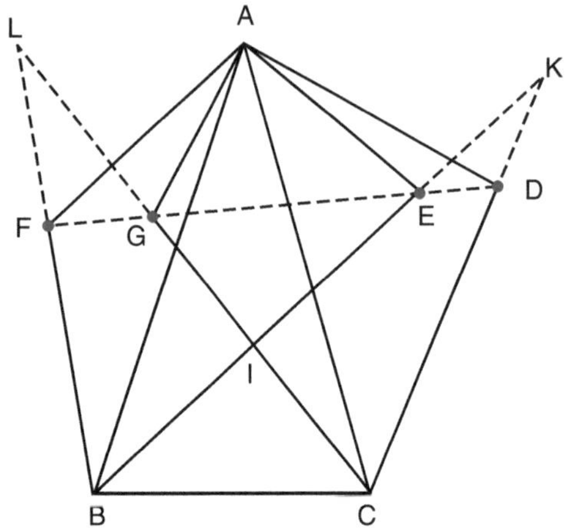

Diagram 3.10 .

A. Then we see that G, E, D three points are collinear. Similarly, for $\triangle BCL$ and the point A, by Simpson line theorem, we know *F*, *G*, *E* are collinear. Hence, the four points *F*, *G*, *E*, *D* are collinear (Diagram 3.10).

Exercise 3.11 (A part of an IMO problem) Let $\triangle ABC$ be an acute-angled triangle. *CD* is the altitude passing through the point C, *M* is the midpoint of the side *AB*. The line passing through the point M meets CA and CB at K and L, respectively with CK = CL. If *S* is the circumcenter of $\triangle CKL$. Prove that $SD = SM$.

We first draw the circumcenter of the $\triangle ABC$. Join the points C, S and extend CS to meet the circumcircle of $\triangle ABC$ at the point T. Join the points T and M.

Construct $TK' \perp AC$ at the point K' and construct $TL' \perp BC$ at L' (Diagram 3.11).

Notice that S is the circumcenter of $\triangle KLC$ and KC = KL. Therefore CS is the angle bisector of $\angle KCL$, and hence T is the mid point of the arc $\widehat{AB}$.

Further, we recall that M is the midpoint of AB. Then $TM \perp AB$. Apply the Simpson line theorem, we know that K', M, L' three points are collinear. Because CT is also the angle bisector $\angle K'L'M$ and the points K', L', M are collinear, then we have $CK' = CL'$. This implies that the line $K'ML'$ is on the perpendicular line passing through the point M to CT.

Likewisely, the line KML is also a perpendicular line passing through the point M to CS. Hence, the points K' and K are coincide, the points L' and L are coincide. That is, $\angle CKT = \angle CLT = 90°$. This means that the four points C, K, T, L are concyclic. (This is an important step).

Diagram 3.11 .

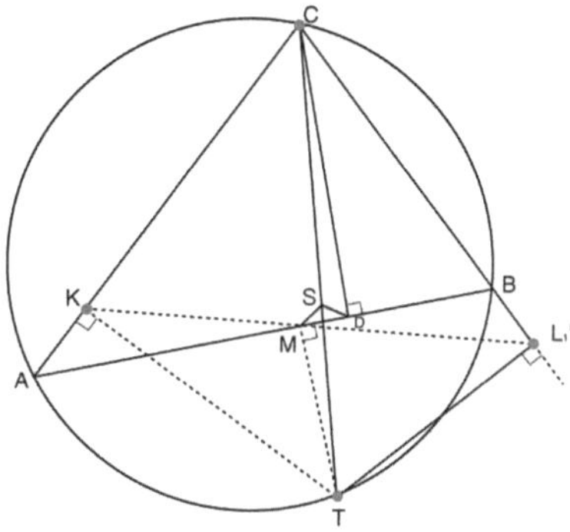

By the above result, we know that S in the center of the circumcircle of the quadrilateral CKTL. We obtain SC = ST. Hence, S is the midpoint of the line TC. As $CD \perp AB$, we have $CD \,//\, MT$. Therefore, we have proved that $SM = SD$. The proof is completed.

3.4 Some Applications of Quadratic Equations

The method of solving quadratic equations is taught in the course of elementary algebra in most high schools. In this section, we introduce the method of solving plane geometry problems by using Vieta's formulas and the discriminants of quadratic equations.

We start with the following example.

Example 4.1 Given a line $l\colon y = 4x$. The point $P\,(6, 4)$ is in the first quadrant of the XY-plane. Through the point P, draw a line meeting the line $l\colon y = 4x$ at the point $Q\ (x_1, y_1)$ in l_2 and the x-axis at M. Then, draw ΔOMQ. When the area of ΔOMQ is minimum, where should be the point Q be located?

Solution: In order to locate the point Q, we need to find the coordinates of the point Q, that is $Q(x_1, y_1)$ (Diagram 3.12).

The area of ΔOMQ varies according to the position of the point Q, therefore we need to express the area of ΔOMQ in terms of the coordinates of Q. Thus, we have to express the area of the ΔOMQ as a function of y, the Y-coordinate of the point Q.

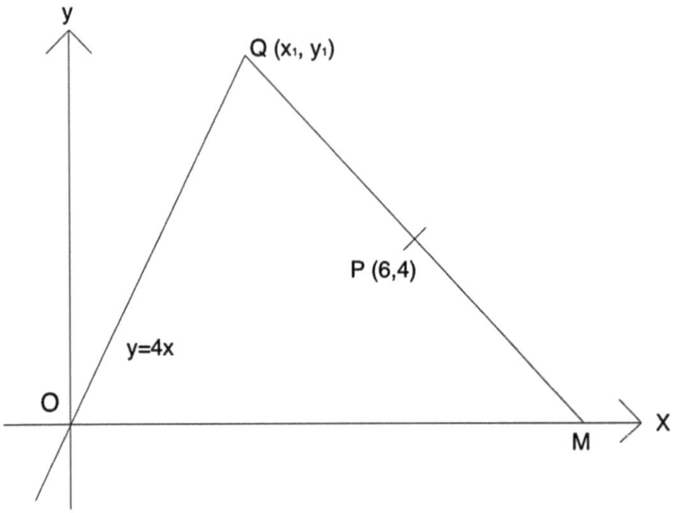

Diagram 3.12 .

Let $Q=(x_1,y_1)$. Since the point Q is on the line $l\colon y=4x$. Therefore, we have $y_1=4x_1$.

For the line QP, the equation of the line passing through the point $P\,(6,4)$ is $(x_1-6)(y-4)=(y_1-4)(x-6)$. Since the point P is in the first quadrant of the XY-plane, $y_1>4$. The intersection point M of line PQ and the x-axis is now M: $\left(\frac{5y_1}{y-4},0\right)$, with $y_1>4$. Therefore, $\frac{5-y_1}{y-4}>0$. Hence the area of ΔOMQ is $S=\frac{1}{2}\cdot\frac{5-y_1^2}{y_1-4}$, that is, $2S(y_1-4)=5y_1^2$, and so we have

$$5y_1^2-2Sy_1+8S=0. \tag{3.1}$$

Clearly, the discriminant of the quadratic Eq. (3.1) is $\Delta=4S^2-160S>0$. Because S > 0, we have $S\geq 40$. Cleraly, when S = 40, the area of ΔOMQ is a minimum. Hence by Eq. (3.1), we get $y_1=8$ which satisfies $y_1>4$. Putting $y_1=8$, we find $x_1=2$.

Thus, the area of ΔOMQ is a minimum at the point $Q\,(2,8)$.

Example 4.2 Let λ_1,λ_2 be the two real roots of the quadratic equation

$$\lambda^2+\lambda x+y=0$$

Suppose that M_1,M_2 are the two real roots of the quadratic equation

Diagram 3.13 See Encyclopedia of the Solutions of Mathematics Problem (1983), problem 168

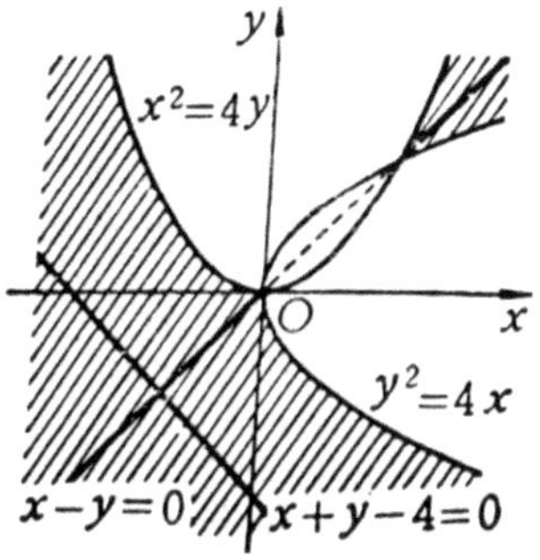

$$M^2+My+x=0$$

If $|\lambda_1-\lambda_2|=|M_1-M_2|$, find the locus of the point (x,y).

Solution: Clearly $|\lambda_1-\lambda_2|=\sqrt{(\lambda_1+\lambda_2)^2-4\lambda_1\lambda_2}=\sqrt{\lambda^2-4y}$;

$$|M_1-M_2|=\sqrt{(M_1+M_2)^2-4M_1M_2}=\sqrt{y^2-4x}.$$

Because we require that λ_1,λ_2 both be real roots, the discriminant of $\lambda^2+\lambda x+y=0$ is $\Delta=yx^2-4y\geq 0$. Similarly, the discriminant of $M^2+My+y=0$ is $\Delta=y^2-4x\geq 0$.

By $|\lambda_1-\lambda_2|=|M_1-M_2|$, we have $x^2-4y=y^2-4x\geq 0$.

Thus, we have $(x-y)(x+y+4)=0$ and so $x-y=0$ or $(x+y+4)=0$.

This means that the locus of the point (x,y) is the straight lines. $x-y=0$ and $x+y+4=0$ with $x+y+4=0$ not in the first quadrant of the XY-plane (Diagram 3.13).

The following example is a typical application of the quadratic equations for the parabola and a straight line.

Example 4.3 Find the area of the triangle bounded by the three straight lines

$$Ax^2+2Bxy+cy^2=0$$
$$lx+my+n=0.$$

Solution: It is well known in plane geometry that the following equation

$$Ax^2+2Bxy+cy^2=0 \tag{3.2}$$

is the expression of two straight lines passing through the point of origin of the XY-plane. Suppose that the line

$$l\colon lx+my+n=0 \tag{3.3}$$

meets these two lines passing through the origin at the points $P(x_1, y_1)$ and $Q=(x_2, y_2)$. When $m \neq 0$, then from (3.3), we have

$$y=-\frac{(lx+n)}{m}. \tag{3.4}$$

Hence, we find $y_1=-\frac{lx_1+n}{m}, y_2=-\frac{lx_2+n}{m}$.
Let the area of the required triangle be denoted by $S_{\Delta PQR}$.
Then, we deduce that

$$\begin{aligned} S_{\Delta PQR} &= \frac{1}{2}\begin{vmatrix} x_1 & y_1 & 0 \\ x_2 & y_2 & 0 \\ 0 & 0 & 1 \end{vmatrix} = \frac{1}{2}(x_1y_2x_2y_2) \\ &= \frac{1}{2}\left(\frac{x_2(lx_1+n)}{m}-\frac{x_1(lx_1+n)}{m}\right)=\frac{1}{2}\left(\frac{n}{m}(x_2-x_1)\right) \end{aligned} \tag{3.5}$$

Put (3.4) in (3.2), after simplification, we get

$$\left(Am^2-2Bml+cl^2\right)x^2-2n(Bm-cl)x+cn^2=0 \tag{3.6}$$

Because (3.6) in a quadratic equation, we can suppose x_1 and x_2 are its two roots.

By Vieta's formulas, we have

$$x_1+x_2=\frac{2n(Bm-cl)}{Am^2-2Bml+cl^2}$$

$$x_1\cdot x_2=\frac{cn^2}{\left(Am^2-2Bml+cl^2\right)^2}$$

Hence, we obtain

$$|x_1-x_2|=\frac{2|mn|\sqrt{H^2-AB}}{|Am^2-2Bml+cl^2|}$$

Putting this result in (3.5), we have

$$S_{\Delta POR} = \frac{n^2\sqrt{B^2 - AC}}{(Am^2 - 2Bml + cl^2)} \tag{3.7}$$

When $m=0$, then $l \neq 0$ and so by (3.3), we have

$$x = -\frac{n}{l} \tag{3.8}$$

$x_1 = x_2 = -\frac{n}{l}$. Thus, the area of ΔPQR is

$$S_{\Delta PQR} = \frac{1}{2}\left(\frac{n}{l}(y_1 - y_2)\right) \tag{3.9}$$

After rearrangement, we obtain

$$Cl^2y^2 - 2Bnly + 4 = 0 \tag{3.10}$$

Suppose the two roots of the quadratic Eq. (3.10) are y_1 and y_2.

Thus, by Vieta's formulas, we have $y_1 + y_2 = \frac{2Bn}{Cl}$, $y_1 \cdot y_2 = \frac{An^2}{Cl^2}$.

Therefore $y_1 - y_2 = \frac{2|n|\sqrt{B^2 - AC}}{|cl|}$

Putting the above result into (3.1), we obtain $S_{\Delta POQ} = \frac{n^2\sqrt{B^2 - AC}}{|cl^2|}$

This formula corresponds to the case $m = 0$ in (3.5). Hence the required area of ΔPOQ is

$$\frac{n^2\sqrt{B^2 - AC}}{(Am^2 - 2Bml + cl^2)}$$

In the following example, we consider parabolas.

Example 4.4 Let the parabola $y^2 = 2px$ meet the line $y = 2x + 1$ at two points $A(x_1, y_1), B(x_2, y_2)$. With $|AB| = \sqrt{15}$. $(2x+1)^2 = 2px$, that is, $4x^2 + 4x + 1 = 2px$.

Hence we have

$$4x^2 + 2(2 - p)x + 1 = 0. \tag{3.11}$$

Because the discriminate of the above quadratic equation is $\Delta = 4(2-p)^2 - 16 > 0$, we must have $p > 4$ or $p < 0$. Using Vieta's formulas, we have

Diagram 3.14 .

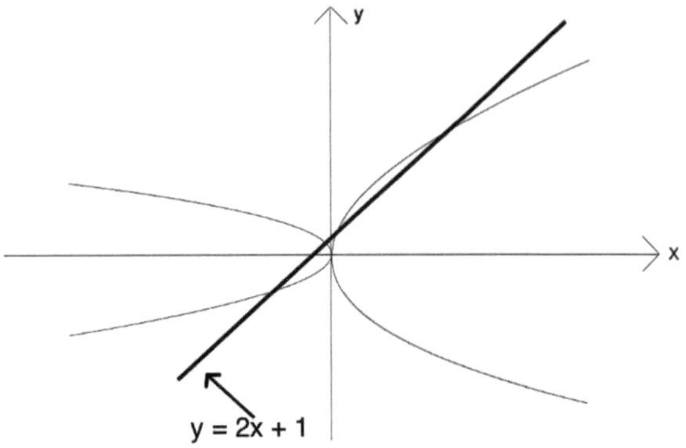

$$x_1 + x_2 = \frac{1}{2}(p-2) \tag{3.12}$$

$$x_1 x_2 = \frac{1}{4}. \tag{3.13}$$

Therefore, we deduce that

$$(x_1 - x_2)^2 = (x_1 + x_2)^2 - 4x_1 x_2 = \frac{1}{4}(p-2)^2 - 1.$$

Because $y_\lambda = 2x_\lambda + 1(\lambda = 1, 2)$, we have $(y_1 - y_2)^2 = 4(x_1 - x_2)^2$.

Also, $\sqrt{(x_1 - x_2)^2 + (y_1 - y_2)^2} = \sqrt{15}$, so that $\sqrt{5(x_1 - x_2)^2} = \sqrt{15}$, that is, $\frac{1}{4}(p-2)^2 - 1 = 3$, the roots of the equation are $p = -2$ or $p = 6$. Therefore, the parabola is $y^2 = -4x$, or $y^2 = 12x$ (Diagram 3.14).

To find the coordinates of the intersection points, we usually first solve a number of parametric equations. If we apply Vieta's formulas for the quadratic equation, we can sometimes simplify the tedious calculations. We present the following example.

Example 4.5 Suppose that the parabola $y^2 = 4ax$ meets the straight line $lx + my + c = 0$ at the points P, Q, and that F is the focus of the parabola. If lines PF and QF meet the parabola $y^2 = 4ax$ at points R and S, respectively, find the equation of the line RS (Diagram 3.15).

Solution:

Because the points P, Q are on the parabola, we may assume that the coordinates of the point P, Q are $P(at_1^2, 2at_1)$, $Q(at_2^2, 2at_2)$. The coordinates

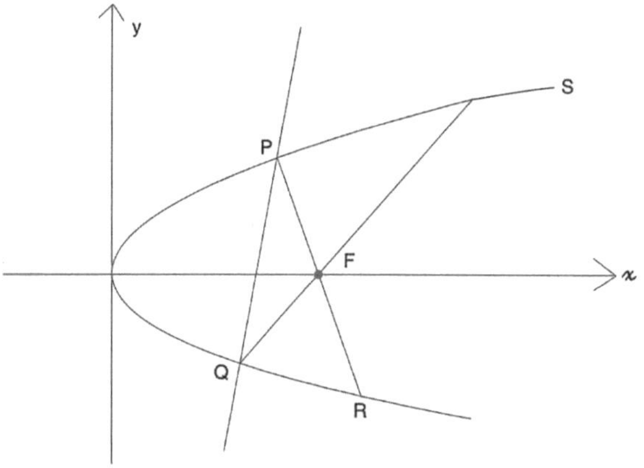

Diagram 3.15 .

of the point R, S be $(at_1'^2, 2at_1')$, $(at_2'^2, 2at_2')$. Then the equation of the line *PR* is

$$y - 2at_1 = \frac{2a(t_1 - t_1')}{a(t_1^2 - t_1'^2)} \cdot (x - at_1^2).$$

That is, $2x - (t_1 + t_2')y + 2at_1t_1' = 0$

Since *PR* passes through the focus point $F(a, 0)$, we have $2a + 2at_1t_1' = 0$, hence $t_1' = -\frac{1}{t_1}$, similarly, $t_2' = -\frac{1}{t_2}$.

Therefore, we find the coordinates of *R*, *S* are $\left(\frac{a}{t_1^2}, \frac{2a}{t_1}\right) \cdot \left(\frac{a}{t_2^2}, \frac{-2a}{t_2}\right)$.

Observe that the points *P*, *Q* are on the straight line $lx + my + n = 0$, hence we have $lat_1^2 + 2amt_1 + n = 0, lat^2 + 2amt_2 + n = 0$.

This means that t_1, t_2 are the two roots of the quadratic equation $al\lambda^2 + 2am\lambda + n = 0$. Apply Vieta Theorem again, we have $t_1 + t_2 = \frac{-2m}{l}$, $t_1t_2 = \frac{n}{al}$. If the equation of the required straight line RS is $Ax + By + c = 0$, then $\frac{Aa}{t_1^2} - \frac{2aB}{t_1} + c = 0, \frac{Aa}{t_2^2} - \frac{2aB}{t_2} + c = 0$. Let t_1, t_2 be the two roots of the equation $cu^2 - 2aBM + Aa = 0$.

Then, we have $t_1+t_2=\frac{2aB}{c}$, $t_1t_2=\frac{Aa}{c}$, and consequently, $\frac{-2m}{l}=\frac{2aB}{c}$, $\frac{n}{al}=\frac{Aa}{c}$, that is, $\frac{B}{c}=-\frac{m}{la}$, $\frac{A}{C}=\frac{n}{la}$.

Thus, the equation of the straight line RS is $\frac{nx}{la^2}-\frac{my}{la}+1=0$, that is $nx-may+la^2=0$.

Now, we mention again the usage of the discriminant of a quadratic equation in the following example.

Example 4.6 Suppose that the length of the chord AB formed by the line $y=2x+k$ cutting the parabola $y^2=4x$ at $3\sqrt{5}$. If P is a point on the x-axis which forms an equilateral triangle PAB and the area of $\Delta PAB=39$, find the coordinates of the point P (Diagram 3.16).

Solution. Let $A=(x_1,y_1), B=(x_2,y_2)$. Then $y_1=2x_1+k, y_2=2x_2+k$.

Then, substituting $y=2x+k$ into $(2x+k)^2=4x$, we get $4x^2+4(k-1)x+k^2=0$.

Now, the discriminant of the above equation is $\Delta=16(k-1)^2-16k^2>0$, that is, $k<\frac{1}{2}$. If x_1,x_2 are the two roots of the quadratic equation, by Vieta's Theorem, we have $x_1+x_2=1-k$, $x_1x_2=\frac{k^2}{4}$ and $y_1-y_2=2(x_1-x_2)$.

Hence $(x_1-x_2)^2+(y_1-y_2)=(3\sqrt{5})^2$. Thus, we have $5(x_1-x_2)^2=45$, so that $1-2k=9, k=-4$.

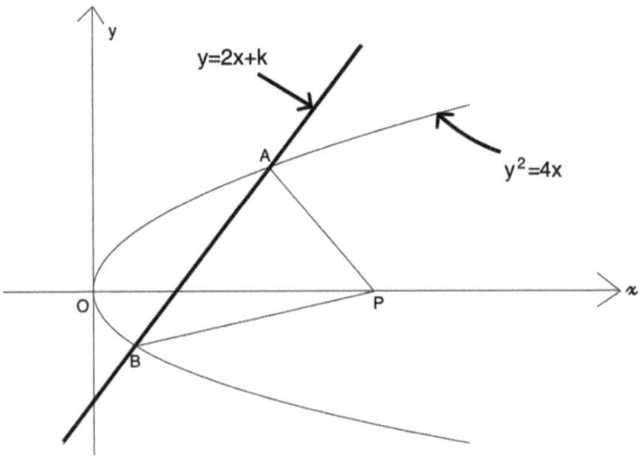

Diagram 3.16 .

Now, let $P=(x,0)$, then the distance from P to AB is $d=\frac{|2x-4|}{\sqrt{5}}$.

Hence $\frac{1}{2}\cdot\frac{|2x-4|}{\sqrt{5}}\cdot 3\sqrt{5}=39$, from which we get $|x-12|=13$, and $x=15$ or $x=-11$. Therefore, $P=(15,0)$ or $(-11,0)$.

Remark In this example, we consider the chord AB formed by the line $y=kx+m$ to meet the quadratic curve $y^2=4x$. Hence the length of the chord AB is

$$l=\sqrt{(x_1-x_2)^2+(y_1-y_2)^2}=\sqrt{(x_1-x_2)^2+k^2(x_1-x_2)^2}$$

$$=\sqrt{Hk^2}\cdot\sqrt{(x_1+x_2)^2-4x_1x_2}=\sqrt{(Hk)^2}\sqrt{\left(\frac{-b}{k}\right)^2-\left(\frac{4c}{a}\right)}=\sqrt{Hk^2}\cdot\frac{\sqrt{\Delta}}{a}$$

Thus, to find the length of the chord AB, we only need to use the coefficient k, the coefficients of the quadratic equation a, b and the value of the discriminant, so there is no need to first find out the coordinates of the intersection points.

3.5 Ceva's Theorem and Its Application

In solving problems in plane geometry, Ceva's Theorem is often used. In this section, we briefly introduce Ceva's Theorem and its application (see Benitez 2007; Grunbaum and Shephard 1955).

Theorem 5.1 *Let A', B', C' be three points on the sides BC, AC, AB, respectively, of ΔABC or on their extension lines. Then*

$$\frac{BA'}{A'C}\cdot\frac{CB'}{B'A}\cdot\frac{AC'}{C'B}=1.$$

Proof There are several situations of the theorem; we first draw the following three diagrams.

In Diagrams 3.17b, If AA', BB', CC' meet at the point P, then we draw a line through A parallel to BC which meets BB', CC' on their extension lines at D, E, respectively. Now, it is clear that $\frac{CB'}{B'A}=\frac{BC}{AD}$ and $\frac{AC'}{C'B}=\frac{EA}{BC}$.

Also, from $\frac{BA'}{AD}=\frac{A'P}{PA}=\frac{A'C}{EA}$, we have $\frac{BA'}{A'C}=\frac{AD}{EA}$.

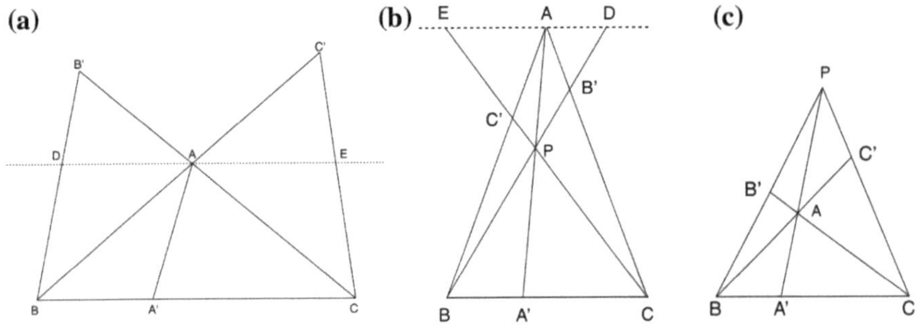

Diagram 3.17 .

Consequently, we have $\frac{BA'}{A'C}\cdot\frac{CB'}{B'A}\cdot\frac{AC'}{C'B}=\frac{AD}{EA}\cdot\frac{BC}{AD}\cdot\frac{EA}{BC}=1.$

If, AA', BB' and CC' are parallel lines, the theorem can be proved similarly.

Remark 5.2 For the Diagram 3.17b and c, Ceva's Theorem can also be proved by using "area". That is by $\frac{BA'}{A'C}\cdot\frac{CB'}{B'A}\cdot\frac{AC'}{C'B}=\frac{S_{\Delta PAB}}{S_{\Delta PCA}}\cdot\frac{S_{\Delta PBC}}{S_{\Delta PAB}}\cdot\frac{S_{\Delta PCA}}{S_{\Delta PBC}}=1.$ We usually call point P the Ceva Point of ΔABC.

Corollary 5.3 *The collinear points of Ceva's Theorem are equivalent to the collinear points of Menelaus' Theorem.*

Proof Consider Diagram 3.17b and c. Then by Menelaus Theorem, the points C', P, C are in the intercept $C'DC$ in $\Delta ABA'$, we have

$$\frac{BA}{CA'}\cdot\frac{A'P}{PA}\cdot\frac{AC'}{C'B}=1 \qquad (a)$$

Secondly, for the intercept $B'PB$ in $\Delta AA'C$, we have

$$\frac{A'B}{BC}\cdot\frac{CB'}{B'A}\cdot\frac{AP}{PA'}=1 \qquad (b)$$

Multiply (a) and (b), we therefore obtain $\frac{BA'}{A'C}\cdot\frac{CB'}{B'A}\cdot\frac{AC'}{C'B}=1.$

Thus, by using Menelaus' Theorem, we can easily prove Ceva's Theorem.

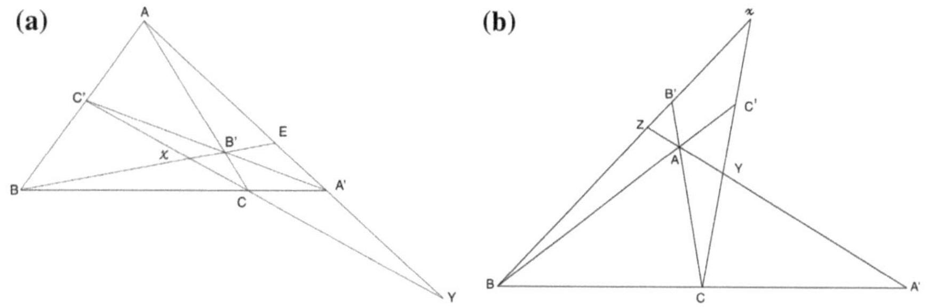

Diagram 3.18 .

Menelaus' Theorem can also be proved by using Ceva's Theorem; we now consider the following diagrams (Diagram 3.18).

In the above two diagrams, we observe the following facts:

i. Let C' be the Ceva point of $\Delta BCB'$. Then by Ceva's Theorem, we have

$$\frac{BA'}{A'C} \cdot \frac{CA}{AB'} \cdot \frac{B'X}{XB} = 1.$$

ii. Let A' be the Ceva point of $\Delta CAC'$. Then by Ceva's Theorem again, we have

$$\frac{CB'}{B'C} \cdot \frac{AB}{BC'} \cdot \frac{C'Y}{YC} = 1.$$

iii. Let B' be the Ceva point of $\Delta ABA'$. Then by Ceva's Theorem, we have

$$\frac{AC'}{B'B} \cdot \frac{BC}{CA'} \cdot \frac{A'Z}{ZA} = 1.$$

iv. Let C be the Ceva point of $\Delta BBC'$. Then by Ceva's Theorem, we have

$$\frac{BX}{XB} \cdot \frac{B'A'}{A'C'} \cdot \frac{C'A}{AB} = 1.$$

v. Let A be the Ceva point of $\Delta CC'A'$. Then by Ceva's Theorem, we have

$$\frac{CY}{YC'} \cdot \frac{C'B'}{B'A'} \cdot \frac{A'B}{BC} = 1.$$

vi. Let B be the Ceva point of $\Delta AA'B$. Then by Ceva's Theorem, we have

$$\frac{AZ}{ZA'} \cdot \frac{A'C'}{C'B} \cdot \frac{B'C}{C'A} = 1.$$

Now, multiplying the above six equations, we get $\left(\frac{BA'}{A'C} \cdot \frac{CB'}{B'A} \cdot \frac{AC'}{C'B}\right)^2 = 1.$

Hence we have proved that $\frac{BA'}{A'C} \cdot \frac{CB'}{B'A} \cdot \frac{AC'}{C'B} = 1$. Thus, Ceva's Theorem is proved. The following example can be regarded as the converse part of Ceva's Theorem.

Remark 5.4 If A', B', C' are three points on the three sides of ΔABC or on their extension lines of ΔABC such that $\frac{BA'}{A'C} \cdot \frac{CB'}{B'A} \cdot \frac{AC'}{C'B} = 1$, then the three lines AA', BB' and CC' are concurrent or mutually parallel.

Proof Suppose that the lines AA' and BB' meet at P. Also, suppose that CP and AB meet at C_1.

Then by Ceva's Theorem, we have $\frac{BA'}{A'C} \cdot \frac{CB'}{C'A} \cdot \frac{AC_1}{C_1B} = 1.$

Now, recall the given condition $\frac{BA'}{A'C} \cdot \frac{CB'}{B'A} \cdot \frac{AC'}{C'B} = 1$, we immediately get that $\frac{AC_1}{C_1B} = \frac{(AC')}{C'B'}$, that is, $\frac{AC_1}{AB} = \frac{AC'}{AB}$. Hence $AC_1 = AC'$. This means that the two points C_1 and C' coincide. Thus, we have proved that the lines AA', BB', CC' are concurrent.

If $AA' // BB'$, then $\frac{CB'}{B'A} = \frac{CB}{B'A}$. Putting this equality into the given condition, we obtain $\frac{AC'}{C'B} = \frac{A'C}{CB}$, and hence $CC' // AA'$, $AA' // BB' // CC'$.

In conclusion, we give the following criterion for concurrent points.

Theorem 5.5 *If A', B', C' are the points in the three sides of ΔABC in BC, CA and AB respectively. Then the three lines AA', BB' and CC' are concurrent or parallel if and only if*

$$\frac{BA'}{A'C} \cdot \frac{CB'}{B'A} \cdot \frac{AC'}{C'B} = 1.$$

In proving that the collinear points on lines are concurrent, Ceva's Theorem is a powerful tool in plane geometry. We now state the following two different forms of Ceva's Theorem (Grunbaum and Shephard 1955*).*

Theorem 5.6 (Ceva's Theorem Form 1) *Let A', B', C' be the three points in the three sides of ΔABC. Then the three lines AA', BB' and CC' are concurrent or parallel if and only if*

$$\frac{\sin\angle BAA'}{\sin\angle A'AC} \cdot \frac{\sin\angle ACC'}{\sin\angle C'C'B} \cdot \frac{\sin\angle CBB'}{\sin\angle B'BA} = 1$$

Proof We first notice that

$$\frac{BA'}{A'C} = \frac{S_{\Delta ABA'}}{S_{\Delta AA'C}} = \frac{AB \cdot \sin\angle BAA'}{AC \cdot \sin\angle B'AC}, \frac{CB'}{B'A} = \frac{BC \cdot \sin\angle CBB'}{AB \cdot \sin\angle B'BA}, \frac{AC'}{C'B} = \frac{AC \cdot \sin\angle ACC'}{BC \cdot \sin\angle C'CB}.$$

By multiplying the above equalities and apply Theorem 5.5, the theorem is proved.

Theorem 5.7 (Ceva's Theorem Form 2) *Let A', B', C' be the points in the three sides of ΔABC and O is a point not in ΔABC. Then the three lines AA', BB' and CC' are concurrent or parallel if and only if*

$$\frac{\sin\angle BOA'}{\sin\angle A'OC} \cdot \frac{\sin\angle AOC'}{\sin\angle C'OB} \cdot \frac{\sin\angle COB'}{\sin\angle B'OA} = 1.$$

Proof Applying the criterion of Ceva's concurrent point Theorem 5.5, we have

$$\begin{aligned} 1 &= \frac{BA'}{A'C} \cdot \frac{CB'}{B'A} \cdot \frac{AC'}{C'B} = \frac{S_{\Delta BOA'}}{S_{\Delta A'OC}} \cdot \frac{S_{\Delta COB'}}{S_{\Delta B'OA}} \cdot \frac{S_{\Delta AOC'}}{S_{\Delta C'OB}} \\ &= \frac{BO \cdot \sin\angle BOA'}{CO \cdot \sin\angle A'OC} \cdot \frac{CO \cdot \sin\angle COB'}{AO \cdot \sin\angle B'OA} \cdot \frac{AO \cdot \sin\angle AOC'}{BO \cdot \sin\angle C'OB'} \end{aligned}$$

Hence, the theorem is proved.

Note In applying Ceva's Theorem above, the reader have to observe whether there are points in the extension lines at the three sides of the ΔABC.

We point out here that Ceva's Theorem has been frequently used to solve many regional and national Olympiad problems. The following are some examples:

Example 5.8 (China MO problem, 1997) In the following Diagram 3.19. The quadrilateral $ABCD$ is inscribed in the circle with the line AB and the extension of the line DC meet at P, and AD and the extension of the line BC meet at the point Q. Construct two tangent lines through the point Q to touch the circle at the points E and F. We prove that the points P, F, E are collinear (Diagram 3.19).

We draw the diagram as shown in Diagram 3.19. Join the point E and F to meet AD, BC at M, N respectively. Also, we let AC to meet BD at K. We first prove the points P, K, M are collinear and the another three points P, N, K are collinear.

Now, we draw the diagram as shown in Diagram 3.19. We need to prove that the lines AC, BD and PM are concurrent. Then by applying the converse of Ceva's Theorem, we need to prove that $\frac{AB}{BP} \cdot \frac{PC}{CD} \cdot \frac{DM}{MA} = 1$.

Because the line QCB cuts ΔPDA, by Menelaus' Theorem, we have $\frac{AB}{BP} \cdot \frac{PC}{CD} \cdot \frac{DQ}{QA} = 1$. We only need to prove that $\frac{DM}{AM} = \frac{DQ}{AQ} = 1$.

Because the line QCB cuts ΔPDA, by Menelaus' Theorem again, we have $\frac{AB}{BP} \cdot \frac{PC}{CD} \cdot \frac{DQ}{QA} = 1$. Thus, we only need to prove that

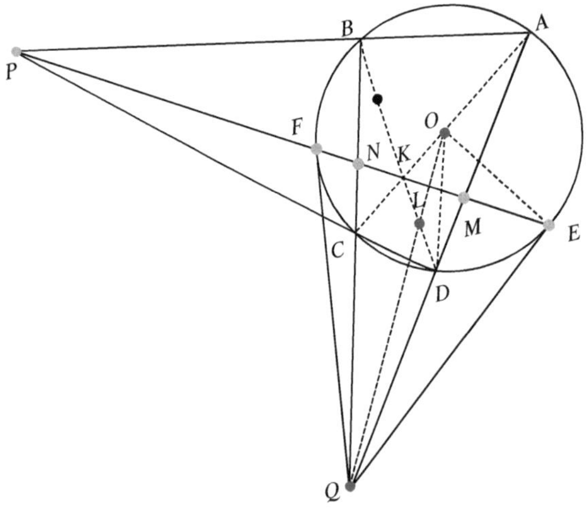

Diagram 3.19 .

$\frac{DM}{AM} \cdot \frac{DM}{MA} = 1.$ Let the center of the circle be O. Join QO to meet EF at L. Also join LD, OD, OA. Then by using the tangent theorem and projection theorem, we immediately have $QD \cdot QA = QE^2 = QL \cdot OD$. Thus, the four points D, L, O, A are concyclic points. Consequently, $\angle DLQ = \angle DAO = \angle ODA = \angle OLA$. This shows that the line QL is the interior angle bisector of $\angle ALD$ and is also the exterior angle bisector of $\angle ALD$ and hence $EF \perp OQ$. Thus, EL is the angle bisector of $\angle ALD$. Thus, we have

$$\frac{DM}{AM} = \frac{DL}{AL} = \frac{DQ}{AQ}.$$

The proof is completed.

Example 5.9 (IMO Problem 1983, Yugoslavia) Let M be a point inside ΔABC such that $\angle MBA = 30°, \angle MAB = 10°$. If $\angle ACB = 80°$ and $AC = BA$, find $\angle AMC$ (Diagram 3.20).

Solution:

Let $\angle ACM = \alpha$. Then $\angle MCB = 80° - \alpha$. By Ceva's theorem Form 1, we immediately have

$$\frac{sin\alpha}{\sin(80° - \alpha)} \cdot \frac{sin10°}{sin40°} \cdot \frac{sin20°}{sin30°} = 1 \tag{3.14}$$

and hence $sin\alpha \cdot sin10° = \sin(80° - \alpha) \cdot \cos20°$.

Therefore, we derive that $2sin\alpha \cdot cos80° = 2\sin(80° - \alpha) \cdot cos20°$.

Thus, we get

$\sin(\alpha + 80°) + \sin(\alpha - 80°) = \sin(100° - \alpha) + \sin(60° - \alpha)$and

$$\begin{aligned}\sin(\alpha - 80°) - \sin(60° - \alpha) &= \sin(100° - \alpha) - \sin(\alpha + 80°)\\ &= 2\cos 90° \cdot \sin(10° - \alpha) = 0.\end{aligned}$$

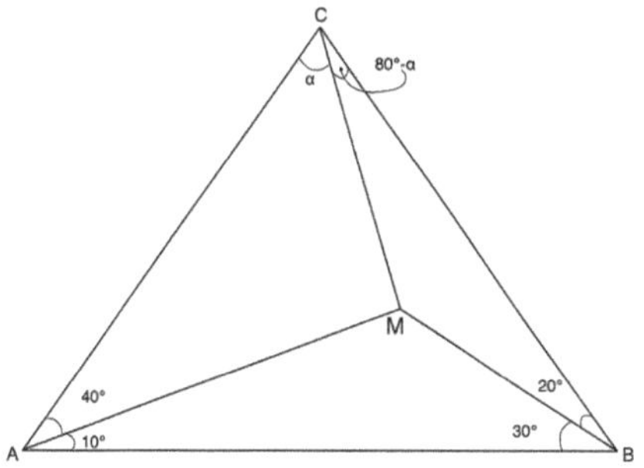

Diagram 3.20 .

We conclude that $\sin(\alpha - 80°) = \sin(60° - \alpha)$.
Now observe that $0 < \alpha < 80°$,
we know that $-80° < \alpha < -80°,\ 60° - \alpha < 60°$.
Hence, $\alpha - 80° = 60° - \alpha$ and so $\alpha = 70°$.
Therefore, we obtain $\angle AMC = 180° - MAC - \angle ACM = 180° - 40° - 70° = 70°$.

Remark This question can be solved directly by using (3.14), that is

$$\frac{\sin\alpha}{\sin 10°} = \frac{\sin 70°}{\sin(80° - \alpha)},$$

where $0 < \alpha,\ 80° - \alpha < 80°$. Then wc obtain $\alpha = 70°$, or by $\frac{sin\alpha}{\sin(80° - \alpha)} \cdot \frac{sin10°}{sin40°} \cdot \frac{sin20°}{sin30°} = 1$, we have

$$\frac{\sin(80° - \alpha)}{\sin\alpha} = \frac{\sin 10°}{\sin 40°} \cdot \frac{\sin 20°}{\sin 30°} = \frac{\sin 10°}{\cos 20°} = \frac{\sin(80° - 70°)}{\sin 70°}$$
$$= \sin 80° \cdot \cos 70° - \cos 80°$$

Because $\frac{\sin(80° - \alpha)}{sin\alpha} = \sin 80° \cdot \cos\alpha - \cos 80°$.
As the function of α is strictly decreasing in the interval $(0°, 180°)$, therefore, $ACM = \alpha = 70$, $\angle AMC = 180° - 40° - 70° = 70°$.
Or for the point C and the ΔMAB, we can apply Ceva's Theorem Form 1 again to get,

$$1 = \frac{sin\angle AMC}{sin\angle CMB} \cdot \frac{sin\angle MBC}{sin\angle CBA} \cdot \frac{sin\angle BAC}{sin\angle CAM} = \frac{sinx}{\sin(220° - x)} \cdot \frac{(sin50°)}{sin50°} \cdot \frac{sin50°}{\sin 40°}.$$

Then,

$$\frac{\sin(220° - x)}{\sin x} = \frac{1}{2cos20°} = \frac{sin(220° - 70°)}{sin70°} = sin220° = cos70° - cos20°$$

Because $\frac{\sin(220° - x)}{\sin x} = sin220° \cdot \cos x - \cos 220° (\sin 220° < 0)$.
As the function of x is strictly decreasing in the interval $(0°, 180°)$, therefore we derive that $\angle AMC = x = 70°$.

Diagram 3.21 .

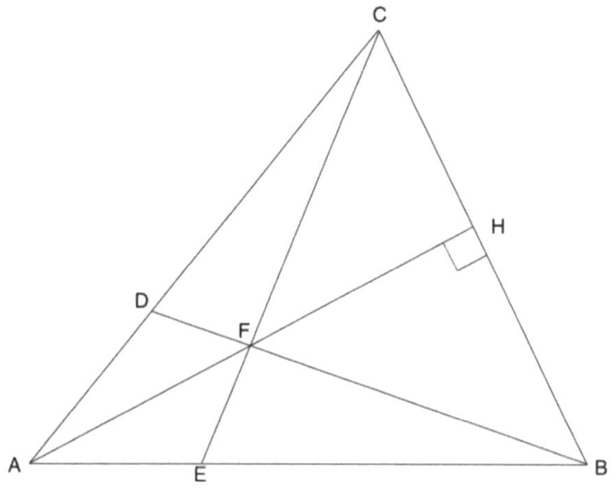

Example 5.10 (A Proposed MO Problem of Canada) In ΔABC, $\angle BAC = 40°$, $\angle ABC = 60°$. D and E are the points on AC and AB such that $\angle CBD = 40°$ *and* $\angle BCE = 70°$. The lines BD and CE meet at F. Prove that $AF \perp BC$.

Proof Let $\angle BAF = \alpha$. Then, $\angle FAC = 40° - \alpha$. Let F be the Ceva point of ΔFAC and apply Ceva's Theorem Form 1, we have $\frac{sin10°}{sin70°} \cdot \frac{sin\alpha}{\sin(40° - \alpha)} \cdot \frac{sin40°}{sin20°} = 1$. Hence, we have $\frac{\sin 10°}{sin70°} \cdot \frac{sin\alpha}{\sin(40° - \alpha)} \cdot \frac{sin40°}{sin20°} = 1$ and so in Diagram 3.21, we have

$$\sin(40° - \alpha) = 2sin\alpha \cdot sin10° = 2sin\alpha \cdot cos80° = \sin(\alpha + 80°) + \tan(\alpha - 80°)$$

It follows that

$$\begin{aligned}\sin(\alpha - 80°) &= \sin(40° - \alpha) - \sin(\alpha + 80°) = 2cos60° \cdot \sin(-20° - \alpha) \\ &= \sin(-20° - \alpha).\end{aligned}$$

Observe that $O < \alpha < 40°$, we know that $-80° < -20° - \alpha$, $-\alpha - 80° < 20°$, so we have $\alpha - 80° = -20° - \alpha$, and $\alpha = 30°$. Extend the line AF to meet BC at H, then $\angle AHB = 180° - \angle FAB - \angle ABH = 180° - 30° - 60° = 90°$.

Thus, we have shown that $AF \perp BC$.

Diagram 3.22 .

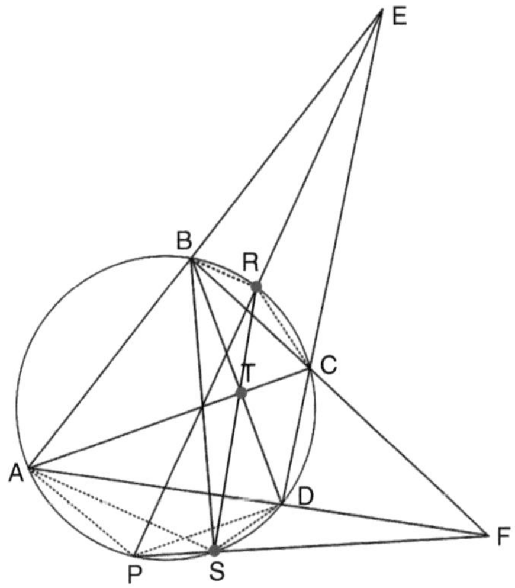

The following example is a Hong Kong IMO preliminary contest problem that required students to know how to apply both the Menelaus' Theorem and Ceva's Theorem.

Example 5.11 Let $ABCD$ be a quadrilateral inscribed in a circle as shown in Diagram 3.22. Prove that R, T, S are collinear.

Hint: By $\Delta EBR \sim \Delta EPA$, $\Delta FDS \sim \Delta FPA$, we have $\frac{BR}{DA} = \frac{EB}{EP}$, $\frac{PA}{DS} = \frac{FP}{FD}$. Multiplying these two equations, we get

$$\frac{BR}{DS} = \frac{EB}{ED} \cdot \frac{FP}{FD}. \tag{3.15}$$

Similarly, by $\Delta ECR \sim \Delta EPD$, $\Delta FPD \sim \Delta FAS$, we derive that

$$\frac{BR}{DS} \cdot \frac{AS}{CR} = \frac{EB}{EC} \cdot \frac{FA}{FD}. \tag{3.16}$$

Multiplying (3.15) and (3.16) by $\frac{DC}{AB}$, we get $\frac{BR}{RC} \cdot \frac{CD}{DS} \cdot \frac{SA}{AB} = \frac{EB}{BA} \cdot \frac{AF}{FD} \cdot \frac{DC}{CE}$.

Now, for the intercept BCF and ΔEAD, applying Menelaus' Theorem, we have $\frac{EB}{BA} \cdot \frac{AF}{FD} \cdot \frac{DC}{CE} = 1$. Therefore, we have $\frac{BR}{RC} \cdot \frac{CD}{DS} \cdot \frac{SA}{AB} = 1$.

Recall Ceva's Theorem Form 1 and its corollary, we know immediately that the three lines *BD*, *RS* and *AC* are concurrent at a point and hence *R*, *T*, *S* are collinear. The proof is completed.

3.6 Ptolemy's Theorem and Stewart's Theorem

Apart from the Simpson line theorem, Menelaus' Theorem, Ceva's Theorem etc., Ptolemy's Theorem, Stewart's theorem and the Erdös-Mordell inequalities are often used as tools in solving problems in plane geometry. In this section, we briefly introduce the proofs of these tools (see Pech 2005; Moise 1990; Some Geometry Problems in Mathematical Olympiad Competitions 2015; Encyclopedia of Solved Problems 2016).

Theorem 6.1 (Ptolemy's Theorem) *If a convex quadrilateral ABCD is inscribed in a circle, then*

$$AB \cdot CD + BC \cdot AD = AC(BP + PD) = AC \cdot BD.$$

Proof It is clear that $\angle PAB = \angle CAD$ and so $\Delta ABP \sim \Delta ACD$.
Therefore

$$\frac{AB}{AC} = \frac{BP}{CD} \Rightarrow AB \cdot CD = AC \cdot BP. \tag{3.17}$$

Similarly since $\Delta ABC \sim \Delta APD$, we have

$$BC \cdot AD = AC \cdot PD. \tag{3.18}$$

Adding (3.17) and (3.18), we have $AB \cdot CD + BC \cdot AD = AC(BP + PD) = AC \cdot BD$

Corollary 6.2 (Euler's Theorem) *A special case of Ptolemy's Theorem is the following. A, B, C, D are four ordered points on a line. Then, by Ptolemy's Theorem, we have* $AB \cdot CD + BC \cdot AD = AC \cdot BD$.

The following example is a typical application of the Ptolemy's Theorem (CMO, China 2000).

Example 6.3 Let ΔABC be an acute-angled triangle. Also, let *E*, *F* be two points on the line *BC* with $\angle BAE = \angle CAF$. Construct the segments $FM \perp AB$ at the point *M* and $FN \perp AC$ at *N*.

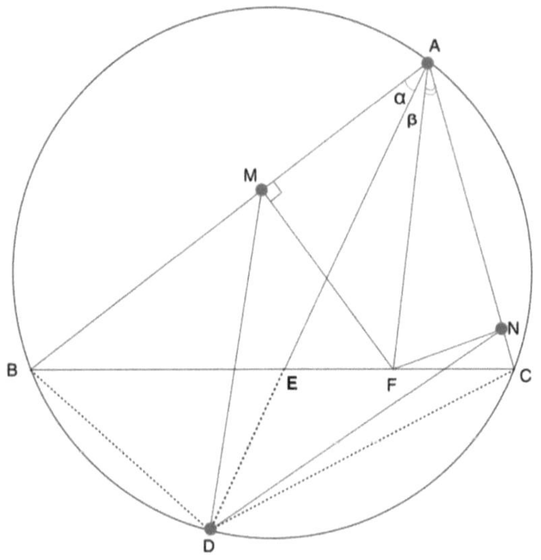

Diagram 3.23 .

The extension of the line AE meets the circumcircle of ΔABC at D (see Diagram 3.23). Prove that the area of $AMDN$ and the area of ΔABC are equal.

Proof Let $\angle BAE = \angle CAF = \alpha,\ \angle EAF = \beta$. Then

$$S_{\Delta ABC} = \frac{1}{2}AB \cdot AF \cdot \sin(\alpha+\beta) + \frac{1}{2}AC \cdot AF \cdot sin\alpha = \frac{AF}{4R}(AB \cdot CD + AC \cdot BD),$$

where R is the radius of the circumcircle.

Now, we also find that

$$\begin{aligned} S_{AMDN} &= \frac{1}{2}AM \cdot AD \cdot sin\alpha + \frac{1}{2}AD \cdot AN \cdot \sin\alpha + \beta \\ &= \frac{1}{2}AD[AF \cdot \cos(\alpha+\beta) \cdot sin\alpha + AF \cdot cos\alpha \cdot \sin(\alpha+\beta)] \\ &= 1/2\, AD \cdot AF \cdot \sin(2\alpha+\beta) = \frac{4F}{4R}AD \cdot BC. \end{aligned}$$

Applying Ptolemy's Theorem, we have $AB \cdot CD + AC \cdot BD = AD \cdot BC$ and hence $S_{AMDE} = S_{\Delta ABC}$.

Theorem 6.4 (Stewart's Theorem) *Let P be an arbitrary point on the side BC of* ΔABC. *If point P is distinct from points A and C, then*

$$AB^2 \cdot PC + AC^2 \cdot BP = AP^2 \cdot BC + BP \cdot BP \cdot BC \tag{3.19}$$

or

$$AP^2 = AB^2 \cdot \frac{PC}{BC} + AC^2 \cdot \frac{BP}{BC} - BC^2 \cdot \frac{BP}{BC} \cdot \frac{PC}{BC}. \tag{3.20}$$

For the sake of convenience, let a, b, c be the three sides of a ΔABC *opposite to* $\angle A, \angle B, \angle C$, *respectively. Also, we let d be the Ceva point lying in the side BC which divides a into the ratio* $m{:}\,n$ with $b^2m^2 + c^2n = a\,(d^2 + mn)$.

[After rearrangement, we can write the above expression as $man + dad = bmb + cnc$. "A man and his dad put a bomb into the sink", a form which invites mnemonic memorization. This expression yields a relation between the lengths of the sides of the triangle].

Proof Without loss of generality, we may assume that $\angle APC < 90°$. Then by using the Cosine Law, we have (Diagram 3.24)

$$\begin{aligned} AC^2 &= AP^2 + PC^2 - 2AP \cdot PC \cdot cos\angle APC, \\ AB^2 &= AP^2 + BP^2 - 2AP \cdot BP \cdot cos(180° - \angle APC) \\ &= AP^2 + BP^2 + 2AP \cdot BP \cdot cos\angle APC \end{aligned}$$

Multiplying (3.19) and (3.20) by *BP*, *PC* and adding up, we have proved the theorem.

Diagram 3.24 .

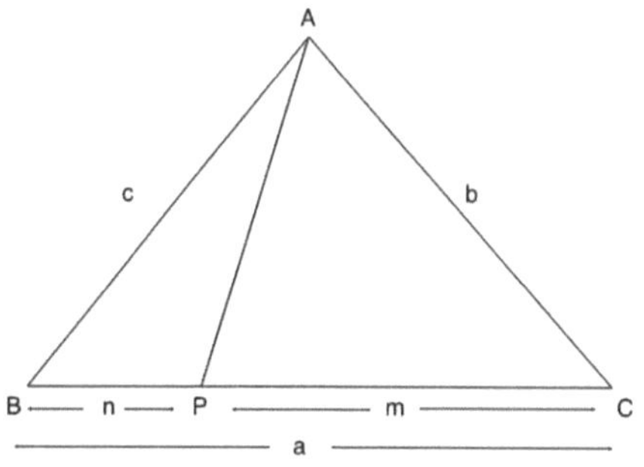

As the converse of Stewart Theorem, we let B, P, C be points on the projective lines AB, AP, AC. Then we claim that if

$$AB \cdot PC + AC^2 \cdot BP = AP^2 \cdot BC + BP \cdot PC \cdot BC$$

or

$$AP^2 = AB^2 \cdot \frac{PC}{BC} + AC^2 \cdot \frac{BP}{BC} - BC^2 \cdot \frac{BP}{BC} \cdot \frac{PC}{BC},$$

then the points B, P, C are collinear.

Corollary 6.5 (Stewart's Theorem)

(i) *If* P *is a point on the extension line of* BC, *then*

$$AP^2 = -AB^2 \cdot \frac{PC}{BC} + AC^2 \cdot \frac{BP}{BC} + BC^2 \cdot \frac{PC}{BC} \cdot \frac{BP}{BC}.$$

(ii) *If* P *is a point in the opposite extension line of* BC, *then*

$$AP^2 = AB^2 \cdot \frac{PC}{BC} - AC^2 \cdot \frac{BP}{BC} + BC^2 \cdot \frac{PC}{BC} \cdot \frac{BP}{BC}$$

Corollary 6.6 (see Pech 2005)

a. If $\triangle ABC$ *is an isosceles triangle and* P *is on* BC, *then* $AP^2 = AB^2 - BP \cdot PC$.
b. *If* AP *is a median of side* BC, *then* $AP^2 = \frac{1}{2}AB^2 + \frac{1}{2}AC^2 - \frac{1}{4}BC^2$.
c. *If* AP *is the interior angel bisector of* $\angle A$, *then* $AP^2 = AB \cdot AC - BP \cdot PC$.
d. *If* AP *is the exterior angle bisector of* $\angle A$, *then* $AP^2 = -AB \cdot AC + BP \cdot PC$.
e. *If* P *divides* BC *such that* $\frac{BP}{BC} = \lambda$, *then*

$$AP^2 = \lambda(\lambda - 1)BC^2 + (1 - \lambda)AB^2 + \lambda AC^2.$$

f. *If* $\frac{BP}{PC} = k$, then $AP^2 = \frac{1}{1+k} \cdot AB^2 + \frac{k}{1+k}AC^2 - \frac{k}{(1+k)^2} \cdot BC^2$.

We give some examples to show the application of Stewart's Theorem.

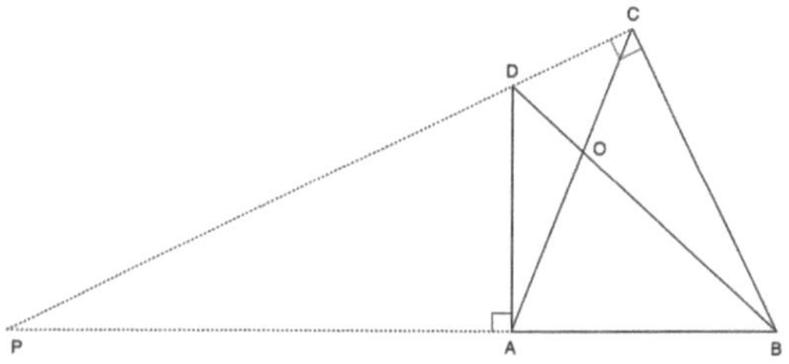

Diagram 3.25 .

Example 6.7 (High school Math competition problem 1996 in Beijing) In the convex quadrilateral $ABCD$, $\angle ABC = 60°$, $\angle BAD = 90°$, $AB = 2$, $CD = 1$, AC and BD meet at O. Find $\angle AOB$.

Solution:

Extend BA, CD to meet at the point P. Let $BC = x$. Then $PB = 2x$, $PC = \sqrt{3}x$. For $\triangle PBC$ and the point A on PB, by Stewart's Theorem, we have (Diagram 3.25)

$$CA^2 = PC^2 \cdot \frac{AB}{PB} + BC^2 \frac{PA}{PB} - AB \cdot PA$$
$$= \left(\sqrt{3}x\right)^2 \cdot \frac{2}{2x} + x^2 \cdot \frac{2x-2}{2x} - 2(2x-2) = x^2 - 2x + 4$$

Now, since $\triangle ADP \sim \triangle CBP$ are right-angled $\triangle$, we have $PD \cdot PC = PA \cdot PB$; that is, $\left(\sqrt{3}x - 1\right) \cdot \sqrt{3}x = (2x-2) \cdot 2x$, and $BC = x = 4 - \sqrt{3}$. Therefore, $CA^2 = 15 - 6\sqrt{3}$. Recall that $\angle BCD$ is a right angle, $BD^2 = x^2 + 1 = 20 - 8\sqrt{3}$. Hence $BD \cdot AC = \sqrt{4(5-2\sqrt{3})} \cdot \sqrt{3(5-2\sqrt{3})} = 10\sqrt{3} - 12$. Moreover, we have $S_{ABCD} = S_{\Delta ABD} + S_{\Delta BCD} = (2\sqrt{3} - 2) + \frac{1}{2}(4 - \sqrt{3}) = \frac{3\sqrt{3}}{2}$.

Therefore, $\frac{1}{2}\left(10\sqrt{3} - 12\right) \cdot sin\angle AOB = \frac{3\sqrt{3}}{2}$, whence we find $sin\angle AOB = \frac{15 + 6\sqrt{3}}{26}$.

Example 6.8 (A summer camp MO problem, China, 2001) In the following diagram, prove that $\frac{2}{PC} = \frac{1}{PA} + \frac{1}{PB}$ (Diagram 3.26).

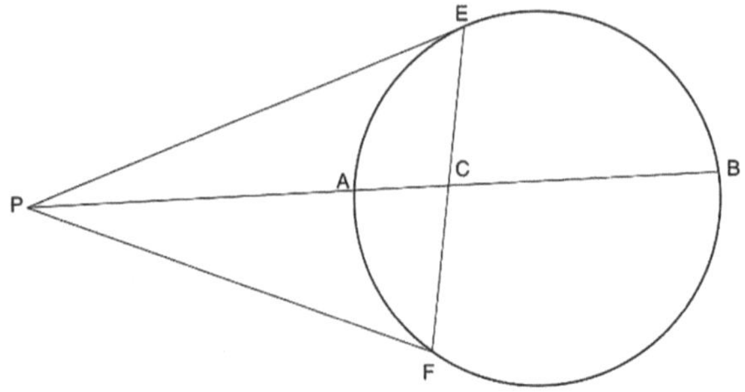

Diagram 3.26 .

Proof Because the lines EF and PB meet at the point C, we have $EC \cdot CF = AC \cdot CB$.

Because $PE = PF$, applying Corollary 6.6(a) of Stewart's Theorem, we have

$$PC^2 = PE^2 - EC \cdot CF,$$

that is,

$$\begin{aligned} PE^2 &= PC^2 + EC \cdot CF = PC^2 + AC \cdot CB \\ &= PC^2 + (PC - PA) \cdot (PB - PC) \\ &= PC^2 - PC^2 - PA \cdot PB + PC \cdot PB + PC \cdot PA. \end{aligned}$$

Hence, $PE^2 = PA \cdot PB$, and consequently we have $2PA \cdot PB = PA \cdot PC + PB \cdot PC$.

Therefore, we deduce that $\dfrac{2}{PC} = \dfrac{1}{PA} + \dfrac{1}{PB}$ and the proof is completed.

3.7 Erdős-Mordell Inequality

We first state the following theorem.

Theorem 7.1 *Let P be a point inside or on a side of $\triangle ABC$. Let the distances from the point P to the three sides be* $|PD|$, $|PE|$ and $|PF|$. *Then*

$$|PA| + |PB| + |PC \geq 2|PD| + |PE| + |PF|.$$

Proof We use polar coordinates to prove this theorem. Let P be the polar point and $|PA|$ be denoted by polar coordinates. Then, $A = (\varphi_1, 0)$, $B = (\varphi_2, 2\theta_1)$, $C = (\varphi_3, 2\theta_1 + \theta_2)$. Thus $\angle CPA = 2\theta_3 = 2\pi - (2\theta_1 + 2\theta_2)$, $|PA| = \varphi_1$, $|PB| = \varphi_2$, $|PC| = \varphi_3$. Let PQ, PR and PS be the bisectors of $\angle APB$, $\angle BPC$ and $\angle CPA$, respectively.

Write $|PQ| = t_1$, $|PR = t_2$, $|PS| = t_3$. The polar coordinates of the points θ, R, S are (t_1, θ_1), $(t_2, 2\theta_1 + \theta_2)$, $(t_3, 2\pi - \theta_3)$. Because the points A, Q, B are collinear, by Ceva's Theorem, we have

$$\frac{sin(0 - 2\theta_1)}{t_1} + \frac{sin(2\theta_1 - \theta_1)}{\varphi_1} + \frac{sin(\theta_1 - o)}{\varphi_2} = 0.$$

Hence $t_1 = \frac{2\varphi_1\varphi_2}{t_1} + sin\frac{(2\theta_1 - \theta)}{\varphi_1} + sin\frac{(\theta_1 - 0)}{\varphi_2} = 0$ and therefore, $t_1 = \frac{2\varphi_1\varphi_2}{\varphi_1 + \varphi_2} cos\theta_1 \leq \sqrt{\varphi_1\varphi_2} cos\theta_1$.

Similarly, we have $B(\varphi_2, 2\theta_1)$ and $t_2 \leq \sqrt{\varphi_2\varphi_3} cos\theta_2$, $t_3 \leq \sqrt{\varphi_3\varphi_1} cos\theta_3$ (Diagram 3.27).

By the triangle inequality, if $\alpha + \beta + \gamma = \pi$, and x, y, z are any real numbers, then

$$x^2 + y^2 + z^2 \geq 2xy\ cos\alpha + 2yz\ cos\beta + 2zx\ cos\gamma.$$

Since we have shown that

$$t_1 \leq \sqrt{\varphi_1\varphi_2} cos\theta_1,\ t_2 \leq \sqrt{\varphi_1\varphi_3} cos\theta_2,\ t_3 \leq \sqrt{\varphi_3\varphi_1} cos\theta_3,$$

we can immediately verify that

$$\begin{aligned}\varphi_1 + \varphi_2 + \varphi_3 &\geq 2\left(\sqrt{\varphi_1\varphi_2} cos\theta_1 + \sqrt{\varphi_2\varphi_3} cos\theta_2 + \sqrt{\varphi_2\varphi_1} cos\theta_3\right)\\ &\geq 2(t_1 + t_2 + t_3) \geq 2(|PD| + |DE| + |PF|).\end{aligned}$$

We can also verify that the equality holds if and only if ΔABC is an equilateral triangle and P is its centroid.

Erdös-Mordell inequality

We illustrate the Erdös-Mordell inequality by using diagram.

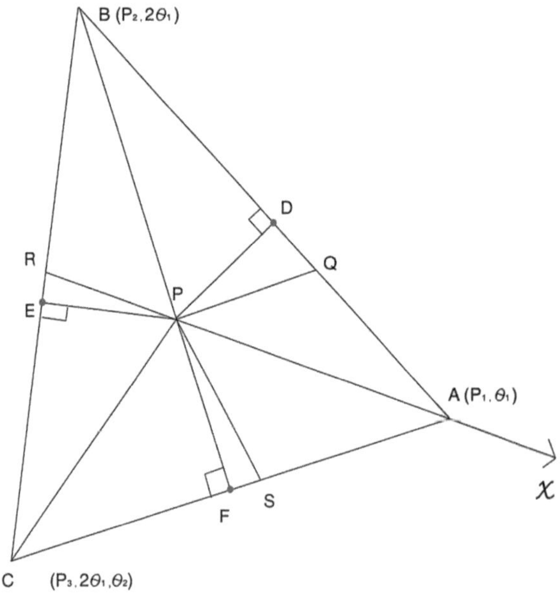

Diagram 3.27 .

Consider the diagram:

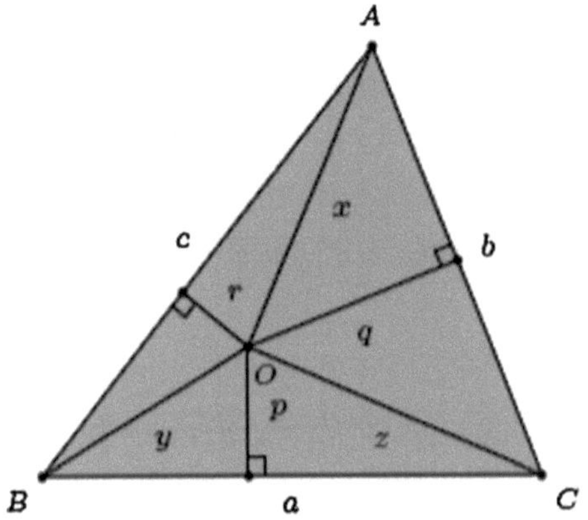

Erdös-Mordell inequality corresponds to proving that

$$x+y+z \geq 2\,(b+q+z)$$

We rearrange the above diagram into the following diagram.

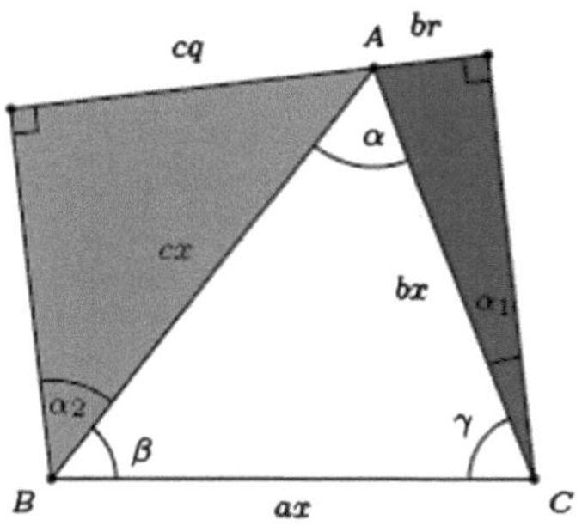

In the above trapezoid, we observe that $ax \geq br + cq$. Similarly, we have $by \geq ar + cp$ and $cz \geq aq + bp$. Adding these inequalities yields

$$x + y + z \geq \left(\frac{b}{a} + \frac{a}{b}\right)r + \left(\frac{c}{a} + \frac{a}{c}\right)q + \left(\frac{c}{b} + \frac{b}{c}\right)p.$$

By using the well-known inequality $AM \geq GM$, the Erdös-Mordell inequality is proved. If we apply $AM \geq GM$ again to the Erdös-Mordell inequality, we will derive that $xyz \geq 8pqr$.

Many inequalities can be solved by using points and lines of a triangle as inspired by the proof of the Erdös-Mordell inequality.

Example 7.2 Suppose that a, b, c, x, y, z are positive numbers satisfying $a + x = b + y = c + z = k$. Prove that $ay + bz + cx \leq k^2$.

Proof Construct an equilateral triangle PQK with side length k. On the three sides of the triangle, let the points N, M, L be such that $QL = x$, $LK = a$, $KM = y$, $MP = b$, $PN = z$, and $NQ = c$. Then it is clear that $S_{\Delta LKM} + S_{\Delta MPN} + S_{\Delta NRL} < S_{\Delta PQK}$ (Diagram 3.28).

Therefore, $\frac{\sqrt{3}}{4}ay + \frac{\sqrt{3}}{4}bz + \frac{\sqrt{3}}{4}cx < \frac{\sqrt{3}}{4}k^2$ and $ay + bz + cx < k^2$.

Example 7.3 Prove the inequality
$\sqrt{a^2 + b^2 - ab} + \sqrt{b^2 + c^2 - bc} \geq \sqrt{a^2 + c^2 + ac}$ for any positive real numbers a, b, c.

Proof Draw lines OA, OB, OC, where O is a point such that

Diagram 3.28 .

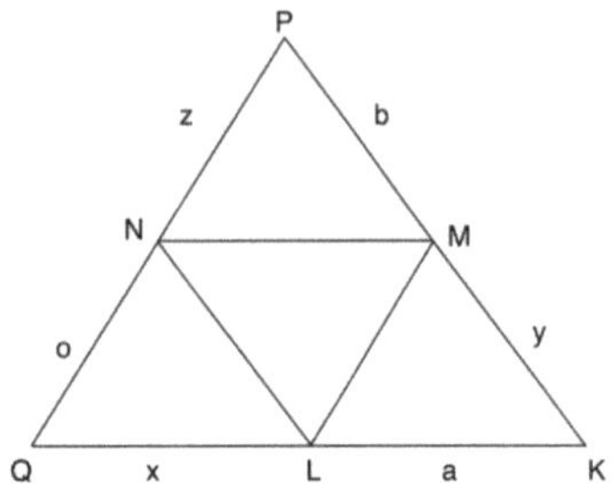

$$|OA| = a,\ |OB| = b,\ |OC| = c,\ \angle AOB = \frac{\pi}{3},\ \angle BOC = \frac{\pi}{3},\ \angle AOC = \frac{2\pi}{3}.$$

In ΔAOB, by the law of Cosines, we have

$$|AB| = \sqrt{a^2 + b^2 - 2ab\frac{cos\pi}{3}} = \sqrt{a^2 + b^2 - ab}.$$

Similarly, in ΔAOC and ΔBOC, use the law of Cosines to get

$$|AC| = \sqrt{a^2 + c^2 - ac} \text{ and } |BC| = \sqrt{b^2 + c^2 - bc}$$

Because in ΔABC, $|AB| + |AC| > |A|$, we have

$$\sqrt{a^2 + b^2 - ab} + \sqrt{b^2 + c^2 - bc} > \sqrt{a^2 + c^2 - ac},$$

and the inequality is proved. If points A, B, C are collinear, $|AB| + |BC| = |AC|$. Thus, the area sum of ΔAOB and ΔBOC equals the area of ΔAOC.

This means that $\frac{1}{2}ab \sin 60° + \frac{1}{2}bc \sin 60° = \frac{1}{2}ac \sin 60°$, that is, $ab + bc = ac$; or, in other words, $\frac{1}{a} + \frac{1}{c} = \frac{1}{b}$ when equality holds.

In closing this Chapter, we cite an interesting problem concerning five points that are concyclic. This problem was proposed by Mr. Jiang Zemin, the former chief Party Secretary of China, on 6th April 2000.

His problem asks to prove the five intersection points N, M, K, L, O, of circles in the diagram are concyclic. This problem would be a challenge problem for IMO.

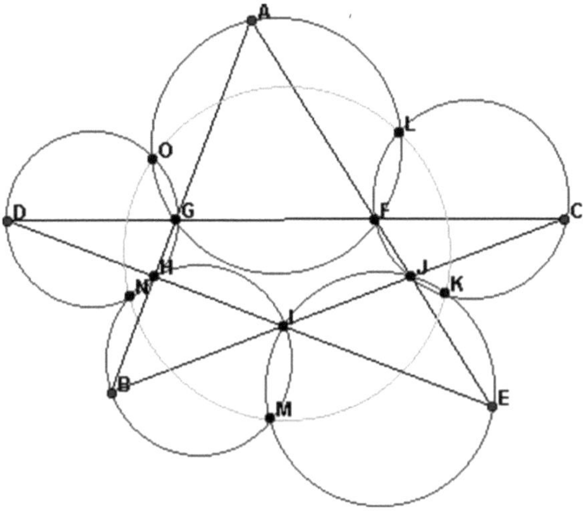

Hint:

(1) Because the four points F, C, K, J are concyclic, $\angle DCK = \angle FCK = 180° - \angle JKJ = \angle KJE$.
(2) Because the four points I, J, K, E are concyclic, therefore $\angle KJE = \angle KIE$.
(3) From (1) and (2), we know that $\angle DCK = \angle KIE$, and therefore the four points D, C, K, I are concyclic.
(4) Similarly, D, C, I, N are four concyclic points.
(5) From (3) and (4), we know that the four points D, C, K, I are concyclic.
(6) From (5), we know $\angle KNI = \angle KCI$.
(7) Because K, C, L, J are four concyclic points, we know that $\angle KCI = \angle KCJ = \angle KLJ$.
(8) In a similar way as in steps (1)–(5), we can prove that B, M, J, L are four concyclic points.
(9) In particular, by (8), we know that B, M, J, L are four concyclic points so that $\angle IBM = \angle JBM = \angle JLM$.
(10) Since the four points N, B, M, I are concyclic, we have $\angle INM = \angle IBM$.
(11) From steps (6), (7), (9) and (10), we know that

$$\angle KNM = \angle KNI + \angle INM = \angle KLJ + \angle MLJ = \angle KLM.$$

(12) From (11), we know that N, M, K, L are four concyclic points.
(13) Similarly, we can prove that N, M, K, O are four concyclic points.

(14) From (12) and (13), we conclude that the five points N, M, K, L, O are concyclic.

Acknowledgements The author would like to thank the Referee, Professor Alexander Soifer and Mr. Yip Choyui for helping to improve and modify the final version of the article.

References

Benitez, J. (2007). A unified proof of Ceva and Menelaus theorem using projective geometry. *Journal of Geometry and Graphics, 11*(1), 39–44.

Encyclopedia of the Solutions of Mathematics Problem (1983). S.-W. Tong (Ed.), sheng hai encyclopedia publication. 數學題解辭典：平面解析幾何, 唐秀穎 主編, 上海辭書出版社.

Encyclopedia of Solved Problems (2016). W.-W. Zhu (Ed.), Guangdong Education Press. 高中數學解題辭典, 朱華偉 主編, 廣東教育出版社.

Grunbaum, B., & Shephard, G. C. (1955). Ceva, Menelaus and the area principle. *Mathematics Magazine, 68*, 254–268.

Klamkin, M. S., & Liu, A. (1992). Simultaneous generalization of the theorems of Ceva and Menelaus. *Mathematics Magazine, 65*(1), 48–82.

Liu, A. C. F. (1979). Lattice points are pick's theorem. *Mathematics Magazine, 52*(4), 232–235.

Moise, E. W. (1990). *Elementary geometry from an advanced standpoint* (edn. 3). Amazon, Science and Mathematics.

Pech, P. (2005). On Simpson Wallace theorem and its generalizations. *Journal of geometry and graphics, 9*(2), 141–156.

Some Geometry Problems in Mathematical Olympiad Competitions. (2015). W. S. Shen, Y, Zhang & K. S. Leng (Eds.), Wu Nan Normal University Press. 數學中的幾何問題(奧林匹克叢書), 張堯、冷崗松、沈文選 編著, 湖南師範大學出版社.

Part III
Combinatorics for Competitions

Chapter 4
Arrangements and Transformations of Numbers on a Circle: An Essay Inspired by Problems of Mathematics Competitions

Kiril Bankov

Abstract There are contest problems dealing with the following situation: several numbers are arranged on a circle and a certain admissible operation can be consecutively done finite number of times; the task is to find conditions under which a specific final arrangement of the numbers can be obtained. The variety of these problems is determined by different initial and final arrangements of the numbers and by the admissible operations with them. The change of some of these elements often leads to interesting generalizations. This chapter discusses several such examples. It also presents some other contest problems dealing with arrangements of numbers on a circle. Didactical approaches to teaching how to solve such problem are also considered.

Keywords Combinatorial situation · Admissible operation · Optimal arrangements of numbers · Problems of mathematics competitions

4.1 Introduction

The intellectual treasure of every mathematics competition is the set of the problems given to the participants. Competitions present variety of problems: from these that are closely connected to the school curriculum to those

K. Bankov (✉)
Faculty of Mathematics and Informatics, University of Sofia, James Boucher bul. 5, 1164 Sofia, Bulgaria
e-mail: kbankov@fmi.uni-sofia.bg

K. Bankov
Institute of Mathematics and Informatics, Bulgarian Academy of Sciences, Acad. G. Bonchev str. bl. 8, 1113 Sofia, Bulgaria

A. Soifer (ed.), *Competitions for Young Mathematicians*,
ICME-13 Monographs, DOI 10.1007/978-3-319-56585-9_4

that deal with "non-standard" situations. The latter usually stimulate creative thinking and remain in the minds for a long time. Finding their solutions develops mathematical abilities. Many of these problems give rise to numerous mathematical ideas. This chapter discusses such problems: some are taken from mathematics competitions, others are inspired by competition problems. In both cases, as some of the best examples of beauty in mathematics, they provoke an interest in mathematics that often begins with the consideration of attractive problems.

4.2 Examples with Admissible Operations

The life is full of operations. Many times in a day we make decisions about series of operations that have to be done in order to obtain a particular result. The correctness of these decisions depends on the ability to estimate the final results. Mathematics helps in modelling this reality by tasks using a particular admissible operation to transform a given situation to a different one. These problems lead to interesting generalizations by changing either the admissible operation or the initial/final situations. This part presents such examples taken from mathematics competitions in the context of arrangements of numbers on a circle.

4.2.1 First Situation

Let $n \geq 3$ cells be arranged into a circle. Each cell can be occupied by 0 or 1. The following operation is admissible: choose any cell C occupied by a 1 and reverse the entries in the two cells adjacent to C (so that x, y become $1-x$, $1-y$).

In order to create problems using the above situation, it is useful first to understand some of the properties of the admissible operation.

Property 1 *The operation does not change the parity of 1s. Certainly, the admissible operation either does not change the number of 1s (if the entries of the two cells adjacent to C are different, 0 and 1), or changes it by 2 (two more 1s if both cells adjacent to C contain 0s, or two less 1s if they both contain 1s).*

Property 2 *Let an even number (say 2k) of consecutive cells be occupied by 1s and 0s elsewhere. If the admissible operation is consecutively performed clockwise (or anticlockwise) on the 1-st, 3-th, 5-th, ..., (2k–1)-th 1s (or on* *the 2-d, 4-th, 6-th, ..., (2k)-th 1s), the number of 1s does not change. The group of these 1s moves one cell anticlockwise (or clockwise).*

This property is demonstrated in Fig. 4.1. The operation starts clockwise from the leftmost 1 and moves the group of the four 1s one cell anticlockwise.

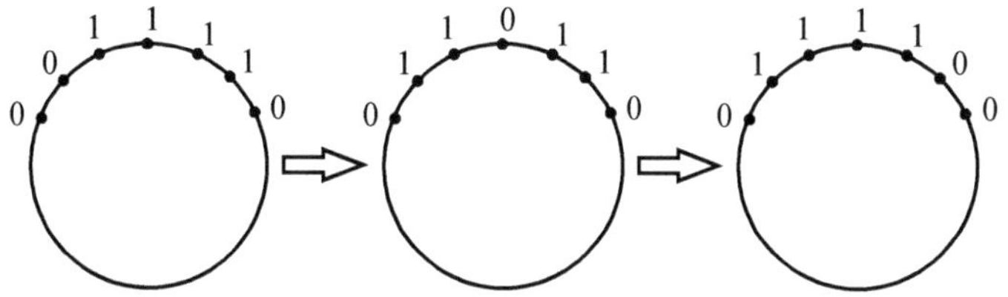

Fig. 4.1 Demonstration of property 2

Property 3 *Let an odd number (say 2k+1) of consecutive cells be occupied by 1s and 0s elsewhere. If the admissible operation is consecutively performed clockwise (or anticlockwise) on the 1-st, 3-th, 5-th, ..., (2k+1)-th 1s, the number of 1s increases by 2 with two 1s at the both ends of the initial group of 1s.*

This property is demonstrated in Fig. 4.2. The operation starts clockwise from the leftmost 1s and increases the group of five 1 to a group of seven 1s.

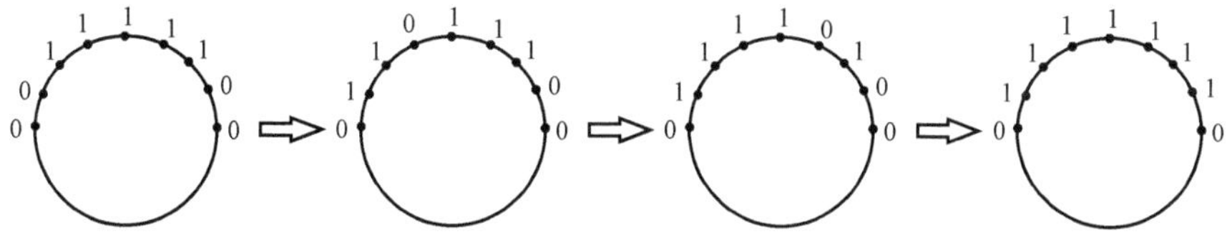

Fig. 4.2 Demonstration of property 3

Property 4 *Let an odd number (say 2k+1) of consecutive cells be occupied by 1s and 0s elsewhere. If the admissible operation is consecutively performed clockwise (or anticlockwise) on the 2-d, 4-th, 6-th, ..., (2k)-th 1s, the number of 1s decreases by 2 by transferring to 0 the 1s at the both ends of the initial group.*

This property is demonstrated in Fig. 4.3. The operation starts clockwise from the second to the left 1 and decreases the group of five 1s to a group of three 1s.

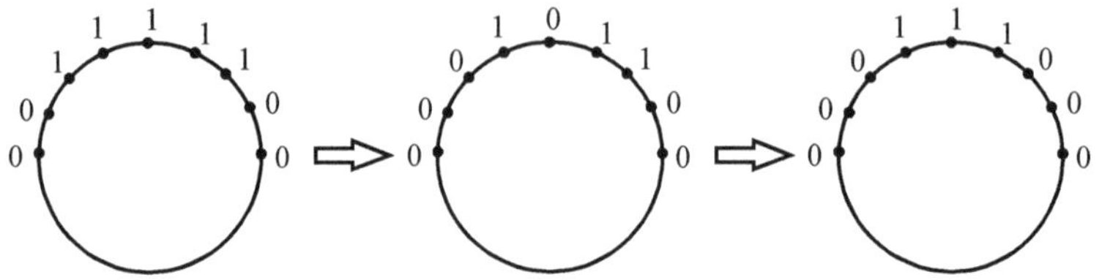

Fig. 4.3 Demonstration of property 4

It is now time to pose several problems. For each of them (Problems from 2.1 to 2.6) the First Situation is used.

Problem 2.1 (*National competition in Bulgaria, 2016*) Let $n = 20$.

(A) Initially, there is a 1 in one cell and 0s elsewhere. Is it possible to obtain 1s in all cells in a finite number of admissible steps?
(B) Initially, there are 1s in two consecutive cells and 0s elsewhere. Is it possible to obtain 1s in all cells in a finite number of admissible steps?
(C) Initially, there are 1s in two cells that stand in one and 0s elsewhere. Is it possible to obtain 1s in all cells in a finite number of admissible steps?
(D) Is it possible to choose two cells in such a way so that if initially there are 1s in both of them and 0s elsewhere to obtain 1s in all cells in a finite number of admissible steps?

Solution

(A) No. According to Property 1, the number of 1s will always be odd and cannot be 20.
(B) No. According to Property 2 the number of 1s will always be equal to 2.
(C) No. After the first performance of the operation there will be four consecutive 1s. According to Property 2, the number of 1s will always be equal to 4.
(D) Let initially there are 1s in two non-adjacent cells and 0s elsewhere. Start the operation from one of the 1s (Fig. 4.4) and continue adding 1s until the second 1 is attached to the group of 1s (Property 3). We have now even number of consecutive 1s. According to Property 2 their number cannot be changed. In order not to have 0s, the initial cells that contain 1s must be diametrically opposite.

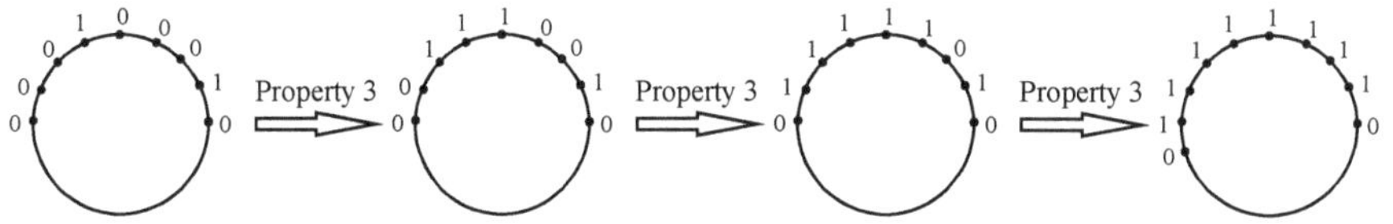

Fig. 4.4 Joining the two 1s

Clearly, for parts (A) and (D) what matters is the parity of n while for parts (B) and (C) the answers are the same for any n.

Problem 2.2 (*National competition in Bulgaria, 2016*) Let $n = 21$. Initially, there is a 1 in one cell and 0s elsewhere. Is it possible to obtain 1s in all cells in a finite number of admissible steps?

Solution Yes. Apply Property 3.

Problem 2.3 Initially, there is a 1 in one cell and 0s elsewhere. For which values of n is it possible to obtain 1s in all cells in a finite number of admissible steps?

Solution Because of Property 1, for even n it is not possible to obtain 1s in all cells. Let now n be an odd number. By applying Property 3, it is possible to obtain 1s in all cells. The required values of n are all odd numbers.

Problem 2.3 in cases $n = 1990$ and $n = 1991$ is given on a national mathematics competition in Bulgaria, 1991 (Rakovska et al. 2007).

Problem 2.4 Initially, there is a 1 in one cell and 0s elsewhere. For which values of n is it possible to reverse the entries in all cells (i.e. to obtain 0 in the cell where the initial 1 was, and 1s elsewhere) in a finite number of admissible steps?

Solution Let n be such a number that it is possible to reverse the entries in all cells in a finite number of admissible steps. Because of Property 1, n must be an even number. By applying Property 3, it is possible to obtain only one 0 that is in the cell diametrically opposite to the cell having contained the initial 1, and 1s everywhere. After the admissible operation is performed, the 0 "jumps" across one cell. Therefore, this single 0 may occupy the cell where the initial 1 was if and only if n is divisible by 4.

Didactical consideration. It is always difficult to teach students solving problems. It is much more difficult to teach them solving competition problems. The difficulty is that these problems usually need a "non-standard" approach. One of the first important steps in solving such problems is to understand the situation. This is the basis for a good start in considering the methods and strategies for moving ahead. In the case of the problems presented in this paper, understanding the properties of the admissible operation is crucial. This is why the recommendation is to first acquaint the students with the situation and the properties of the admissible operation. Then the teacher may pose (or even ask the students to pose) several problems, the ideas for which solutions are (more or less) already explained in the properties of the operation.

The next problem can be considered as a “reverse” of Problem 2.3.

Problem 2.5 Initially, there are 1s in m cell and 0s elsewhere. For which values of m is it possible to obtain a 1 in one cell and 0s elsewhere in a finite number of admissible steps?

Solution Let m be such a number that it is possible to obtain a 1 in only one cell in a finite number of admissible steps. Because of Property 1, m must be an odd number. Let now $m = 2k + 1$ be an odd number. (i) If the m cells containing 1s are consecutive, we apply Property 4 and decrease the number of 1s by 2, i.e. we get $m - 2 = 2k - 1$ consecutive 1s. This way, step by step, it is possible to obtain 1s in only one cell. (ii) Let now the m cells containing 1s are not consecutive. Choose a group A of odd number (say $2s - 1$) 1s. (Such a group exists, since m is an odd number.) Using Property 3 we can get two more 1s to group A. If we continue this way, we can “integrate” another group of 1s with A, i.e. we decrease the groups of 1s at the expense of increasing of the number of the consecutive 1s. Using this procedure, we can obtain one group of odd number consecutive 1s. Then we can proceed as in (i). Therefore, the answer of Problem 2.3 is “all odd numbers”.

A similar situation can be examined in different shapes. For example, consider a figure in shape of “eight”, namely:

Problem 2.6 Let $n \geq 3$ and $(2n - 1)$ cells be arranged into two tangent circles in such a way that one of the cells is in the tangent point (i.e. it is a common cell for the both circles) and each circle has n cells. Initially, there is a 1 in the cell that is common for both circles and 0s elsewhere. The following operation is admissible: choose any cell C occupied by a 1 and reverse the entries in the two cells adjacent to C (so that x, y become $1 - x$, $1 - y$). Figure 4.5 represents the adjacent cells of C_0 (the common cell for the both circles)—they are either the cells C_1 and C_2, or the cells C_3 and C_4. For which values of n is it possible to obtain 1s in all cells of the figure in a finite number of admissible steps?

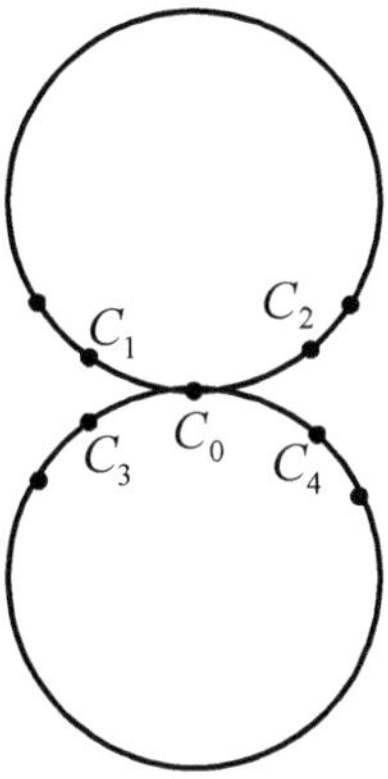

Fig. 4.5 Adjacent cells

Solution The next three cases are considered:

(i) Let n be an odd number. According to Problem 2.3, it is possible to obtain 1s in all cells of each of the circles separately in a finite number of admissible steps. Therefore we can obtain 1s in all cells of the figure.
(ii) Let n be divisible by 4. Following the procedure described in the solution of Problem 2.4 for each of the circles, we may obtain two 0s, one in each of the circles in the cells that are diametrically opposite to C_0, and 1s everywhere. Because each of these 0s may "jump" across one cell, the most left situation in Fig. 4.6 can be obtained. Then we consecutively apply the admissible operation on the cells C_2 and C_4 (Fig. 4.6), and obtain 1s in all cells.

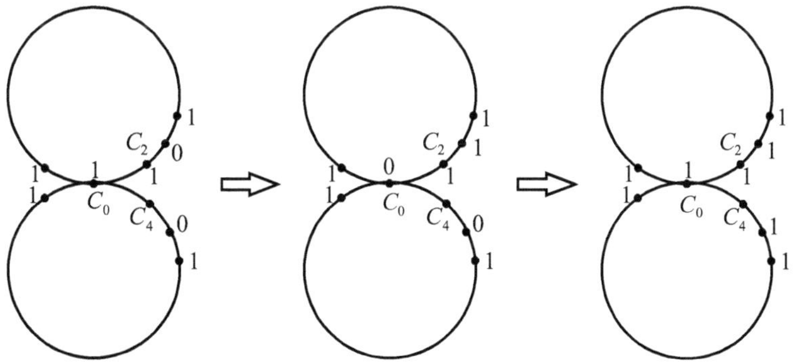

Fig. 4.6 The case n divisible by 4

(iii) Let n be an even number that is not divisible by 4. Assume that it is possible to obtain 1s in all cells in a finite number of admissible steps. Consider the last step: it should be applied on a cell containing 1, and two 0s only in its adjacent cells. Property 1 tells us that if we have only two 0s, they are situated one on each circle. In order to be able to perform the last step, these two 0s should be placed in a way that is similar to what is shown on the central situation in Fig. 4.6, i.e. one of the 0s is in cell C_0. Because after the admissible operation is performed, the 0 "jumps" across one cell, this means that in a finite number of admissible steps the only 0 in one of the circles (the upper circle in Fig. 4.6) may occupy cell C_0, i.e. the cell where the initial 1 was. According to Problem 2.4, n is divisible by 4, which is a contradiction.

Therefore, it is possible to obtain 1s in all cells in a finite number of admissible steps if and only if n is an odd number or n is divisible by 4.

Another possible figure that could be examined with this situation is formed by two intersecting circles with cells containing 1s in their intersecting points and 0s elsewhere. The consideration of different arrangements is left to the reader.

4.2.2 Second Situation

Let $n \geq 3$ cells be arranged into a circle. Each cell can be occupied by 0 or 1. The following operation is admissible: choose any cell C occupied by a 1, change it into a 0 and simultaneously reverse the entries in the two cells adjacent to C (so that x, y become $1-x$, $1-y$).

The difference with the First Situation is that the admissible operation changes also the entry in the chosen cell C. Because of this, the Second Situation has different properties.

Property 1 *Any three consecutive 1s can be transformed into three consecutive 0s.*

Property 2 *Using the admissible operation it is not possible to obtain 1s in all cells.*

This is because each operation is performed in a cell containing 1, which transfers into 0.

Property 3 *If the admissible operation is performed clockwise (or anticlockwise) on a group of consecutive 1s, the following arrangement is obtained: the group of consecutive 1s increases by one, following by one 0 and one 1, then a group of consecutive 0s, decreasing by one.*

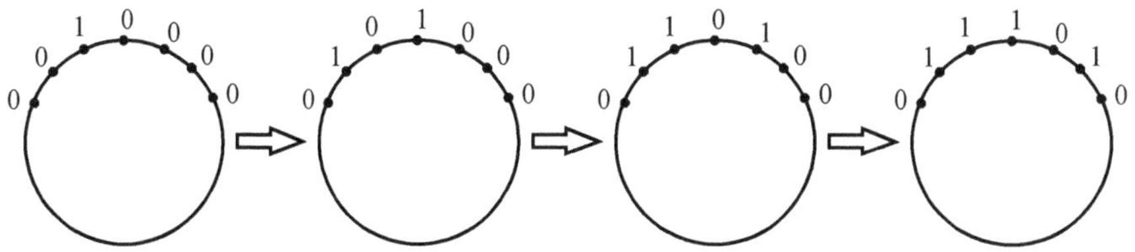

Fig. 4.7 Demonstration of property 3

This property is demonstrated in Fig. 4.7. The operation is performed clockwise on the consecutive 1s.

The next two problems (Problems 2.7 and 2.8) use the Second Situation.

Problem 2.7 (*4-th Austrian–Polish Mathematical Competition* 1981) Initially, there is a 1 in one cell and 0s elsewhere. For which values of n is it possible to obtain 0s in all cells in a finite number of admissible steps?

Solution Denote the cells by $C_1, C_2, \ldots, C_n$. Let initially C_1 be occupied by 1 and n be such a number that after a certain number of admissible steps all cells are occupied by 0. Denote by s_i the number of operations performed in C_i and by a_i the number of the changes in C_i (i.e. when the admissible operation is performed in C_i or any of its neighbor cells C_{i-1} or C_{i+1}). Since in the final arrangement all cells are occupied by 0, then $a_1 \equiv 1(\text{mod}2)$ and $a_i \equiv 0(\text{mod}2)$ for $i \neq 1$. It is clear that $a_i \equiv s_{i-1} + s_i + s_{i+1}(\text{mod}2)$ for $i = 1, 2, \ldots, n$. (We assume that $s_0 = s_n$ and $s_{n+1} = s_1$.)

(i) If n is divisible by 3, then $1 \equiv a_1 + a_4 + a_7 + \cdots + a_{n-2} \equiv \sum_{i-1}^{n} s_i(\text{mod}2)$ and $0 \equiv a_2 + a_5 + a_8 + \cdots + a_{n-1} \equiv \sum_{i-1}^{n} s_i(\text{mod}2)$, which is not possible.

(ii) Let $n \equiv 1(\text{mod}3)$. Because of Property 3, after the execution of the admissible operation clockwise on the consecutive cells, the arrangement presented in Fig. 4.8 can be obtained. Cell C_{n-2} contains 0 and 1s

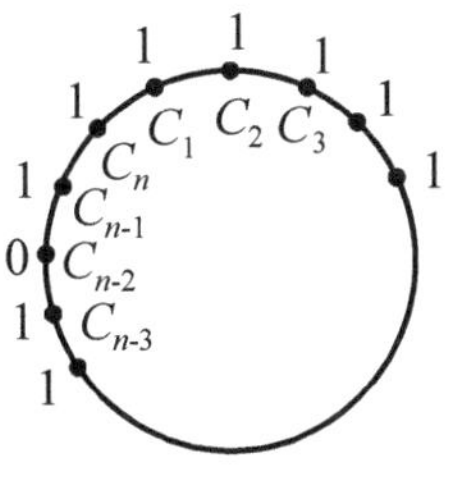

Fig. 4.8 Arrangement obtained in case (ii)

elsewhere. There are $3k$ consecutive 1s. Using Property 1, they can be grouped in k groups by 3 and 0s can be obtained everywhere.

(iii) Let $n \equiv 2(\text{mod}3)$. Because of Property 3, after the execution of the admissible operation clockwise on the consecutive cells, the arrangement presented in Fig. 4.9 can be obtained. Cells C_{n-1} and C_{n-2} contain 0 and 1s elsewhere. There are $3k$ consecutive 1s. Using Property 1, they can be grouped in k groups by 3 and 0s can be obtained everywhere.

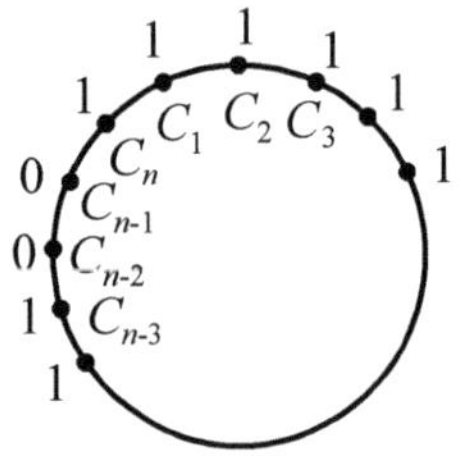

Fig. 4.9 Arrangement obtained in case (iii)

Therefore, it is possible to obtain 0s in all cells in a finite number of admissible steps if and only if n is not divisible by 3.

A possible change is to ask the same question under a different initial arrangement. Here is an example.

Problem 2.8 Initially, all cells are occupied by 1. For which values of n is it possible to obtain 0s in all cells in a finite number of admissible steps?

Solution The answer is that for any n it is possible to obtain 0s in all cells in a finite number of admissible steps. Indeed, this is obvious if $n \equiv 0(\text{mod}3)$ (Property 1). If $n \equiv 1(\text{mod}3)$, we may arrange several groups of three ones and this way to obtain a situation in which there is a 1 in one cell and 0s elsewhere; now we may apply (ii) in problem 2.7. If $n \equiv 2(\text{mod}3)$, we may arrange several groups of three 1s and this way to obtain a situation in which there is only two neighbor 1s and 0s elsewhere; in the next step we will get a 1 in one cell and 0s elsewhere; now we may apply (iii) in problem 2.7.

4.2.3 *Third Situation*

Let $n \geq 3$ cells be arranged into a circle. Each cell can be occupied by 1 or 0. The following operation is admissible: draw another n cells—one between any two of the existing cells; in each of these new cells write 0 if the numbers in the two neighbor existing cells are equal, and 1 if these numbers are different; then delete the existing cells.

Property 1 *If n is an odd number and there is at least one 1, it is not possible to obtain 0s in all cells in a finite number of admissible steps.*

Certainly, assume that in a finite number of admissible steps all cells contain 0s. Then in the second to the last arrangement all cells must contain 1s. Therefore, in the previous arrangement any two neighbor cells contain different numbers, which is impossible, having an odd number of cells.

Property 2 *Considering the modulo-2 arithmetic, the admissible operation can be reworded the following way: draw another n cells—one between any two of the existing cells; in each of these new cells write the modulo-2 sum of the numbers in the two neighbor existing cells; then delete the existing cells.*

This is because in modulo-2 arithmetic, $1+1=0+0=0$ and $1+0=0+1=1$.

Property 3 *After the execution of the admissible operation k times, the numbers in the cells are obtained as a sum of* $k+1$ *consecutive numbers among the initially written numbers with coefficients that are the numbers in the k-th row of the Pascal's triangle modulo-2, known also as Sierpinski's triangle* (Fig. 4.10).

```
0                                 1
1                                1 1
2                               1 0 1
3                              1 1 1 1
4                             1 0 0 0 1
5                            1 1 0 0 1 1
6                           1 0 1 0 1 0 1
7                          1 1 1 1 1 1 1 1
8                         1 0 0 0 0 0 0 0 1
9                        1 1 0 0 0 0 0 0 1 1
10                      1 0 1 0 0 0 0 0 1 0 1
11                     1 1 1 1 0 0 0 0 1 1 1 1
12                    1 0 0 0 1 0 0 0 1 0 0 0 1
13                   1 1 0 0 1 1 0 0 1 1 0 0 1 1
14                  1 0 1 0 1 0 1 0 1 0 1 0 1 0 1
15                 1 1 1 1 1 1 1 1 1 1 1 1 1 1 1 1
16                1 0 0 0 0 0 0 0 0 0 0 0 0 0 0 0 1
17               1 1 0 0 0 0 0 0 0 0 0 0 0 0 0 0 1 1
18              1 0 1 0 0 0 0 0 0 0 0 0 0 0 0 0 1 0 1
19             1 1 1 1 0 0 0 0 0 0 0 0 0 0 0 0 1 1 1 1
20            1 0 0 0 1 0 0 0 0 0 0 0 0 0 0 0 1 0 0 0 1
21           1 1 0 0 1 1 0 0 0 0 0 0 0 0 0 0 1 1 0 0 1 1
22          1 0 1 0 1 0 1 0 0 0 0 0 0 0 0 0 1 0 1 0 1 0 1
23         1 1 1 1 1 1 1 1 0 0 0 0 0 0 0 0 1 1 1 1 1 1 1 1
24        1 0 0 0 0 0 0 0 1 0 0 0 0 0 0 0 1 0 0 0 0 0 0 0 1
25       1 1 0 0 0 0 0 0 1 1 0 0 0 0 0 0 1 1 0 0 0 0 0 0 1 1
26      1 0 1 0 0 0 0 0 1 0 1 0 0 0 0 0 1 0 1 0 0 0 0 0 1 0 1
27     1 1 1 1 0 0 0 0 1 1 1 1 0 0 0 0 1 1 1 1 0 0 0 0 1 1 1 1
28    1 0 0 0 1 0 0 0 1 0 0 0 1 0 0 0 1 0 0 0 1 0 0 0 1 0 0 0 1
29   1 1 0 0 1 1 0 0 1 1 0 0 1 1 0 0 1 1 0 0 1 1 0 0 1 1 0 0 1 1
30  1 0 1 0 1 0 1 0 1 0 1 0 1 0 1 0 1 0 1 0 1 0 1 0 1 0 1 0 1 0 1
31 1 1 1 1 1 1 1 1 1 1 1 1 1 1 1 1 1 1 1 1 1 1 1 1 1 1 1 1 1 1 1 1
32 1 0 0 0 0 0 0 0 0 0 0 0 0 0 0 0 0 0 0 0 0 0 0 0 0 0 0 0 0 0 0 0 1
```

Fig. 4.10 Pascal's triangle modulo-2 (or Sierpinski's triangle)

This is because of the way the numbers are obtained described in Property 2.

Problems 2.9 and 2.10 use the Third Situation.

Problem 2.9 (*Mathematics competition in former Yugoslavia, 1975 Sergeev* 1987) Let $n = 9$ and four of the cells be occupied by 1, the other five be occupied by 0. Is it possible to obtain 0s in all nine cells in a finite number of admissible steps?

Solution No, because of Property 1.

This relatively simple problem gives birth to variety of generalizations. There are different variations of the initial arrangements, depending on the number of the cells and on the number and the positions of the initial 1s. Here is an example.

Problem 2.10 Initially there is a 1 in one cell and 0s elsewhere. For which values of n is it possible to obtain 0s in all cells in a finite number of admissible steps?

Let's call an *eligible value* a value of n for which it is possible to obtain 0s in all cells in a finite number of admissible steps. Property 1 tells us that an odd number cannot be an eligible value. This is why we will consider only even values of n. We will show that the eligible values are all powers of 2.

Because in the initial arrangement there is only one 1 and 0s elsewhere, it follows from Property 3 that the numbers in the cells of the circle after the execution of the admissible operation k times for $k < n$ can be obtained by consecutively writing the numbers in the k-th row of the Sierpinski's triangle and completing the remaining cells with 0s (if necessary). For $k \geq n$ these numbers can be obtained the following way: consider the numbers in the k-th row of the Sierpinski's triangle and roll them up around the cells of the circle; then add modulo-2 the numbers that go into one and the same cell.

Let n be a power of 2, i.e. $n = 2^m$ for any natural number m. If $k < n$, after the execution of the admissible operation k times, the number 1 appears with coefficient 1 modulo 2 at least once (i.e. the first coefficient in the k-th row of the Sierpinski's triangle). This means that it is not possible to obtain 0s in all cells in a number of admissible steps that is less than n. Since the row number $n = 2^m$ of the Sierpinski's triangle consists of two 1s (at the both ends) and 0s elsewhere (see, for example, Proposition 4.1.11 on page 230 of Gross 2008), after the execution of the admissible operation $n = 2^m$ times, the number 1 appears with coefficient 0 everywhere. This is because the two 1s at the both ends in the n-th row go into one and the same cell and their sum modulo 2 is 0. Therefore after the execution of the admissible operation $n = 2^m$ times the number in each cell is 0. This means that all powers of 2 are eligible values.

We will prove now that if n is not a power of 2, it is not an eligible value. We will make use of the following

Theorem Let n and k be non-negative integers. Then

$$\binom{n}{k} \equiv \begin{cases} 0 \bmod 2 & \text{if } n \text{ is even and } k \text{ is odd} \\ \binom{\lfloor \frac{n}{2} \rfloor}{\lfloor \frac{k}{2} \rfloor} \bmod 2 & \text{otherwise} \end{cases}$$

The proof of the theorem can be found on pp. 228–230 of Gross 2008.

The above theorem allows comparison of the entries in the s-th and $(2s)$-th rows of the Sierpinski's triangle, where s is a natural number. Because $\binom{2s}{2k+1} \equiv 0 \bmod 2$ and $\binom{2s}{2k} \equiv \binom{s}{k} \bmod 2$, it follows that the entries in the $(2s)$-th rows of the Sierpinski's triangle can be obtained by writing 0s between any two of the numbers in the s-th row.

Lemma 2.1 *Let $n > 2$ cells be arranged into a circle A. Each cell can be occupied by 1 or 0. Initially there is a 1 in one cell and 0s elsewhere. Let $2n$ cells be arranged into another circle B. Each cell can also be occupied by 1 or 0. Initially there is a 1 in one cell and 0s elsewhere. The admissible operation described in the Third Situation is performed $s \geq 1$ times on circle A and $2s$ times on circle B. Then the numbers on circle B can be obtained by writing 0s between any two numbers on circle A.*

Lemma 2.1 follows from the comparison of the entries of the s-th and $(2s)$-th rows of the Sierpinski's triangle and rolling up the numbers of the corresponding rows around the cells of the circles.

Lemma 2.2 *Let $m > 2$ be an eligible value. Then $\frac{m}{2}$ is also an eligible value.*

Proof Let $m > 2$ cells be arranged into a circle B. Each cell can be occupied by 1 or 0. Initially there is a 1 in one cell and 0s elsewhere. Let after s steps all cells contain 0s in a finite number of admissible steps. Then after $(s-1)$ steps all cells contain 1s. According to Lemma 2.1, $(s-1)$ cannot be an even number, i.e. s is an even number. Let now $\frac{m}{2}$ cells be arranged into a circle A. Each cell can be occupied by 1 or 0. Initially there is a 1 in one cell and 0s elsewhere. Lemma 2.1 tells us that all 0s in circle B can be obtained by writing 0s between any two numbers on circle A after $\frac{s}{2}$ steps. This means that all cells on circle A after $\frac{s}{2}$ steps also contain 0s, i.e. $\frac{m}{2}$ is also an eligible value.

We are now ready to show that if n is not a power of 2, it is not an eligible value. Certainly, let $n = q2^r$, where q is an odd number, $q > 1$ and r is a natural number. Suppose that n is an eligible value. Apply Lemma 2.2 r times. The result is that q is also an eligible value. This is not possible, since q is an odd number. Therefore, n is not an eligible number.

Problem 2.10 gives rise to another issue that needs exploration. Let n is not a power of 2. Since n is not an eligible number, the execution of the admissible operation described in the Third Situation will never end. On the other hand, there are finite number of arrangements of 0 and 1 in the cells of the circle. This means that after a certain number of steps the arrangements of the numbers will cyclically repeat. The least number of the steps in this repetition is called a period. The reader may try to explore how the period depends on n.

4.2.4 Fourth Situation

Let $n \geq 3$ cells be arranged into a circle. Each cell contains either 1 or (–1). The numbers in any two neighbor cells are multiplied, so that n products are obtained and S is the sum of these products.

Denote the numbers in the cells consecutively by $a_1, a_2, \ldots, a_n$. The obtained products are $p_1 = a_1a_2, p_2 = a_2a_3, \ldots, p_n = a_na_1$. Then $S = p_1 + p_2 + \cdots + p_n$.

Property 1 *Each of the products $p_1 = a_1a_2, p_2 = a_2a_3, \ldots, p_n = a_na_1$ is either 1 or (–1).*

This is because each of the numbers $a_1, a_2, \ldots, a_n$ is either 1 or (–1).

Property 2 *There is an even number of (–1)s among the products $p_1, p_2, \ldots, p_n$.*

This follows from the equation $P = p_1p_2 \ldots p_n = a_1^2a_2^2 \ldots a_n^2 = 1$.

Problem 2.11 and the discussion after it use the Fourth Situation.

Problem 2.11 (*Regional competition in Bulgaria*) Prove that if $S = 0$, then n is divisible by 4.

Solution Each term of the sum $S = p_1 + p_2 + \cdots + p_n$ is either 1 or (–1). Because $S = 0$, the number of (–1)s is equal to the number of 1s. According to Property 2 there is an even number (say $2k$) of (–1)s among the products $a_1a_2, a_2a_3, \ldots, a_na_1$ and the same number ($2k$) are the 1s. Therefore, n, which is the number of the terms in S, is the sum of one and the same even number ($2k + 2k = 4k$) and is divisible by 4.

An interesting phenomena is that the reverse statement is also true, namely *if n is divisible by 4, there is an arrangement of 1s and (–1)s into the* *cells, so that* $S = 0$. Certainly, consecutively write the four $\{1, 1, 1, -1\}$ several times and this gives $S = 0$.

The next step toward a possible extension is the observation that *if S is divisible by 4, then n is also divisible by 4*. This is because if S is divisible by 4, there is an even number of (–1)s and also an even number of 1s among the products $a_1a_2, a_2a_3, \ldots, a_na_1$. If we allocate one and the same number (say $2k$) of 1s and (–1)s to obtain a sum of 0, the sum of the remaining (–1)s (or 1s) is divisible by 4, therefore their number is also divisible by 4—let this number be $4m$. Then $n = 2k + 2k + 4m = 4(k + m)$ is divisible by 4. This way we conclude that *n is divisible by 4, if and only if S is divisible by 4*.

Similar reasoning can be applied when n is not divisible by 4. The conclusion is that n and S have one and the same remainder modulo 4. Therefore the following assertion is true:

Statement. In the notation of the Forth Situation, $n \equiv S(\text{mod} 4)$.

In line with the previous considerations, it seems worth mentioning one more problem that uses the Fourth Situation. Let $n \geq 3$ cells be arranged into a circle. Each cell can be occupied by 1 or (–1). The following operation is admissible: draw another n cells—one between any two of the existing cells; in each of these new cells write the product of the numbers in the two neighbor existing cells; then delete the existing cells. Initially there is a (–1) in one cell and 1s elsewhere. For which values of n is it possible to obtain 1s in all cells in a finite number of admissible steps? By itself the task is interesting, but it is actually analogous to problem 2.10.

4.3 Static Arrangements

Contest problems sometimes present interesting situations involving arrangements of numbers on a circle that satisfy particular properties. Even though this looks like "static arrangements", some operations are also involved in the justification of the properties of the arrangements. Two examples are considered below.

4.3.1 Example 1

One of the aims of this paper is to give some guidance of how to teach solving "non-standard" problems from mathematics competitions. For this purpose we refer to Polya and his works on problem solving that are considered as masterpieces in the area. The discussion below is guided by the

Polya's phases for problem solving (Polya 1946). They are: (i) understanding the problem; (ii) designing a plan; (iii) carrying out the plan; (iv) looking back. It is strongly recommended to consider them when teaching problem solving.

Problem 3.1 Let $2n+1,\ n \geq 2$, cells be arranged into a circle. Each cell is occupied by one of the numbers $1, 2, 3, \ldots, 2n+1$ and all these numbers are used. The numbers in any two neighbor cells are multiplied, so that $2n+1$ products are obtained. Denote by S the sum of these products. Find an arrangement of the numbers into the cells such that S has the largest possible value.

Solution As promised, the solution follows the Polya's phases for problem solving.

(i) *Understanding the problem.* Larger values of S can be obtained if the values of the terms of S are larger. Since each term is a product of two natural numbers, larger values means that "smaller" numbers has to be multiplied by "small" numbers; "larger" numbers—by "large" numbers. Therefore, the neighbors of 1 have to be 2 and 3. Then, 4 and 5 should go next to 2 and 3. Figure 4.11 shows the correct positions of 4 and 5—the one that is presented on the right circle.

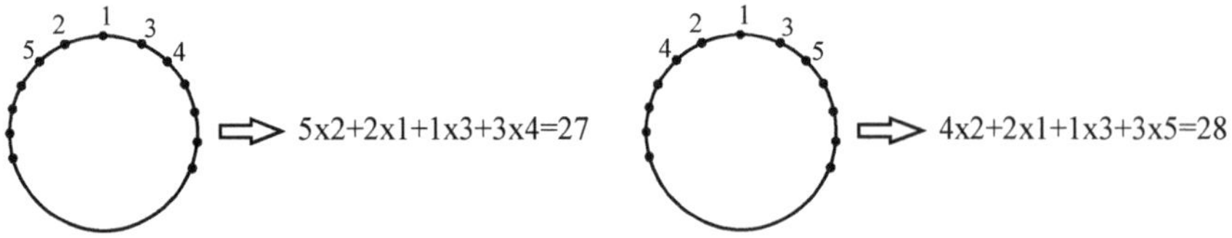

Fig. 4.11 Positions of 4 and 5

(ii) *Designing a plan.* The plan is to prove that the required optimal arrangement is the one presented in Fig. 4.12.

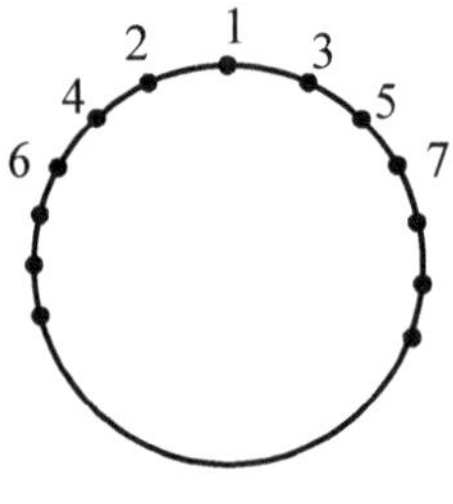

Fig. 4.12 The optimal arrangement

(iii) *Carrying out the plan.* We will use the Lemma 3.1 (see below, after the completion of the solution of the problem). We may assume that the cells divide the circle into $2n+1$ equal arcs. Let the numbers $1, 2, 3, \ldots, 2n+1$ be arranged in an optimal way, i.e. S has the largest possible value. Consider a diameter from one of the cells. According to the Lemma 3.1, for each pair of numbers that are symmetrical against this diameter, the smaller numbers are in one and the same of the semicircles and the larger numbers are in the other semicircle. It follows that numbers 1 and 2 are adjacent. This is because if they are not, there is a diameter separating 1 and 2, such that 1 and 2 are not symmetrical against this diameter. Denote by A and B the numbers symmetrical to 1 and 2 respectively (Fig. 4.13). Then $A > 1$ and $2 < B$,

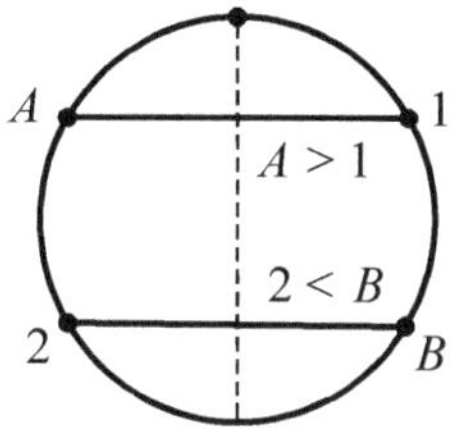

Fig. 4.13 Arrangements of numbers 1 and 2

which is a contradiction, because 1 and B are in one of the semicircles but 2 and A are in the other semicircle. Using mathematical induction we will prove that the required optimal arrangement is the one presented in Fig. 4.12. Assume that for any k, $k < n$, the numbers $2k, 2k-2, \ldots 4, 2, 1, 3, 5, \ldots, 2k-1$ are adjacent, as shown in Fig. 4.14. Let A and B be the adjacent numbers to $2k$ and $2k-1$ as in Fig. 4.14. Suppose that $A > 2k+1$ and $B > 2k+1$. Denote by C the

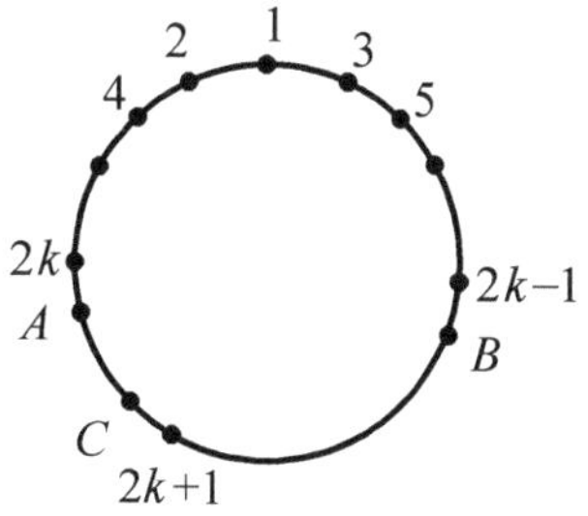

Fig. 4.14 Inductive process

number adjacent to $2k+1$ as shown in Fig. 4.14. There is a diameter such that the pairs $(C, 2k-1)$ and $(2k+1, B)$ are symmetrical. Since $C > 2k-1$ and $2k+1 < B$, this is a contradiction. Therefore either $A = 2k+1$ or $B = 2k+1$. The same way we may prove that either $A = 2k+2$ or $B = 2k+2$. Because of the Lemma 3.1, the only possibility is that $A = 2k+2$ and $B = 2k+1$. This completes the proof of the optimal arrangement.

(iv) *Looking back.* Some people miss this phase by thinking that when the solution is completed there is not a need to pay attention to the problem. "Looking back" phase gives conviction that the solution is correct. It also stimulate looking for "better" solutions (easier, more effective, smarter, etc.). This phase considers questions like "What did we do?", "Which methods did we use?", "What is important in the solution?", "Can something be improved?", etc. Guided by the last two questions we may realize that in the case of Problem 3.1 it is not important that the terms of *S* are the products of two neighbor cells. The reasoning is similar if the terms of *S* are the products of any given number of neighbor cells. This way we approach a problem given in the Moscow Mathematical Olympiad, 1999 (Fedorov et al. 2011).

Problem 3.2 Let 1999 cells be arranged into a circle. Each cell is occupied by one of the numbers 1, 2, 3, …, 1999 and all these numbers are used. Denote by *S* the sum of the products of the numbers in all sets of 10 consecutive numbers. Find an arrangement of the numbers into the cells such that *S* has the largest possible value.

Its solution is very similar to what was already done.

Here is the Lemma that was already used in the solution of Problem 3.1.

Lemma 3.1 *Let* $2n+1$, $n \geq 2$ *cells be arranged into a circle. These cells are occupied by different natural numbers* $a_1, a_2, a_3, \ldots, a_{2n+1}$ *such that* $a_1 > a_{2n}$. *The numbers in any two neighbor cells are multiplied, so that* $2n+1$ *products are obtained. Denote by S the sum of these products. Consider the following operation: for each* $i = 1, 2, \ldots n$, *numbers* a_i *and* a_{2n+1-i} *change their places if* $a_i < a_{2n+1-i}$, *and do not change the places otherwise* (Fig. 4.15). *Prove that if at least one pair of numbers change their places, the value of S gets larger.*

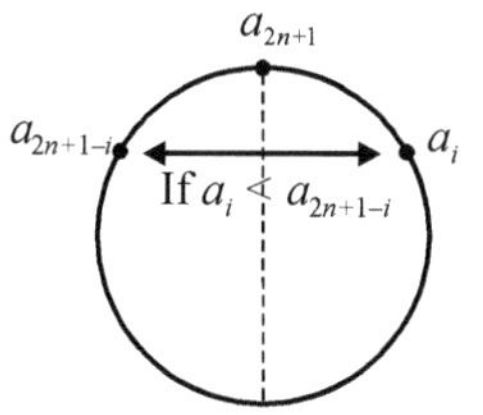

Fig. 4.15 The operation

Proof Consider the diameter from the cell containing the number a_{2n+1}. The meaning of the operation is to arrange the numbers in such a way that for each pair of numbers that are symmetrical against this diameter, the smaller numbers are in one and the same of the semicircles and the larger numbers are in the other semicircle. (For example, in Fig. 4.12, the smaller numbers are on the left semicircle and the larger—on the right semicircle.) Let the operation be completed and numbers a_i and a_{2n+1-i} change their places. This means that $a_i < a_{2n+1-i}$. The sum of the products that contain a_i and a_{2n+1-i} before the operation is

$$S_1 = a_{i-1}a_i + a_ia_{i+1} + a_{2n-i}a_{2n+1-i} + a_{2n+1-i}a_{2n+2-i}$$

and after the operation it is

$$S_2 = a_{i-1}a_{2n+1-i} + a_{2n+1-i}a_{i+1} + a_{2n-i}a_i + a_ia_{2n+2-i}.$$

Therefore,

$$S_2 - S_1 = (a_{2n+1-i} - a_i)(a_{i-1} - a_{2n+2-i})(a_{i+1} - a_{2n-i}).$$

The first factor is positive (because $a_i < a_{2n+1-i}$). The other two factors are also positive, since theirs first terms are from the right semicircle (Fig. 4.15) and the second terms are from the left semicircle. This is why $S_2 > S_1$.

4.3.2 Example 2

The solution to the last problem in this chapter uses the famous Unique Factorization Theorem. It states that every integer greater than 1 either is prime itself or is the product of prime numbers, and this product is unique, up to the order of the factors.

Problem 3.3 Let $2n+1$, $n \geq 2$ cells be arranged into a circle. Each cell is occupied by a natural number. For the numbers in any two neighbor cells the ratio of the greater to the smaller is calculated, so that $2n+1$ quotients are obtained. Is it possible to occupy these cells by natural numbers in such a way that each of these quotients is a power of a prime number?

Solution Assume that the natural numbers $a_1, a_2, a_3, \ldots, a_{2n+1}$ are written into the cells in the same order in such a way that each quotient $\frac{a_i}{a_{i+1}}$, $i=1,2,\ldots,2n+1$, $(a_{2n+2}=a_1)$ is either a power of a prime number or its reciprocal. Let m of these quotients are powers of prime numbers and they are $p_1^{\alpha_1}, p_2^{\alpha_2}, \ldots, p_m^{\alpha_m}$, and the other $2n+1-m$ are reciprocal to the powers of prime numbers and they are $\frac{1}{q_1^{\beta_1}}, \frac{1}{q_2^{\beta_2}}, \ldots, \frac{1}{q_{2n+1-m}^{\beta_{2n+1-m}}}$. Then

$$1=\frac{a_1}{a_2}\cdot\frac{a_2}{a_3}\cdot\ldots\cdot\frac{a_i}{a_{i+1}}\cdot\ldots\cdot\frac{a_{2n+1}}{a_1}=p_1^{\alpha_1}.p_2^{\alpha_2}\ldots p_m^{\alpha_m}\cdot\frac{1}{q_1^{\beta_1}}\cdot\frac{1}{q_2^{\beta_2}}\cdot\ldots\cdot\frac{1}{q_{2n+1-m}^{\beta_{2n+1-m}}}.$$

Therefore $A=p_1^{\alpha_1}.p_2^{\alpha_2}\ldots p_m^{\alpha_m}=q_1^{\beta_1}.q_2^{\beta_2}\ldots q_{2n+1-m}^{\beta_{2n+1-m}}$. Because of the unique factorization of number A, it follows that $m=2n+1-m$, i.e. $2m=2n+1$, which is impossible. The contradiction shows that such arrangement is not possible.

4.4 Problems to the Reader

Readers may try to find solutions to the following two problems. The first of them is about an admissible operation, the other one deals with a static arrangement.

Problem 4.1 Let 6 cells be arranged into a circle. Each cell is occupied by one number. The following operation is admissible: add 1 to any three neighbor numbers or subtract 1 from any three numbers that stand in one number (i.e. stand on the places numbered 1, 3, and 5; or numbered 2, 4, and 6). Initially, numbers 1, 2, 3, 4, 5, and 6 are written in the cells in this order. Is it possible to obtain one and the same number written in each cell in a finite number of admissible steps?

Problem 4.2 Let $n\geq 3$ cells be arranged into a circle. Each cell is occupied by one digit neither of which is 0. Albert writes a $(n-1)$—digit number by copying clockwise consecutively $(n-1)$ of the digits. Betty also writes $(n-1)$—digit number the same way. Even if they both start from different cells of the circle, their numbers are equal. Prove that the circle can be divided into several arcs in such a way that the numbers written by copying clockwise consecutively all digits of each of the arcs are all equal.

4.5 Conclusion

The circle is an amazing geometric figure. It has been known since before the beginning of recorded history. Natural circles would have been observed, such as the Moon, Sun, and a short plant stalk blowing in the wind on sand, which forms a circle shape in the sand. The circle is the basis for the wheel, which, with related inventions such as gears, makes much of modern machinery possible.

In mathematics, the study of the circle has helped inspire the development of geometry, astronomy, and calculus. Even in 1700 BC year the Rhind papyrus (the best example of Egyptian mathematics) gives a method to find the area of a circular field. The result corresponds to $\frac{256}{81} \approx 3.16049\ldots$ as an approximate value of π. In 300 BC years Book 3 of Euclid's Elements deals with the properties of circles.

Besides the many interesting geometrical properties, the circle is also a close curve. This makes it possible to consider the problems in this paper. An interesting feature of these problems is that they are not closely connected with the curriculum usually taught at school. Actually, they are not connected to any curricula because to understand the problems one does not need to possess particular mathematical knowledge. However, finding solutions needs a lot of mathematical reasoning, experience, and intuition. In this respect these problems are one of the best examples of the beauty of mathematics.

The German psychologist Karl Duncker said: "Problem arises when someone has a goal for which he/she does not know a path for its achievement". From this point of view this chapter presents excellent examples of "problems". They are in contrast with what is usually taught at school: routine problems and exercises that are purposeful activities with known path for its achievement.

Thanks to mathematics competitions such problems become known to students, teachers and many others that are interested in mathematics. The acquaintance and discovering of their solutions is the best way for students to get involved in sensible mathematical activities. This is also a way to present the students the beauty of mathematical ideas and to attract them to mathematics.

References

4-th Austrian—Polish Mathematical Competition (1981). Retrieved November 31, 2016, from http://imomath.com/othercomp/Apo/ApoMC81.pdf.

Fedorov, R., Belov, A., Kovaldzhi, A., & Yashchenko, I. (2011). *Moscow Mathematical Olympiads, 1993–1999*. AMS: MSRI.

Gross, J. (2008). *Combinatorial methods with computer applications*. New York: Taylor & Francis Group.

Polya, G. (1946). *How to solve it*. Princeton University Press.

Rakovska, D., Tonov, I., Bankov, K., Vitanov, T., Tonova, T., & Lozanov, Ch. (2007). *Mathematics competitions, 4–7 grades*. Sofia, Regalia 6. (In Bulgarian), p. 79.

Sergeev, I. (Eds.). (1987). *Foreign mathematical Olympiad*. Nauka, Moskow. (In Russian), p. 74.

Chapter 5
Combinatorial Problems in the Mathematical Olympiad of Central America and the Caribbean

L.F. Cáceres-Duque, J.H. Nieto Said and R.J. Sánchez Lamoneda

Abstract In this article we analyze the combinatorial problems proposed at the Mathematical Olympiad of Central America and the Caribbean, during its eighteen years of existence. The different types of combinatorial problems (counting, existence, strategy games, etc.) are explained and illustrated with various examples. Some original problems, submitted to the olympiad but not selected in the papers, are also discussed.

Keywords Olympiad · Combinatorics · Problem · Solution · Central America · Caribbean

5.1 Introduction

The Mathematical Olympiad of Central America and the Caribbean (Centro) was created in 1999 with the aim of encouraging the participation of countries of the region in international mathematical competitions, goal that has been successfully achieved. The Centro is an International Mathematical

L.F. Cáceres-Duque (✉)
University of Puerto Rico, Mayagüez Campus, Mayagüez, USA
e-mail: luis.caceres1@upr.edu

J.H. Nieto Said
Universidad Del Zulia, Maracaibo, Venezuela
e-mail: jhnieto@gmail.com

R.J. Sánchez Lamoneda
Universidad Antonio Nariño, Bogotá, Colombia
e-mail: lamonedar@gmail.com

A. Soifer (ed.), *Competitions for Young Mathematicians*,
ICME-13 Monographs, DOI 10.1007/978-3-319-56585-9_5

Olympiad (IMO) type competition: the exam is applied in two consecutive days; each day the contestants are allowed four and a half hours for solving three problems, each one with a seven points value. Each country participates with a delegation consisting of a Leader, a Deputy Leader and at most three students. The contest is addressed to young high school students, less than 16 years old the year previous to the contest.

More information about the Centro and statistical data may be found in Cáceres et al. (2016), Nieto and Sánchez (2005, 2009).

The main areas for the contest problems are algebra, combinatorics, geometry and number theory. The proposed problems from 1999 to 2014, with complete solutions, are found in Nieto (2015). In this article we focus on combinatorial problems, for the following reasons:

1. Combinatorial problems are around one third of all the problems proposed so far.
2. Combinatorics is the area less understood and developed in the high school's mathematical curricula in the region.

Our purpose is to analyze the set of combinatorial problems proposed in the Centro, their types (counting problems, existence problems, strategy games, etc.) and the main ideas involved. We will also include some original problems from the bank of problems which were not selected for the competition.

5.2 Contest Problems

The contest problems in combinatorics, proposed from 1999 to 2016, may be classified in five categories: strategy games, configuration problems, extremal problems, counting problems and miscellaneous problems. In the following pages we examine typical problems in each of these classes. As we mentioned above, each year's competition has six problems. We name the problems specifying the year and the position of the problem on the exam. We should mention that the difficulty of the problems is intended to be in the order 1, 4, 2, 5, 3, 6, from the easiest to the most difficult.

5.2.1 Counting Problems

These problems ask to count the number of some kind of configurations. This is one of the main concerns of combinatorics. To succeed in this task it is necessary to know the basic counting principles: the addition principle, the product principle and the inclusion-exclusion principle. In addition the students should know how to count several typical configurations: subsets, functions, sequences, multisets, permutations, etc. A basic knowledge of recurrence relations is also important. This material may be found in many books, for example in Andreescu and Feng (2004).

Proposed problems in this area during the 18 years of the competition: 2003-5, 2004-6, 2008-4, 2011-1, 2014-4.

Problem 2003-5. An 8×8 square board is divided into 64 square cells, each cell of side 1cm. Each cell can be colored either white or black. Find the total number of ways to color the board such that each 2×2 square formed by four cells with a common vertex contains two white cells and two black cells.

Solution: We start by painting the first row arbitrarily and then we try to extend the coloring to the whole board satisfying the given conditions. If two consecutive cells in the first row have the same color, we paint the two cells below them with the opposite color; then it is easy to see that the only admissible way to paint the second row is to color each cell with the opposite color that the cell above it. We repeat this way to color the third, fourth, and so on, up to the last row. On the other hand, if in the first row there are not two consecutive cells with the same color, this means the cells are painted in an alternating pattern, WBWBWBWB or BWBWBWBW; then the second row can be coloured either WBWBWBWB or BWBWBWBW. The same is true for the remaining rows. Hence any of the two alternating ways to color the first row can be extended to 2^7 ways to color the whole board, and each one of the $2^8 - 2$ non alternating ways to color the first row can be extended in an unique way. Therefore we will have $2 \cdot 2^7 + 2^8 - 2 = 2^9 - 2 = 510$ different ways to paint the board.

Problem 2011-1. In each vertex of a cube there is a fly. When a whistle sounds, each fly moves to a vertex located on some of the faces to which its vertex belongs, but diagonally opposed to it. In how many ways can the flies move, in such a way that no vertex ends with two or more flies?

Solution: Note first that the set of vertices of the cube can be partitioned into two disjoint subsets: $\{A, C, F, H\}$ (represented with white dots in Fig. 5.1)

Fig. 5.1 Problem 2011-1

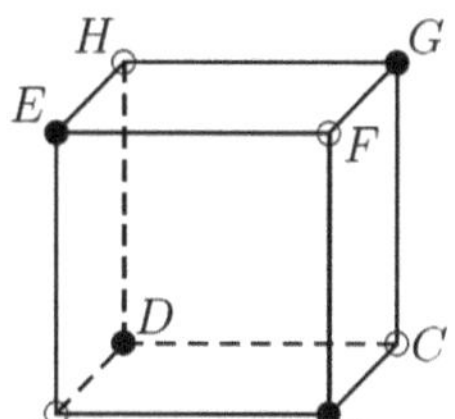

and $\{B, D, E, G\}$ (represented with black dots in Fig. 5.1), and each fly can only move to a vertex belonging to the same set of its initial vertex.

Therefore it suffices to compute in how many ways the flies located in each subset can move, and then multiply both numbers. A fly located in A has 3 possible destinations (C, F and H). Without loss of generality let us consider that it moves to C. If the fly in C moves to A, then the flies in F and H also have to interchange positions. If the fly in C moves to F, the fly in F must move to H and this one to A (if the fly in F moves to A then the one in H would not have a place to go). Similarly if the fly in C moves to H, the fly in H must move to F and this one to A. Then there are 3 possibilities if A moves to C, and of course another 3 if it moves to F and another 3 if it moves to H, making a total of 9 possibilities. The flies in B, D, E and G can also move in 9 different ways and therefore by the product principle the answer is $9 \times 9 = 81$.

Note: The number of different ways that the flies located in A, C, F and H can fly are the *derangements* of $\{A, C, F, H\}$, namely the number of permutations of 4 elements without fixed points, whose number is $D_4 = 4!(\frac{1}{2!} - \frac{1}{3!} + \frac{1}{4!}) = 9$.

5.2.2 *Strategy Games*

We have noticed that this type of problem is clearly a favorite in the Centro. The games are mostly bipersonal, finite, complete information and without tie. The statements ask to find a winning strategy for some of the players (by Zermelo's theorem, one of the players has a winning strategy). There are two exceptions: 2010-3, which is a solitaire, and 2013-4, a game in which no player has a winning strategy (Zermelo's theorem does not apply because the game is not finite).

To solve these problems a useful strategy is to examine simple cases and look for a pattern. In some problems it is also useful to look for invariants.

Proposed problems in this area during the 18 years of the competition: 1999-3, 2001-1, 2003-1, 2004-1, 2005-4, 2008-3, 2009-3, 2010-3, 2013-4, 2015-4 and 2016-4.

Problem 2010-3. A player places a tile (a 1×1 square) on a cell of an $m \times n$ board divided into squares of size 1×1. The player moves the tile according to the following rules:

- In each move, the tile goes from the occupied cell to another cell having one side in common with it.
- The tile cannot move to a previously occupied cell.
- Two consecutive movements cannot have the same direction.

The game ends when the player cannot move the tile. Determine all values of m and n for which the player can place the tile on the board and move it in such a way that the tile has occupied all the cells before finishing the game.

Solution: It will be shown that the possible values of (m, n) are $(1, 1)$, $(2, k)$ and $(k, 2)$, where k is any positive integer.

The case $(m, n) = (1, 1)$ is clear, since by placing a tile on the board, automatically the game ends and the tile has occupied all the cells.

For $2 \times k$ boards, consider the moving pattern shown in Fig. 5.2, where the $\circ$ represents the initial location of the tile. This pattern will eventually fill the $2 \times k$ board, ending the game. We proceed analogously for a $k \times 2$ board.

It remains to show that other board sizes are not possible. If one side of the board is of length 1 and the other is of length greater or equal than 3, it is clear that the third condition (two consecutive movements cannot have the same direction) makes impossible to visit all the cells in the board, because after the first movement a turn is required, which is impossible.

If the lengths of the two sides of the board are greater than 2, it is possible to isolate at least one corner as that shown in Fig. 5.3, which satisfies the following conditions:

(a) The tile has not started in cells 1, 2 or 3.
(b) The route of the tile will not end in cell 1.

According to condition (b), cell 1 must be reached from another cell and from this cell you must go into another cell. Assume, without loss of generality, that the way around cell 1 is made in the order $2 \to 1 \to 3$.

Now, as the tile has not started in cell 2, it must come from another cell, and due to the prohibition of two consecutive moves in the same direction,

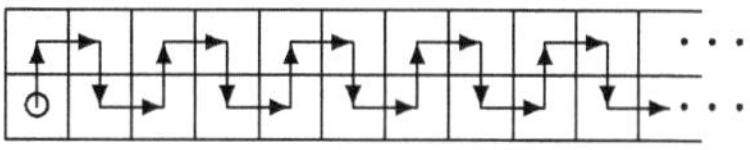

Fig. 5.2 Problem 2010-3 moving pattern

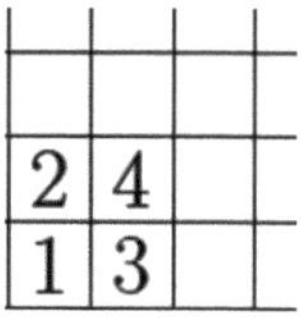

Fig. 5.3 Special case for Problem 2010-3

it is necessary that the tile comes from the cell marked as 4. Thus, the route can be traced back to $4 \to 2 \to 1 \to 3$.

Note that cell 4 is already part of the route, so that the journey should end in cell 3, since it is impossible to continue from cell 3 without continuing in the same direction or repeating a cell.

The above paragraphs show that, of the four corners of the board, at least two meet condition (a) and at least one meets both conditions. As shown, that cell should be the penultimate cell of the route, and the last cell of the route cannot be another corner. Thus, there is another corner that satisfies the conditions, and we would have two corners of the board being simultaneously the penultimate corner of the route, which is absurd. This concludes the proof.

Problem 2013-4. Ana and Beatriz alternate turns in a game that starts with a square of side 1 drawn on an infinite board. A move is to draw a square that does not overlap with the figure already drawn, so that one of the sides is a (full) side of the rectangle which is already drawn. The player who draws a figure whose area is a multiple of 5 is the winner. If Ana makes the first move, is there a winning strategy for any player?

Solution: There is no winning strategy for any player. Suppose that at some moment the figure is an $a \times b$ rectangle, where a, b and $a + b$ are not multiples of 5. The player X who has the turn in that moment can convert the rectangle into an $(a + b) \times b$ rectangle or into an $a \times (a + b)$ rectangle. None of these moves wins immediately, but at least one of them avoids losing in the next turn. Indeed, if $5 \mid 2a + b$ and $5 \mid a + 2b$, then $5 \mid 3(a + b)$ and therefore $5 \mid a + b$, contrary to the assumption. Thus, if $5 \nmid a + 2b$, X avoids losing making up a rectangle of $(a + b) \times b$; otherwise they will have that $5 \nmid 2a + b$ and X avoids losing making up an $a \times (a + b)$ rectangle. Since each player can make a play that avoids losing in the next turn, neither of them has a winning strategy.

Alternative solution: If at some moment the figure made is an $a \times b$ rectangle, where a and b are not multiples of 5, the player with the turn can convert it into an $(a + b) \times b$ rectangle or into an $a \times (a + b)$ rectangle. If $5 \mid a + b$ she wins, otherwise she must choose a move that avoids losing in the next round. The first moves, if nobody wants to lose, are unique:

$$(1,1) \to (1,2) \to (1,3) \to (4,3) \to (4,7) \to (11,7) \to (11,18) \to (29,18) \to \cdots$$

Since we only care if a or b are multiples of 5, we can work modulo 5 to obtain

$$(1,1) \to (1,2) \to (1,3) \to (4,3) \to (4,2) \to (1,2) \to (1,3) \to (4,3) \to \cdots$$

We see that after the second move the cycle $(1, 2) \to (1, 3) \to (4, 3) \to (4, 2)$ repeats, therefore this sequence is periodic and there cannot be a winning strategy.

Problem 2016-4. On the board the number 3 is written. Ana and Bernardo play alternately, starting with Ana, as follows: if the number n is written on the board, the player with the turn must replace it by any integer m that is coprime with n and such that $n < m < n^2$. The first player to write a number greater or equal than 2016 looses the game. Determine which player has a winning strategy and describe it.

Solution: Let us say that a number is a *winner* if any player who writes it can ensure victory, otherwise it is a *loser*. Obviously $2015 = 5 \cdot 13 \cdot 31$ is a winner. The number $2005 = 5 \cdot 401$ is also a winner, because any possible answer between 2006 and 2014 can be replied with 2015.

Let us show now that Ana has a winning strategy. In her first move she writes 5. Then she can proceed in several ways, writing multiples of 5 up to 2005 or 2015. One way to do this is as follows: Bernardo must respond to 5 with an m_1 coprime with 5 and such that $5 < m_1 < 25$. Then Ana writes $25 = 5^2$. Bernardo must respond with an m_2 coprime with 5 and such that $25 < m_2 < 625$. Then Ana writes $625 = 5^4$. Bernardo must respond with an m_3 coprime with 5 and such that $625 < m_3 < 5^8$. If $m_3 \geq 2016$, Ana wins. Obviously m_3 cannot be 2015. If $2005 < m_3 < 2015$, Ana writes 2015 that is a winner. If m_3 is $802 = 2 \cdot 401$, $1203 = 3 \cdot 401$ or $1604 = 4 \cdot 401$, Ana writes 2015 that is a winner. If $625 < m_3 < 2005$ and m_3 is coprime with 5 and 401, Ana writes $2005 = 5 \cdot 401$ that is a winner.

5.2.3 Configuration Problems

In these problems one has to find (or prove the existence of) discrete configurations with certain properties, or to prove some property of a class of configurations.

Useful tools for these problems are existence results such as the *pigeonhole principle* (if more than n objects are distributed among n boxes, then some box contains more than one object) and its generalization, the Dirichlet's principle (if more than nk objects are distributed among n boxes, then some box contains more than k objects). However, these problems often require hard work to construct the desired configurations.

Proposed problems in this area during the 18 years of the competition: 2000-4, 2001-6, 2002-1, 2002-6, 2005-6, 2009-4 and 2015-1.

Problem 2002-1. For which integers $n \geq 3$ is it possible to accommodate, in some order, the numbers 1, 2, ..., n in a circle so that each number divides the sum of the two following numbers in the clockwise direction?

Solution: Clearly it is possible for $n = 3$, and we will see that this is the only possible case. If 1, 2,..., n may be accommodated satisfying the requested condition, then there cannot be two consecutive even numbers, otherwise the next one would also be even, and then all of them would be even. Also it is not possible to have two even numbers with just one odd number in between. Therefore after each even number there must be at least two odd numbers. This implies that the amount of odd numbers is at least twice the amount of even numbers, which only happens if $n = 3$.

Problem 2005-6. There are n cards numbered from 1 to n and p boxes to store them, with p prime. Determine the possible values of n for which it is possible to store all the cards so that the sum of the cards in each box is the same.

Solution: We will say that a set A of integers is r-decomposable if there exists a partition of A in r disjoint blocks, such that the sums of the elements of each block are all equal. It is easy to see that any set of $2kr$ consecutive integers $\{a, a+1, \ldots, a+2kr-1\}$ is r-decomposable since, as each pair $\{a+j, a+2kr-1-j\}$ (for $j = 0, 1, \ldots, kr-1$) has the same sum $2a+2kr-1$, joining these pairs in groups of k we obtain r blocks of equal sum. It is also clear that the union of r-decomposable disjoint sets is also r-decomposable.

The objective here is to determine for which positive integers n the set $A_n = \{1, 2, \ldots, n\}$ is p-decomposable.

If $p = 2$, the solution is the positive integers n congruent with 0 or 3 modulo 4. Indeed, if $n \equiv 1 \pmod 4$ or $n \equiv 2 \pmod 4$ then $1+2+\cdots+n = n(n+1)/2$ is odd and A_n is not 2-decomposable. As shown above A_{4k} is 2-decomposable. Since $A_3 = \{1, 2\} \cup \{3\}$ is 2-decomposable and so is $\{4, 5, \ldots, 4k+3\}$ (because they are $4k$ consecutive integers), then their union A_{4k+3} is 2-decomposable.

If $p > 2$, the integers n such that A_n is p-decomposable are those $n \geq 2p-1$ congruent with 0 or -1 modulo p. Indeed, since $1+2+\cdots+n = n(n+1)/2$, if A_n is p-decomposable then $p \mid n(n+1)/2$, i.e., $p \mid n$ or $p \mid (n+1)$, or equivalently $n \equiv 0 \pmod p$ or $n \equiv -1 \pmod p$.

But evidently neither A_{p-1} nor A_p are p-decomposables. However A_{2kp} is p-decomposable. Also A_{2p-1} is p-decomposable because $A_{2p-1} = \{1, 2p-2\} \cup \{2, 2p-3\} \cup \cdots \cup \{p-1, p\} \cup \{2p-1\}$, so A_{2kp-1} is p-decomposable since it is the union of A_{2p-1} and $\{2p, 2p+1, \ldots, 2kp-1\}$ (which are

$2(k-1)p$ consecutive integers). It only remains to see that A_{kp} and A_{kp-1} are p-decomposable for $k \geq 3$ odd. Let us begin with $k = 1$ (that is to say $n = 3p$).

A_{3p} is partitioned into the blocks:

$\{i, (3p-1)/2 + i, 3p + 2 - 2i\}$ for $1 \leq i \leq (p+1)/2$ and

$\{i, (p-1)/2 + i, 4p + 2 - 2i\}$ for $(p+3)/2 \leq i \leq p$.

A_{3p-1} is partitioned into the blocks:

$\{(3p-1)/2, 3p-1\}$,

$\{i-1, (3p-1)/2 + i - 1, 3p + 1 - 2i\}$ for $2 \leq i \leq (p+1)/2$,

and $\{i-1, (p-3)/2 + i, 4p + 1 - 2i\}$ for $(p+3)/2 \leq i \leq p$.

Now for $k = 3 + 2t$, as $A_{kp} = A_{3p} \cup \{3p+1, \ldots, 3p+2tp\}$ and $A_{kp-1} = A_{3p-1} \cup \{3p, \ldots, 3p+2tp-1\}$, then A_{kp} and A_{kp-1} are p-decomposable, being disjoint unions of p-decomposable sets.

If $p \equiv 1 \pmod 4$, A_{3p} can be partitioned into three different types of blocks: $(3p-3)/4$ blocks with two elements each, $(p-1)/4$ blocks with six elements each and an additional block with the remaining elements. Each block has sum $(9p+3)/2$. The blocks are:

$\{(3p+3)/2, 3p\}, \{(3p+5)/2, 3p-1\}, \ldots, \{(9p-1)/4, (9p+7)/4\}$,

$\{1, 2, 3, (3p-5)/2, (3p-3)/2, (3p-1)/2\}$,

$\{4, 5, 6, (3p-11)/2, (3p-9)/2, (3p-7)/2\}, \ldots$,

$\{(3p-11)/4, (3p-7)/4, (3p-3)/4, (3p+5)/4, (3p+9)/4, (3p+13)/4\}$

and $\{(3p+1)/4, (3p+1)/2, (9p+3)/4\}$.

For $p = 3$ we have $A_9 = \{1, 2, 3, 4, 5\} \cup \{6, 9\} \cup \{7, 8\}$.

If $p \equiv 3 \pmod 4$ and $p \geq 7$, A_{3p} can be partitioned into the following blocks:

$\{(3p+3)/2, 3p\}, \{(3p+5)/2, 3p-1\}, \ldots, \{(9p+1)/4, (9p+5)/4\}$,

$\{1, 2, 3, (3p-5)/2, (3p-3)/2, (3p-1)/2\}$,

$\{4, 5, 6, (3p-11)/2, (3p-9)/2, (3p-7)/2\}, \ldots$,

$\{(3p-17)/4, (3p-13)/4, (3p-9)/4, (3p+11)/4, (3p+15)/4, (3p+19)/4\}$

and $\{(3p-5)/4, (3p-1)/4, (3p+3)/4, (3p+7)/4, (3p+1)/2\}$.

Similarly for A_{3p-1}. If $p \equiv 1 \pmod 4$, the blocks are:

$\{(3p-1)/2, 3p-1\}, \{(3p+1)/2, 3p-2\}, \ldots, \{(9p-5)/4, (9p-1)/4\}$,

$\{1, 2, 3, (3p-7)/2, (3p-5)/2, (3p-3)/2\}$,

$\{4, 5, 6, (3p-13)/2, (3p-11)/2, (3p-9)/2\}, \ldots$,

$\{(3p-11)/4, (3p-7)/4, (3p-3)/4, (3p+1)/4, (3p+5)/4, (3p+9)/4\}$.

For $p = 3$ we have $A_8 = \{1, 2, 3, 6\} \cup \{4, 8\} \cup \{5, 7\}$.

If $p \equiv 3 \pmod 4$ and $p \geq 7$, the blocks are:

$\{(3p-1)/2, 3p-1\}, \{(3p+1)/2, 3p-2\}, \ldots, \{(9p-7)/4, (9p+1)/4\}$,

$\{1, 2, 3, (3p-7)/2, (3p-5)/2, (3p-3)/2\}$,

$\{4, 5, 6, (3p-13)/2, (3p-11)/2, (3p-9)/2\}, \ldots$,

$\{(3p-17)/4, (3p-13)/4, (3p-9)/4, (3p+7)/4, (3p+11)/4, (3p+15)/4\}$

and $\{(3p-5)/4, (3p-1)/4, (3p+3)/4, (9p-3)/4\}$.

Problem 2009-4. We want to place natural numbers around a circle satisfying the following property: the differences between each pair of neighboring numbers, in absolute value, are all different.

(a) Is it possible to place the numbers from 1 to 2009 satisfying the property?
(b) Is it possible to remove one of the numbers from 1 to 2009, so that the remaining 2008 numbers can be placed satisfying the property?

Solution: (a) It is not possible. There should be 2009 differences, but the lowest possible difference is 1 and the highest possible difference is $|2009 - 1| = 2008$, hence by the pigeonhole principle some difference should appear more than once.

(b) Yes. Suppose that a is withdrawn. If the remaining numbers are placed in the order $1, 2009, 2, 2008, \ldots, a-1, 2011-a, a+1, 2010-a, \ldots, 1005, 1006$, the differences between neighbors are:

$2008, 2007, \ldots, 2012-2a, 2010-2a, 2009-2a, \ldots, 2, 1,$

and the difference between the first and the last numbers is 1005. In order to have all of them different it suffices that $2011 - 2a = 1005$, this is $a = 503$, and we have the order

$1, 2009, 2, 2008, \ldots, 502, 1508, 504, 1507, 505, 1506, \ldots, 1005, 1006.$

Problem 2015-1. We want to write n different real numbers, with $n \geq 3$, around a circle, so that each of them is equal to the product of his neighbor on the right by his neighbor on the left. Determine all values of n for which this is possible.

Solution: The only possible value is $n = 6$. Let $a_1, a_2, \ldots, a_n$ be the numbers ordered clockwise.

If $n = 3$, then $a_1 = a_2a_3$ (1) and $a_2 = a_1a_3$ (2). Replacing a_2 from (2) into (1) we obtain $a_1 = (a_1a_3)a_3$, then $1 = a_3^2$ and a_3 is 1 or -1. Similarly $a_1 = \pm 1$ and $a_2 = \pm 1$. But by the pigeonhole principle at least two of these numbers must be equal, therefore for $n = 3$ it is not possible.

If $n = 4$, we have $a_2 = a_1a_3 = a_4$, which is not possible.

If $n = 5$, we have $a_2 = a_1a_3$ (1), $a_3 = a_2a_4$ (2) and $a_4 = a_3a_5$ (3). By replacing a_2 from (1) and a_4 from (3) in (2) we obtain $a_3 = (a_1a_3)(a_3a_5)$, therefore $a_1a_3a_5 = 1$. Similarly, starting with a_3 instead of a_1 we obtain $a_3a_5a_2 = 1$. Then $a_1a_3a_5 = a_3a_5a_2$, hence $a_1 = a_2$, which is not possible.

For $n = 6$ there are solutions, for example $2, 6, 3, \frac{1}{2}, \frac{1}{6}, \frac{1}{3}$.

If $n \geq 7$ then $a_1a_3a_5 = 1$ (as seen in the $n = 5$ case) and similarly $a_3a_5a_7 = 1$, then $a_1a_3a_5 = a_3a_5a_7$ and $a_1 = a_7$, therefore it is not possible.

5.2.4 *Extremal Problems*

These problems may be viewed as a subclass of the preceding one, where one is asked to find configurations which maximize or minimize a certain function.

Proposed problems in this area during the 18 years of the competition: 1999-6, 2004-4, 2012-4, 2012-5 and 2015-6.

Problem 1999-6. Let S be a subset of $\{1, 2, 3, \ldots, 1000\}$ such that the sum of every two different elements of S does not belong to S. Find the maximum number of elements of S.

Solution: Let S be a set which satisfies the condition given in the problem and let m be its maximum element. If m is odd, the set $\{1, 2, \ldots, m-1\}$ can be partitioned in pairs $\{x, m-x\}$, with $1 \leq x \leq (m-1)/2$, each one of them can contain at most one element of S. Therefore $|S| \leq (m-1)/2 + 1 \leq 499 + 1 = 500$. Similarly if m is even, the set $\{1, 2, \ldots, m-1\}$ is partitioned by the pairs $\{x, m-x\}$, with $1 \leq x \leq m/2 - 1$, and the single set $\{m/2\}$. In this case $|S| \leq (m/2 - 1) + 1 + 1 = m/2 + 1 \leq 501$. Since the set $\{500, 501, \ldots, 1000\}$ has the desired property, the maximum we are looking for is 501. In fact, $\{500, 501, \ldots, 1000\}$ is the only set with 501 elements which has the property.

Problem 2004-4. We have a 10×10 board and each cell is painted either white or black. Half of the cells are painted in white and the other half in black. A common side between two neighbouring cells is called a *border side* if these two cells have different colors. Determine the minimum and the maximum numbers of border sides that the table can have. Justify your answer.

Solution: The maximum is 180 and it can be obtained when we color the table as a checkerboard. Indeed, the segments that can be border sides are the interior segments (those who are not on the edge of the board) and all of them are border sides when we color the board as a checkerboard.

The minimum number is 10 and we get it in the following way: let us consider one of the lines which joins the middle points of two opposite sides of the board. Let us paint in white all the cells in one of the sides of this line and in black the cells on the other side. In order to show that 10 is the minimum, we note that the number of vertical border sides between two adjacent columns cannot be less than the difference (in absolute value) between the amount of black cells in each column. Hence, if we modify each column moving all the white cells to the top and the black cells to the bottom, the number of border

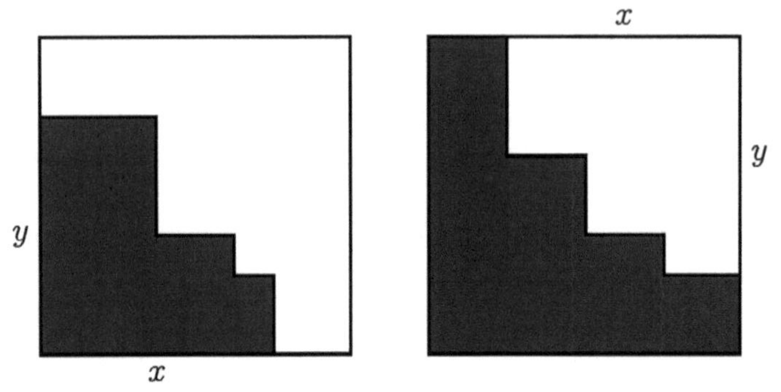

Fig. 5.4 Problem 2004-4

segments does not increase. Repeating this procedure on the rows we obtain a new coloring with the number of border segments less than or equal to the number of border segments of the original. Even more, in this new coloring if a cell is black, then the cells below it and to its left are also black. If there is a whole row whose cells are white and another row with all the cells black (or a whole column white and another one black), it is clear that there are at least 10 border segments. Otherwise we will have one of the cases shown in Fig. 5.4. In each one of them the number of border segments is $x + y$, and in both cases we have

$$x + y \geq 2\sqrt{xy} \geq 2\sqrt{50} > 14.$$

Problem 2015-6. In a Mathematical Olympiad 39 students participated. They had to solve 6 problems and each problem was worth 1 point if it was correct or 0 otherwise. For any three students there is at most one problem that was not solved by any of them. Let B be the sum of the points obtained by the 39 students. Find the smallest possible value for B.

Solution: Let S_i be the set of all students that did not solve problem i. By hypothesis $|S_i \cap S_j| \leq 2$ for every $1 \leq i, j \leq 6$. Since $B = 39 \cdot 6 - \sum |S_i|$, minimizing B is the same as maximizing $\sum |S_i|$. By the inclusion-exclusion principle and Bonferroni's inequality we have

$$\begin{aligned} 39 \geq |\cup S_i| &= \sum |S_i| - \sum |S_i \cap S_j| + \cdots \geq \sum |S_i| - \sum |S_i \cap S_j| \\ &\geq \sum |S_i| - \binom{6}{2} \cdot 2. \end{aligned}$$

Then $\sum |S_i| \leq 39 + 30 = 69$ and $B = 39 \cdot 6 - \sum |S_i| \geq 234 - 69 = 165$.

Now let us see that it is possible to get 165 points. Let P be the 6 problems set. For every subset $Q \subset P$ with $|Q| = 4$, let us take 2 students who solved, each one of them, the problems in Q and no other problem. Since there are $\binom{6}{4} = 15$ subsets Q we have 30 students. Let us say that each one of the other 9 students solved 5 problems. This configuration satisfies the problem conditions and the total number of points is $30 \cdot 4 + 9 \cdot 5 = 165$.

In conclusion the least possible value for B es 165.

5.2.5 *Miscelaneous Problems*

These problems include: (a) problems with a graph-theoretic flavor (1999-1, 2006-5); (b) tesselations (2000-2, 2010-4); (c) production rules in a formal language (2007-4); (d) dynamic process (2013-2).

Graph Theory has not appeared formally in the Centro, with its language and concepts. However there are some problems which may be conveniently represented by graphs. For that reason we think that some familiarity with the basic graph theoretic ideas would be an asset for any competitor. Tesselations is a theme that appears frequently among the submitted problems. Coloring techniques are usually useful for this type of problems.

Problem 2006-5. Olympia is a country formed by n islands. Every island has a different number of inhabitants and Panacentro is the island with the biggest population of all. They wish to build up bridges between the islands such that each bridge can be travelled in both directions and, for every two islands, there will be not more than one bridge connecting them. The following conditions have to be satisfied:

- It is always possible to go from Panacentro to each other island, travelling across the bridges.
- If one goes from Panacentro to any other island, crossing each bridge no more than once, then the numbers of inhabitants in the visited islands must be strictly decreasing.

Determine the number of ways to build up the bridges.

Solution: Let I_1, I_2,..., I_n be the islands of Olympia in decreasing order of inhabitants (I_1 is Panacentro). We claim that for every island I_k, with $2 \leq k \leq n$, there exists a unique $j < k$ such that I_j and I_k are connected by a bridge. Indeed, by hypothesis we know that there is a path c from I_1 to I_k, and without loss of generality we may assume that it does not pass more than once by the same bridge. If the last island visited before arriving to I_k is I_j, by the second condition of the problem it must be $j < k$. Let us assume now, by contradiction, that there is a bridge from another island I_i to I_k, with $i < k$. If c does not pass by I_i, then we could prolong c up to I_i, but this is a contradiction because I_i has more inhabitants than I_k. Otherwise if c passes by I_i it must be $i < j < k$ and the path c must be of the form $I_1 \dots I_i \dots I_j I_k$. But then we could build a new path touring the initial section $I_1 \dots I_i$ of c, going from there directly to I_k and then to I_j, and again it is a contradiction.

Hence I_2 must be directly joined to I_1. I_3 must be linked by a bridge either to I_1 or I_2. I_4 must be linked by a bridge either to I_1, I_2 or I_3, and so on. By the multiplication principle the total number of possible ways to build the bridges is given by $1 \times 2 \times 3 \times \cdots \times (n-1) = (n-1)!$.

Problem 2007-4. The inhabitants of an island speak a language in which every word can be written with the following letters: a, b, c, d, e, f and g. A word *produces* another word if it is possible to arrive from the first word to the second one applying at least one of the following rules:

1. Change one letter for two letters according to the following rules:

$$a \to bc, b \to cd, c \to de, d \to ef, e \to fg, f \to ga, g \to ab.$$

2. If two equal letters surround another one you can eliminate the equal letters. Example: $dfd \to f$.

   For example, $cafcd$ produce $bfcd$, because

$$cafcd \to cbcfcd \to bfcd.$$

Show that in this island every word produces any other word.

Solution: Let us use the notation $\Rightarrow$ to denote that one word produces another word. We see that $a \to bc \to cdc \to d$ therefore $a \Rightarrow d$. Analogously $d \Rightarrow g$, $g \Rightarrow c$, $c \Rightarrow f$, $f \Rightarrow b$, $b \Rightarrow e$ and $e \Rightarrow a$. Hence, since $\Rightarrow$ is transitive, each letter produces any other letter. This implies that any word with n letters, just changing orderly each one of its letters by a, produces a word formed by n letters a. Then, if n is odd, we can apply the second rule repeatedly to obtain a word formed by just one a. If n is even, we can obtain the word aa. But a produce g, so we have $aa \Rightarrow ga \to aba \to b \Rightarrow a$, and also in this case we can obtain the word a.

Now observe that the first rule is reversible, in the sense that $bc \Rightarrow a$, $cd \Rightarrow b, \dots, ab \Rightarrow g$. Indeed, from $bc, cd, \dots, ga, ab$ we can produce a, and from a we can produce $b, c, \dots, g$. Hence all instances of the first rule are reversible.

To reverse the second rule we first observe that $a \to bc \to cdc$. Now, given two letters x and y, since any letter produces any other we can go from x to a, from a to cdc and finally from cdc to yxy. Hence all instances of the second rule are reversible.

Therefore, since any word produces a, we can reverse the process to produce any other word from a.

Problem 2010-4. We wish to tile an $N \times N$ square yard, with N a positive integer. We have two kinds of tiles: square tiles with side equal to 5 and rectangular tiles of size 1×3. Determine all the values of N for which it is possible to tile the yard.
Note: The yard should be completely covered, without overlapping tiles.

Solution: It is possible to tile the yard for every natural number N except 1, 2 and 4. Obviously we cannot tile neither 1×1 nor 2×2 yards, because we will not have space for the given tiles. It is also impossible for $N = 4$, because the only tiles we could use are the rectangular 1×3 tiles, and the yard area should be a multiple of 3, but 16 is not. Clearly it is possible to tile the yard for $N = 3$ and $N = 5$. Then we can tile any rectangular yard with integer sides such that at least one side is a multiple of 3, in particular the 6×6 yard.

For $N = 7$, Fig. 5.5 shows how to tile the yard:

Given any $N \geq 8$ we always can write $N = 5k + 3h$, with k and h non negative integers. Indeed, we can write $N = 5q + r$, con $0 \leq r < 5$. But $N \geq 8$, hence $q \geq 1$, and if $r \leq 2$, then $q \geq 2$.

If $r = 0$ or $r = 3$, we are done.

If $r = 1$, then $N = 5q + 1 = 5(q - 1) + 3 \cdot 2$.

If $r = 2$, then $N = 5q + 2 = 5(q - 2) + 3 \cdot 4$.

If $r = 4$, then $N = 5q + 4 = 5(q - 1) + 3 \cdot 3$.

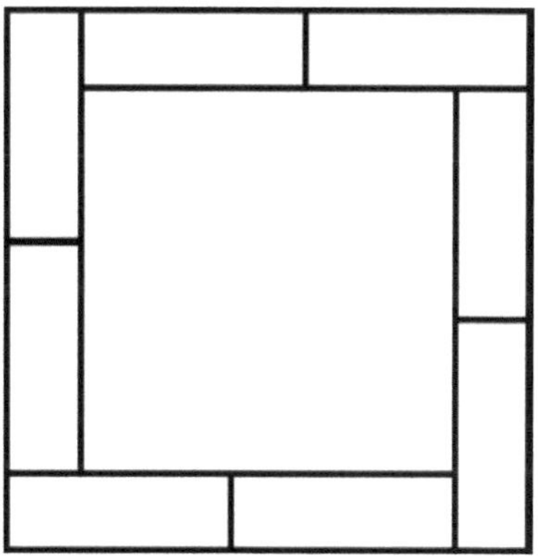

Fig. 5.5 Problem 2010-4

At this point we see that the $N \times N$ square, can be decomposed in a $5k \times 5k$ square, another $3h \times 3h$ square and two $5k \times 3h$ rectangles. We tile the $5k \times 5k$ square with k^2 tiles of size 5×5, and since the other figures have at least one side multiple of 3, we can tile them with tiles of size 1×3.

5.3 Shortlisted Problems

We present now some problems that were in the short lists of some of the competitions, but for some reason they were not selected for the corresponding papers.

Problem SL1. (Shortlist 2014) In a certain country there are 9 towns. For each pair of towns there is one and only one connecting flight, operated by one of two airlines: AirSun and AirMoon. It is known that, given any three towns, at least one of the flights between them is operated by AirMoon. Prove that there are four towns such that all the flights between them are operated by AirMoon.

Solution: Consider the complete graph K_9 whose vertices are the towns and whose edges represent connecting flights. Color an edge red if it is run by AirSun or blue if it is run by AirMoon. Then the problem is equivalent to prove that, if there is no blue triangle (K_3) then there is at least a red K_4. But this is a known result: the Ramsey number $R(3, 4)$ is 9. For that reason this problem was not suitable for the contest.

Problem SL2. (Shortlist 2007) In Mathland's market each merchant had an item to sell. But some of them were unhappy with their item and wanted to exchange it. Each item was desired by one and only one merchant. Every day, they could exchange their product with another, but only once. Find the minimum number of days after which all merchants may become satisfied.

Solution: This problem may be modeled with a directed graph (G, E), where G is the set of merchants and $uv \in E$ if and only if u wants v's item. Each vertex has outdegree and indegree 1. Thus the merchants may be grouped in one or more disjoint cycles $(u_1, u_2, \ldots, u_k)$ such that u_i wants u_{i+1} item (indexes are taken modulo k).

If all the cycles have length 1 then everyone is satisfied and no day is needed. If the cycles have length 2 or 1 then clearly one day is enough.

If there is a 3-cycle (u, v, w), since each day only two of u, v, w may exchange, we need at least 2 days. For example if u and v exchange their items the first day and v and w the second day, the problem is solved.

A 4-cycle (u, v, w, x) may be solved in 2 days too, with the exchanges $v \rightleftarrows x$ the first day and $u \rightleftarrows x$, $v \rightleftarrows w$ the second day.

A 5-cycle (u, v, w, x, y) may also be solved in 2 days, with the exchanges $u \rightleftarrows y$ and $v \rightleftarrows x$ the first day and $u \rightleftarrows x$, $v \rightleftarrows w$ the second day.

Now we will show by induction on n that any n-cycle $(u_1, u_2, \ldots, u_n)$ with $n > 4$ may be solved in two days, in such a way that u_3 does not exchange the first day. The first day we make the exchanges $u_1 \rightleftarrows u_5$ and $u_2 \rightleftarrows u_4$, leaving for the second day $u_1 \rightleftarrows u_4$ and $u_2 \rightleftarrows u_3$. We are left with the cycle $(u_5, u_6, \ldots, u_n)$. But this cycle may be written as $(u_{n-1}, u_n, u_5, u_6, \ldots, u_{n-2})$, which by induction hypothesis may be solved in 2 days without using u_5 the first day. Hence this cycle may be solved in parallel with the exchanges between u_1, u_2, u_3, u_4 and u_5.

Problem SL3. (Shortlist 2011) Two cardboard regular decagons have their vertices numbered from 1 to 10, but in any order (the order may be different for each decagon). The first decagon is superposed over the second one so that each vertex of one decagon is in contact with a vertex of the other one. The numbers of the vertices in contact are multiplied, then the ten products are added. Show that it is possible to superpose the decagons in such a way that this sum is at least 303.

Solution. Select a vertex v in the first decagon and let $a_1, a_2, \ldots, a_n$ be the numbers found when the vertices are traversed clockwise, beginning with k. Analogously let $b_1, b_2, \ldots, b_n$ denote the numbers found on the second decagon when their vertices are traversed clockwise, beginning with one of them. If vertex 1 in the first decagon is over vertex j in the second, then the sum of products is

$$S_j = a_1 b_j + a_2 b_{j+1} + \cdots + a_{10} b_{j+9},$$

where indexes are taken modulo 10. Hence

$$\sum_{j=1}^{10} S_j = \sum_{j=1}^{10} \sum_{i=1}^{10} a_i b_{i+j-1} = \sum_{i=1}^{10} a_i \sum_{j=1}^{10} b_{i+j-1} = \Big(\sum_{i=1}^{10} a_i\Big)\Big(\sum_{j=1}^{10} b_{i+j-1}\Big).$$

But $\sum_{i=1}^{10} a_i = \sum_{j=1}^{10} b_{i+j-1} = \sum_{i=1}^{10} i = 55$, therefore $\sum_{j=1}^{10} S_j = 55^2$. The ten numbers S_j cannot be all of them less than $55^2/10$ (otherwise their sum would be less than 55^2), thus $S_j \geq 55^2/10 = 605/2$ at least for one j, hence $S_j \geq 303$.

Problem SL4. (Shortlist 2011) Find all positive integers n such that it is possible to tile an $n \times n$ square board with 1×2 rectangular tiles (without overlapping) in such a way that the board may not be divided in two parts with a straight line cut, without breaking any tile.

Solution. The answer is: all even integers $n \geq 8$. Clearly if an $n \times n$ board may be tiled, its area is even and n must be even. Let us call a tiling *irreducible* if it is not possible to divide the board with a straight line without breaking any tile. Let us call an integer *nice* if the $n \times n$ board admits an irreducible tiling. Suppose that n is nice (hence even). Let $(0, 0)$, $(n, 0)$, $(0, n)$ and (n, n) be the coordinates of the board's vertices. We affirm that the line $y = k$, with $k \in \{1, 2, \ldots, n-1\}$, must go through an even number of tiles. Indeed, if it crosses j tiles, the area occupied by the tiles which are completely below $y = k$ is $kn - j$, but that number is even, as kn, hence j is even. Thus the $n-1$ lines $y = k$ go through at least $2(n-1)$ tiles. The same thing happens with vertical lines $x = k$. Since each tile is crossed by one and only one of those lines, and there are $n^2/2$ tiles, we have $n^2/2 \geq 4(n-1)$, or equivalently $n^2 \geq 8(n-1)$. Hence $n = 2, 4$ and 6 are not nice. The first n which satisfies the inequality is 8, and indeed 8 is nice, as the irreducible tiling in Fig. 5.6 shows.

Now we will show that if n is nice then $n + 4$ is nice too. It suffices to enlarge an irreducible tiling of the $n \times n$ board surrounding it as Fig. 5.7 shows.

Since 8 is nice, so are the numbers $8 + 4k$ for $k \geq 0$. But 10 is also nice, as shows the tiling in Fig. 5.8.

Thus the numbers $10 + 4k$ for $k \geq 0$ are nice too. Therefore the nice numbers are all the even integers $n \geq 8$.

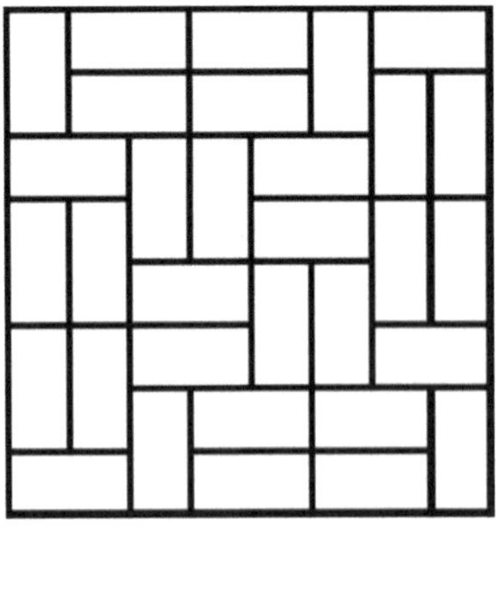

Fig. 5.6 Problem SL4 with $n = 8$

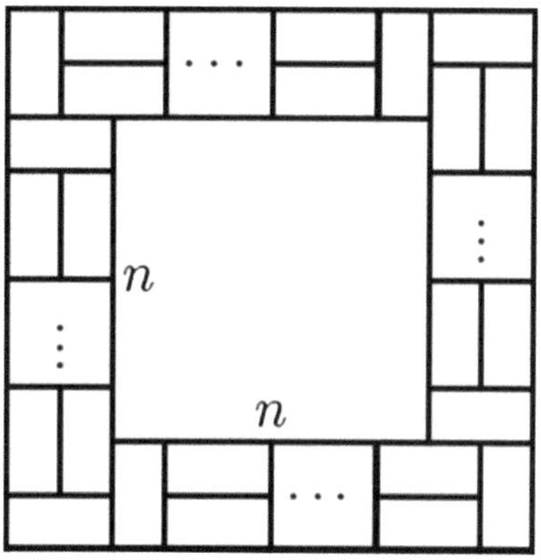

Fig. 5.7 Problem SL4 for $n+4$

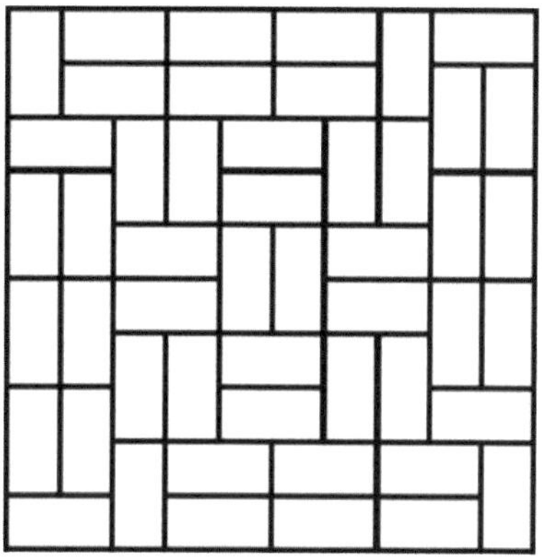

Fig. 5.8 Problem SL4 with $n = 10$

5.4 Conclusions

The mathematics areas with more representation in the papers of the Centro are combinatorics and geometry, each of them with 32.4% of the proposed problems. Number theory problems are the 20.4%, and algebra problems come just to 14.8%.

In the case of combinatorics, the basic concepts needed to solve the problems have not changed that much, but the difficulty of the problems has increased over the years.

During the last 18 years the Centro has proved to be an excellent way to initiate the high school students of the region in international mathematical competitions, preparing them for more demanding events such as the Iberoamerican, Asian-Pacific and International Mathematical Olympiads, among others. In some cases the Centro has been the only opportunity for a country to expose their students to an international mathematics contest.

The Centro has fostered friendly relationships among students and teachers of the participating countries, creating many opportunities for the exchange of information and experiences on the teaching of mathematics in a region with similar culture and common problems.

References

Andreescu, T., & Feng, Z. (2004). *A path to combinatorics for undergraduates—Counting strategies*. Boston: Birkhäuser.

Cáceres, L. F., Nieto, J. H. & Sánchez, R. J. (2016). *The mathematical olympiad of Central America and the Caribbean: 18 years supporting math contests in the region*. Hamburg: ICME-13.

Nieto, J. H. (2015). *Olimpiadas Matemáticas de Centroamérica y el Caribe (1999-2014)*. Ed. Astro Data. Maracaibo. ISBN 978-980-12-7980-8.

Nieto, J. H., & Sánchez, R. J. (2005). The mathematical olympiad of Central America and the Caribbean. *Mathematics Competitions*, *18*(1), 16–38.

Nieto, J. H., & Sánchez, R. J. (2009). Ten years of the mathematical olympiad of Central America and the Caribbean. *Mathematics Competitions*, *22*(1), 19–37.

Part IV
Role of Competitions in the Classroom

Chapter 6
The Rainbow of Mathematics—Teaching the Complete Spectrum and the Role Mathematics Competitions Can Play

Robert Geretschläger

Abstract Although it is clear to all of us with some stake in the teaching of mathematics, that it is an important, valuable and fascinating pursuit, there does not seem to be any real agreement concerning where its central value lies with respect to what is taught in school. The core values of the subject present themselves differently to teachers, math education researchers, professional mathematicians and engineers, and this fact makes it difficult to speak with a common vocabulary about what should be taught and how it should be taught. In this paper, a model for the various aspects of mathematics, ranging from "recreational" through "school" to "applied" is presented, and the role of mathematics competitions in the continuum of this model is discussed. The various points raised in this model are then illustrated by a concrete example.

Keywords Mathematics competitions • Secondary schools • Recreational mathematics • Applications of mathematics • History of mathematics

6.1 Introduction

When people from heterogeneous backgrounds get together to think about the role of mathematics in schools, it is important to have some kind of common starting point for the discussion. As things stand, it has been my experience that such a common starting point does not necessarily exist.

R. Geretschläger (✉)
BRG Kepler, Keplerstraße 1, 8020 Graz, Austria
e-mail: robert.geretschlaeger@brgkepler.at

A. Soifer (ed.), *Competitions for Young Mathematicians*,
ICME-13 Monographs, DOI 10.1007/978-3-319-56585-9_6

(Note that much of what is being said in this paper is derived from my personal experience. I am therefore taking the liberty of putting some things in the first person. I am fully aware that this goes against common practice in such papers, but it is my hope that it will be helpful for further discussion if my personal opinions and experiences are clearly recognizable as such.)

In dealing with people involved with the didactics of mathematics and curriculum development in Austria, there is currently a wide consensus to the effect that the important central aspect of school mathematics lies more or less exclusively in the application of mathematics to the "real world" (whatever that may be; a point I will be getting back to in a moment). It is taken as given to this group, that "pure" math is not really worthy of deep consideration in the classroom, other than what is needed to be able to deal with the most elementary of everyday applications. From this, a commonly derived opinion states that any calculations or algorithmic aspects of mathematics in the school context can and should be left completely to calculators or computers, and not be done by actual human thought at all, except in the most trivial of circumstances.

Speaking to people in the math competition community on the other hand, we have an utterly different view of what is important (and fundamental) in mathematics. While there is certainly also disagreement within this group, it is clear for people involved in competitions that the fundamentals of mathematics are represented by that part which is commonly called "elementary" mathematics. The term "advanced elementary mathematics" is often bandied about among the members of this community, despite the fact that the concept is utterly foreign to the application-oriented group. Members of this group also tend to be in full agreement that enjoyment of the study of mathematics is of central importance. The internal disagreement here often manifests itself with respect to the question of whether or not the specific skills obtained in preparing for competitions will transfer to actual research mathematics. There is, however, definitely agreement concerning the fact that subjects in pure mathematics, which for the purposes of mathematics competitions normally include combinatorics, Euclidean geometry, algebra and number theory, are the most important things for students to learn about and study in order to form a useful base of mathematical knowledge and competence. The logical skills acquired in the somewhat deeper study of these elementary topics are considered most vital in students' academic development.

Finally, speaking to teachers at the university level, the expectations of math skills that students should bring along from secondary school are different again. Furthermore, they are quite dependent on the specific academic discipline. Students in economics are expected to have quite deep knowledge of statistical methods, for example, while students in the natural sciences or engineering are expected to have some knowledge of things like differential equations or matrices; topics that go beyond what both of the previously mentioned groups generally consider fundamental.

Obviously, we are dealing with a huge disconnect here. The object of this paper is to shed some light on this disconnect, and to offer a fresh perspective. My hope is that this will make it possible to reflect better on the somewhat contradictory viewpoints held dear by the various groups of players in this corner of academia, and to ultimately improve the discussion to the benefit of the students in our secondary schools. It is my firm belief that the viewpoint offered by the universe of mathematics competitions has a great deal to offer in this respect.

In order to find a common denominator for fruitful discussion, we first need to achieve some basic agreement on what it is exactly that we are trying to decide. We must therefore find common answers, acceptable to all, to some quite fundamental questions.

The first of these is existential. Why do we think that mathematics is an important subject in school? What are our fundamental reasons for teaching mathematics as a core discipline in secondary education? As mentioned, the answer to this question seems to depend greatly on the circumstances of the person formulating an answer, and it seems clear that the concerns of each of these groups should addressed seriously.

A second important question to be answered in this context has to do with methodology. How do we best get students interested in the type of mathematics we want them to learn? Answering this question depends to a great extent on the individual tastes of the students in question. Different students have utterly different ideas of what is interesting and what is not. Relating my own personal experience in this matter, I can certainly state that my own interests have always been defined by pure mathematics, and geometry in particular. On the other hand, I have good friends, who also happen to be mathematicians, whose interests lie almost purely in applications, and their original impetus for becoming mathematicians was not derived from interest in pure math at all. For them, the gateway into mathematical research resulted from the applications first and foremost, and the idea of discovering mathematical ideas was always totally dependent on these ideas being useful to solve concrete problems. They might consider my own deep interest in the subject as being not much more than the enjoyment of mathematical puzzles, and not really worthy of total academic commitment. (Of course, since they are my friends, they are willing to allow me this luxury.)

The third problem to be addressed is purpose oriented. What are we ultimately preparing students for in their mathematics classes? To which extent are we teaching them mathematics for their own enjoyment? To which extent is this even appropriate? Are we teaching them primarily to prepare for a specific role in society? Are we primarily training their capacity for systematic logical thought? Are we preparing them for university entrance, for mathematical capability that will allow them to study technical subjects, natural sciences, or finance? Do we want to prepare them in a deep manner for what is known in German as "Allgemeinbildung"? (The term is, of course, derived from the Humboldtian ideal of higher education. The concept does not translate very easily into English, and has, in fact, mutated a bit in common understanding over the decades. It certainly goes quite a bit beyond the literal translation of "general knowledge". Some thoughts on this can be found, for instance, in Skovsmose 1994.)

6.2 Defining the Rainbow

In order to illustrate some of the ideas in this paper, let us take a look at the following picture (Fig. 6.1).

We first note that the central shaded block is composed of three sections, carrying the labels Recreational Mathematics, School Mathematics, and Applications of Mathematics. Above these, there floats a cloud containing the word History, and underneath, we see a box containing some tools alongside the word Didactics. We can often find a rainbow underneath a cloudy sky, and it is certainly possible to consider the three sections in the center as aspects in a continuous rainbow, just as the full spectrum of a rainbow can be represented in a basic way by red—yellow—blue. (Unfortunately, we will have to make do with grayscale representation here, but we can let our imaginations fill in the colors.) So, what could I mean by this in the context of learning mathematics?

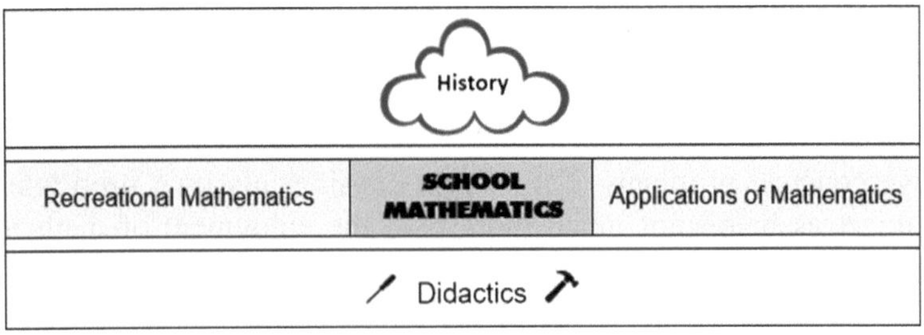

Fig. 6.1 .

Having the box with school mathematics in the center of the diagram (in all capitals for extra emphasis) is meant to illustrate the fact that we are talking about the teaching of mathematics as the central core of our discourse. We are debating mathematical subjects that can and should be talked about in the school context as well as methods that can best be used to engage the interests of students in them. The box with Recreational Mathematics on the left is meant to illustrate the aspects of mathematics that are done primarily for fun. Of course, it is quite possible that there are aspects of mathematics taught in school that students can find quite enjoyable. In fact, if the teaching process is to be successful, we would hope that such topics would be quite common. There are many aspects of so-called recreational mathematics that are not normally dealt with in school. (A very common example of such an aspect is the daily newspaper Sudoku that many people cannot imagine living without. Sudokus are certainly not commonly taught in school, but this is a perfect example of a mathematical topic that many people happily spend their leisure time on, without thought to any external usefulness. We shall be discussing the meaning of Sudoku in this context in greater depth later on.) Still, in an emotionally positive learning environment, we would hope that aspects from this side would spill over into the center.

On the right, we have a box labelled Applications of Mathematics. Many topics commonly covered in school mathematics are taught with a view toward practical applications either in everyday life (as is the case for percentages, for instance), or as a necessary base for higher level applications as can occur in scientific, technical or economic applications. As was the case on the left side however, there are many applications that are certainly never taught in school. Again, we would hope that some ideas from this side seep into the central core of school mathematics, even if higher level applications are almost certainly too sophisticated for consideration at a secondary level.

History hovering above the central boxes is meant to symbolize the fact that all mathematical ideas have a past, and this past can and should have a presence in school, at least up to a certain extent. Some mathematical concepts had their historical start in physical applications (think of differential calculus, for example), while some that may seem very applied from a modern standpoint may have originated in a recreational context (like probability theory, which started from considerations of gambling games). An awareness of this overarching historical aspect of a topic can and should make it easier for the learner to grasp the context of what is being learned. Furthermore, we can hope that an understanding of the historical context of a topic can give many students the necessary motivation toward grappling with its intricacies.

Finally, having the tool-box (represented by the hammer and screwdriver) of practical didactics as the underlying foundation is meant to represent the idea that the entire building of the academic discipline Mathematics rests upon the nuts and bolts of how it is taught. (Sorry about the mixed metaphors. Maybe we need to think of the rainbow as being painted on the side of a grand building.)

In the sections that follow, I will attempt to be a bit more precise about how this model of thinking about mathematical ideas can be useful in thinking about the learning process. More specifically, I will attempt to place mathematics competitions in their appropriate slot in this framework, and illustrate how they can show the path to a more fruitful synthesis of mathematics for enjoyment and useful application. I hope to be able to give a good argument in favor of using mathematics competitions as a tool both for popularizing mathematics as a discipline, and for preparing students for many important aspects that relate to the reason we have the subject in such prominence in the school curriculum.

6.3 Math Is Fun

We are so used to the popular notion of mathematics being called a dry, boring and incomprehensible pursuit in popular discourse that a lot of people outside the math community cannot even conceive of the truth of this heading. But, as we in the community all know, math is indeed fun. And this "fun" aspect of the subject can manifest itself in many different ways.

Why is there even such a thing as the abstract concept of Mathematics? Human nature is such that people have been fascinated by the process of abstraction for at least as long as there has been language. Discovering the fact that there is something highly elementary in the connections between utterly disparate objects exhibiting common traits that can be given a name, like "three" (the leaves on a stalk of clover, or the corners of a triangle, or the more abstract concept of past, present and future) or "circle" (the shape of the sun or the moon in the sky, or a ripple on the surface of a pond when a pebble is tossed in, or the shape you can draw with a stick in the sand by holding one end steady and moving the other) is, simply put, fascinating. And discovering that there are properties that can be found from the definition of such a concept that then turn out to be common to all objects fitting the definition is certainly something wonderful. Realizing this leads us to develop methods of finding such commonalities, resulting in concepts like counting, calculation, axioms and proof. Falling prey to the fascination of such intellectual pursuits is one way in which Math Is Fun.

Another way is well known to all ardent puzzle solvers. There are logical processes involved in solving anything from brainteasers and cryptic crosswords to hidokus and Rubik's Cubes. At first glance, the puzzles seem to be indecipherable, but step-by-step application of logical thought, sometimes combined with some trial and error, lets us inch ever nearer to a solution. Finally, after some effort, the solution presents itself. In a good puzzle, the fact that the result has been found is then completely obvious; there is no doubt that we have succeeded. Most important, a feeling of deep satisfaction results from having found the solution, by application of our own wits, to something that seemed incomprehensible at first glance, but is now utterly clear. This is another way, readily appreciated by any mathematical researcher, of course, in which Math Is Fun.

Another path to enjoyment of mathematics comes from deeper understanding of ways in which mathematical methods allow us to comprehend complex systems. A fine example of this path is the one followed by people involved in high-level financial transactions. The complex mathematical structures that they use make it possible for them to play their high stakes games, and it goes without saying that they have found for themselves a completely different way in which Math Is Fun.

Finally, for some people, simple mental calculation is enjoyable enough, and they may go so far as to cultivate arcane skills involving such things as mental division of huge numbers, memorization of the decimal digits of pi to an incredible number of places or the capacity to manoeuvre freely through hyper-cube cells in four-dimensional space in their minds. Not everyone can appreciate this type of entertainment, but to those who can, they are manifestly yet another way in which Math Is Fun.

Of course, any number of collections of mathematical puzzles is available on the book market, mechanical puzzles are readily available for purchase, and so on. It seems clear that a lot of people are actually quite aware of the fact that mathematics is, indeed, fun.

If there are so many ways in which pure enjoyment of mathematics is possible, isn't it unfortunate that so many people pass through the school system without being able to enjoy the subject in any such a way? In school, we as a society want to help our children speed up the process of abstraction, and expose them to as much as possible of the wealth of knowledge humanity has developed over the millennia. During the course of this process, we present a great deal of that knowledge in a pre-processed way, reducing the elements of individual discovery to a minimum. Of course, this is with good reason. It took humanity many generations to reach the level of sophistication we have now, and it would not be feasible to expect every youngster to figure everything out on his or her own. After all, it took the wisest brains of many generations to come up with what we, as a society, know now. Unfortunately, the accelerated processes typically used in school tend to suck much of the entertainment out of the subject.

Even knowing this, a lot of mathematics remains enjoyable. Sometimes, we may not realize that we are doing mathematics while we are playing with it. Not everyone solving a newspaper number puzzle is cognizant of doing mathematics. Nor did every participant in the great puzzle crazes of the past decades, from the 15-puzzle through Instant Insanity and Soma to the Rubik's Cube necessarily think of their hobbies as intrinsically mathematical, even though they obviously were. It seems clear that any way to introduce this type of enjoyment to the learning process must be advantageous.

Some mathematics competitions offer puzzle problems that give a large number of contestants the opportunity to have some mathematical fun of this type, and the millions of competitors taking part in competitions like the Mathematical Kangaroo, the American Mathematics Competition, or the Australian Mathematics Competitions (just to name a few) show that the enjoyment to be derived from thinking about such questions is well known to many. So, here we have an obvious way in which the math competitions scene is helping to achieve the goals we aim for in regular mathematics education. Helping students to see how enjoyable it can be to solve mathematical problems/puzzles (the distinction becomes quite blurry at times) gives them the impetus to delve deeper into the subject.

Here is the first facet of the Rainbow. One big reason for us to do mathematics is simply because it is fascinating and because it is enjoyable. Next, let us have a look at the opposite end of the spectrum; the other reason we should all be able to agree on for why mathematics is such an important discipline.

6.4 Math Is Useful

On the opposite side of the spectrum of mathematics, we have the utility of mathematical abstraction combined with practical calculation that makes mathematics so useful. Of course, this is also a reason why many people are fascinated by mathematics as a discipline in the first place. Many, for whom mathematics may not have held any particularly high level of fascination in school, become quite enamoured of the pursuit because of the surprising connections it helps to uncover in practical applications. This can be derived from physics, applications in engineering, financial transactions or any number of other things.

Unfortunately even research mathematicians cannot always agree on what exactly is meant by "useful". As was already pointed out, pure mathematicians have a quite different point of view from applied mathematicians, and therefore often find different areas of elementary mathematics to be of elementary importance to their work. Nevertheless, all can agree that things can and should be taught in school because they are, simply, useful. And in any case, the fact remains that mathematics is in some way intrinsic to most any abstract discipline.

For many people, the day-to-day practical aspect of the subject is the central, and perhaps only, justification for its inclusion in the school curriculum in a central role. This is certainly currently the case in the Austrian school system, which I know best from practical experience, and I shall elaborate a bit on in the next section. In my opinion, it is however quite unfortunate if this is considered to be the sole defining justification for the subject. It does seem clear that the things we teach our students in school should have some connections to future applications, of course, but this statement can be interpreted in different ways. We can all agree that school should certainly convey the capability for dealing with everyday calculation to all students. They should learn how to deal with cash transactions in the course of making their daily grocery purchases, calculating the savings involved when something is advertised as being on sale at 10% off, or figuring out how many cans of hi-gloss are required to repaint the garage, and we are certainly all in agreement that the basic intellectual tools needed to solve such problems should be acquired in school.

From the standpoint of preparing secondary school students for the tertiary level, however, there does not seem to be so much common ground. Most would agree that there is a certain amount of higher level mathematics that must be taught in an effective manner, but what does this include? If we want to prepare our high school graduates for studies in the sciences or engineering, we will want them to have some accessible fundamental knowledge of real functions and calculus, algebraic manipulation of polynomials and solving equations, and so on. If we are worried about preparing them for the necessities of anything involving statistical analysis, like medicine, economics or the social sciences, we will want them to have some skills in interpreting statistical tests and working with random distributions. If we are worried about training future mathematicians and computer programmers, we will want them to have some understanding of mathematical proof and algorithms. Or, in the extreme, we can take the position that we do not want to train our students to understand deeply any of this, arguing that they can pick up the necessary knowledge at the tertiary level, and limiting what is taught in secondary school to what is needed for "communication with experts" (see Fischer 2001). This is the current basis for the Austrian school system, and in my opinion this is not at all sufficient.

6.5 Math in School. Connecting the Fun and the Usefulness

I would argue that all aspects of mathematics should be included in an ideal secondary curriculum. In order to keep all students interested and motivated, there should be aspects of recreational mathematics, applications of mathematics, and the history of the subject represented in the classroom. Graduates of our schools should have a reasonably developed feel for numbers, shapes, data and functions. They should understand the value of proof in an axiomatic system and be somewhat schooled in abstract thought. There should be room for the many fascinating aspects and the many uses of the subject, as well as aiming toward achieving the ideal of educated people having a well-grounded understanding of the subject.

Depending on their own point of view, many people think that only one or the other of these aspects is appropriate for schools to worry about. Limiting mathematics in school to practical applicability, however, leaves no room at all for recreational aspects or for the development of pure mathematics as a scientific discipline. Also, the reality of schools often does not allow any kind of deeper insight or any kind of enjoyable work with mathematics because the available time must be used to prepare students for specific types of central exams, which typically test only the ability to deal with highly specific problem formats. This is not good. A good school system will not put undue emphasis on simple calculation, nor will it force the majority of available classroom time to be spent on the study of specific test formats. A good class is one in which the students' minds are challenged in many different ways and in which their individual preferences and interests can find a home, whatever they may be.

Taking a closer look now at the current state of the Austrian school system, we see that there has recently been a shift completely away from anything involving operative mathematics in the secondary schools, and oriented strictly toward applicability.

The opinion of some mathematics educators who feel that all mathematics taught in school should be introduced through "real world" applications now completely dominates the discussion, even if many teachers put up quite a bit of resistance in their classes. (It is worth noting that there is a good reason for the quotation marks here. What is considered the "real world" in mathematical texts is, of necessity, always a stark simplification of reality, with a strong element of pre-digestion having been introduced by the problem authors. The real "real world" is invariably more complex than the highly simplified mathematical models used in the school situation would generally suggest.)

The pure enjoyment of mathematical pursuit is thrown out the window in this educational model, as is the value of abstract thought in a liberal arts education. Both aspects are sacrificed at the altar of applicability.

Furthermore, centralized testing has led to complete dominance of the teaching-to-the-test phenomenon, to the detriment of all else. One can only hope that this state of affairs, which has only come into full force in the last few years, will soon pass, but the plan to move to stronger inclusion of technological aids in mathematics instruction (graphing calculators, CAS and spreadsheets) unfortunately suggests that things will get worse before they get better.

This unfortunate development resulted from an attempt to improve mathematics teaching, of course. Comparing any current textbook to one used, say, in the 1960s, gives an excellent view of what has happened. It is certainly true that there was formerly far too much emphasis placed on calculation for its own sake. Looking at the old textbooks, we find any number of difficult problems involving simplification of quite involved term expressions, for example, and such things can no longer be found in current textbooks. The argument given for the change was that students did not actually gain any real understanding of what they were calculating, and there is a great deal to be said for this. Unfortunately, in the process of reducing this type of rote learning, some topics were eliminated completely, despite the fact that the fascination emanating from them can certainly help a great deal in giving students the motivation to learn.

Different people have different tastes, and while some are readily motivated by pure abstraction and others by the wish to come to grips with practical matters, there cannot be one singular path to motivation equally applicable to all learners. Surely the aim of teaching is to optimize the motivation to learn for as many students as possible, in order to maximize the amount of knowledge students can absorb and develop. Since students can be motivated by quite disparate pathways to such knowledge, it seems quite obvious that all such paths should be reasonably represented.

6.6 Mathematics Competitions: Great at Connecting

One of the main points I would like to get across with this paper is the idea that mathematics competitions are uniquely suited to getting many (though, of course, not all) students more deeply and more actively engaged in mathematical pursuits. Parts of this argument have already been hinted at, but in this section, I would like to present it in a more structured way.

When students get hooked on mathematics competitions, this means that they have developed a feeling for the fascination of problem-solving on an abstract level. Finding solutions to competition problems of progressively higher levels of difficulty leads them on a journey to discovering and writing proofs, and with this they are really learning to be active mathematicians themselves. Compared to what they are confronted with in "regular" math classes, there are some specific qualities to the style of mathematics they encounter in the competitions world.

First of all, there is the feeling of accomplishment that comes from solving a competition problem. This is the same feeling one gets from successfully solving a puzzle or from proving a theorem, but in the context of a competition, it can be reinforced by the fact that points are awarded, and the student may have achieved something that others writing the competition have not. Regular classroom mathematics tends to negatively reinforce not being able to solve a problem (which might even result in failing a test) rather than positively reinforcing the solution of a problem that can be considered at the outset as being optional. It goes without saying that positive reinforcement of this type is preferable from a psychological viewpoint. This positive reinforcement then usually transfers quite well to regular classroom work. (This last claim is what I see quite commonly in my own classrooms, but I am sure that anyone working both directly with students in competition preparation and in a regular classroom setting will agree.)

Essentially, this is part of the argument in favor of using recreational mathematics to get students more actively involved in their classrooms. In the Rainbow, this means that the left-hand box positively influences the central box. The implication is that the participation of students in competitions is therefore quite useful as part of the underlying Didactics tool-box.

Another strong influence of math competitions lies in getting the students to accept the need for logical rigor in their work. If any of their calculations or proofs is logically incomplete, they will simply not score full points, even if they have understood all of the essential parts of an argument. This is disappointing for a student who has become used to the feeling of success that comes with solving a problem. Again, the positive reinforcement that then comes with understanding the need for a logically complete argument in order to get full points in a competition is much better than the negative reinforcement of just being criticized for something incomplete.

While this applies to any kind of mathematical argument, including simple computation, it is especially true for learning to understand the meaning of the axiomatic method in producing proof. Learning this in a normal classroom is quite abstract and involved. In the context of a competition, however, it is very natural (although perhaps not really any easier). It is obvious to all competitors that an argument must be complete if a student hopes to receive full points. It is quite easy to accept this in the context of a competition, as a competitor's more complete argument will obviously be better than mine, if mine is missing some salient points.

For the purpose of learning the axiomatic method and the concept of what constitutes complete proof, classic topics are certainly the best. There is an obvious historical reason why the classic Euclidean topics of geometry and number theory/arithmetic are the areas in which the axiomatic method was developed, and this is certainly also the reason why there is still a wide international consensus that these topics should be included in a central role in competitions. The historical argument is quite strong, not just for intrinsically historical reasons, but because historical development in this area happened for a reason. These topics are basic to human abstract thought, and taking this route during the learning process is as basic and reasonable now as it ever was.

Starting on the right-hand side of the Rainbow, it can also be argued that a similar path from the Applications box is offered by classes in mathematical modelling. In many places around the world, students especially interested in applied mathematical problems are offered participation in such activities that are also competitions of a sort, even if there are generally no "winners" declared. (I refer here specifically to the model of the "Mathematical Modelling Week" as I know it in Styria, in the south of Austria, as this is the one I am most familiar with. Similar programs are, however, offered in many places.) As a path to applied math at a higher level, high school students are invited to work for a week under the tutelage of professional mathematicians on the modelling of some applied problem. These can be from physics, medicine, economics, or any number of other areas, but generally they will be derived from the research specialties of the tutors. While these are not competitions in the traditional sense since there are no winners, it can be argued that all participants in these workshops are "winners" by virtue of their completion of the tasks at hand, and there are simply no "losers". Psychologically, this is certainly a good thing. Otherwise, I would argue that the net positive results of such an activity are the same as those in a more typical competition format. Participants derive the same sense of accomplishment in finding a path toward solving a problem that they could not initially deal with. Through diligent application of logic, they finally arrive at a result that they have every right to be proud of, yielding a strong positive reinforcement.

This can be seen as giving added value to the middle box in the Rainbow from both sides. The problems in such modelling projects can be considered as both Applied and Recreational, at least from the point of view of the active participants.

All told, the argument in favor of mathematical competitions of all types in reinforcing the path to a deeper understanding of mathematics among interested participants is quite strong.

6.7 History on Top; Didactics on the Bottom

Returning briefly now to the picture of the Rainbow (Fig. 6.2), we can concern ourselves a bit more with the top and bottom bars.

The underlying bar labelled "Didactics" is more or less self-explanatory. In school, everything is built up on a base of teaching methodology, and this base is symbolized here by this one term. It includes matters of curriculum, textbooks and worksheets, classroom organization, and so on, and is symbolized here by very elementary tools, namely a screwdriver and a hammer. No matter what we decide to teach in school, we must certainly worry about how we are going to go about teaching it—the nuts and bolts of work in the classroom.

Perhaps a bit more explanation is required for the History cloud. Its floating above all else is meant to imply the fact that all areas of mathematical thought not only have a genesis, but that this genesis is an important intrinsic part of the area.

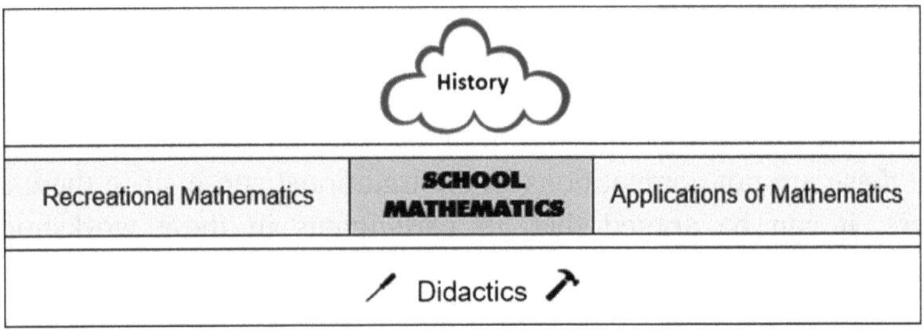

Fig. 6.2 .

No part of mathematics starts in school. Everything starts either as a game like statistics or as an application for further development of something that already existed. Much mathematics is derived from axiomatic interpretation of some aspects of real life. Mathematics is in its core abstraction.

Let us take probability theory as an example. The roots of what we now think of as probability reach back to the 17th century. Some of the biggest thinkers of the day (Cardano, Fermat, Huygens, Pascal) were thinking about games of chance, and the likelihood of winning and losing. While such considerations can certainly have very practical applications for some people, there is an argument to be made for placing these considerations firmly in the realm of recreational mathematics. Throwing dice, flipping coins or playing card games are certainly recreations for all but the most hard-core professional gambler. From this beginning, however, there arose an elaborate theory with applications in such disparate areas as medicine, finance and opinion research.

As has already been alluded to, there are at least two strong arguments to be made for the inclusion of at least some of the history of such a discipline in its teaching.

For one, there is the motivational argument. Getting students interested in a topic gets them invested in the learning process, and the consideration of the historical process that led to the development of a topic can help get students interested in the topic for the same reasons that the scientists that originally developed the theories were interested in them. This is completely independent of the question of applicability of the whole logical structure once it has been developed. The a posteriori uses of a mathematical method are generally not clear at the historical outset of its development.

Furthermore, there is also the methodological argument that a topic can be better understood if it is learned at least in part by following the train of thought that led historically to our current understanding of it. Skipping over the history by reducing mathematics to a system of definition-theorem-proof (which certainly has its place in the university) deprives the student of an important level of understanding.

6.8 An Example from the Rainbow: Sudoku to Graph Coloring

Let us now take a look at a specific topic, how various aspects of it are represented in the different boxes of the Rainbow, and the role that mathematics competitions can play in developing understanding of it.

		1				7		4
	6							
3			8	4	5			
								2
	9	2		7	6	4	5	
8								
			5	9	2			7
							8	
4		7				1		

Fig. 6.3 .

8a: A Popular Pastime: the Daily Sudoku

In the last ten years or so, sudokus have assumed a prominent place in the public consciousness by their ubiquity in the daily papers and in puzzle books available at any book store or news agents'. As is well known, the idea is to fill in a grid of numbers satisfying certain constraints. In a classic sudoku, the numbers from 1 through 9 must be placed in each row and in each column of a 9×9 square grid, and each number must be present in each of the nine 3×3 squares the 9×9 square is composed of. Several numbers are already given in the grid, and the point of the puzzle is to find the unique way to fill in the rest. An example of such a problem grid is shown in Fig. 6.3.

There is no doubt that this is an incredibly popular pastime, and the fact that there is at least a bit of mathematical content involved is already obvious from the fact that numbers are used in the squares. There are many related puzzle types that have found their way into some daily papers and the public consciousness along with them, like Kakuro, Hidoku, Fillomino, and so on.

The reason that such puzzles are so popular lies in the fact that solving them gives the solver a distinct feeling of accomplishment. While we are aware of the fact that we are doing something that isn't really of any immediate use to us (or anyone else for that matter), there is an intrinsic joy in finding the solution. This is the basis for all so-called "recreational mathematics". If the only argument for doing it lies in the recreational aspect, the external value of the actual mathematical content becomes completely irrelevant for the time we spend on the problem.

Of course, this is an aspect of competition mathematics. When students are solving problems in a competition, they are not worried about applicability. They are simply solving the problems for their own sake. The problems themselves are considered interesting, independent of any meaning they may take on in the "real world", and finding the solution (and then possibly being awarded points for it) is the reward they are seeking.

Notably, this is also often the main motivation behind more serious mathematical research. Certainly, some research problems must just be solved in order for a specific application to work, or to guarantee funding for yet another financial period in some research institution. In general, however, anyone involved in any reasonably abstract mathematical research is searching for the solutions because of an intrinsic interest in the problem itself and the deep sense of achievement that comes with finding a solution to a difficult problem.

8b: Mathematical Research and Applications related to Sudoku
Starting from the highly elementary content of Sudokus, there are several different directions our thoughts can take in order to derive mathematical research problems.

Perhaps the most obvious concerns itself with the internal mathematics of the puzzles themselves. There are many questions that can be posed concerning the statement of a sudoku problem or its solution. Some of these are the following:

- What is the smallest number of numbers that can be given in a sudoku grid, such that the solution is unique?
- What is the largest number of numbers that can be given in a sudoku grid, such that the solution is not unique?
- How many minimal sudokus exist? (A "minimal" sudoku is one in which the solution is unique from the given numbers, but in which no given number can be deleted with the resulting sudoku remaining unique.)

Such questions are the focus of a certain strand of mathematical research, and some prove much easier to answer than others. (Interested readers are invited to find out the current state of knowledge concerning such questions by checking Wikipedia (https://en.wikipedia.org/wiki/Mathematics_of_Sudoku) or other easily accessible internet sources.) Solving this type of problem, however, does not stray far from the mathematical content of the Sudokus themselves.

Taking a closer look at the sudoku concept, we see that there is another path to abstraction we can take, that will lead us right into the heart of research mathematics.

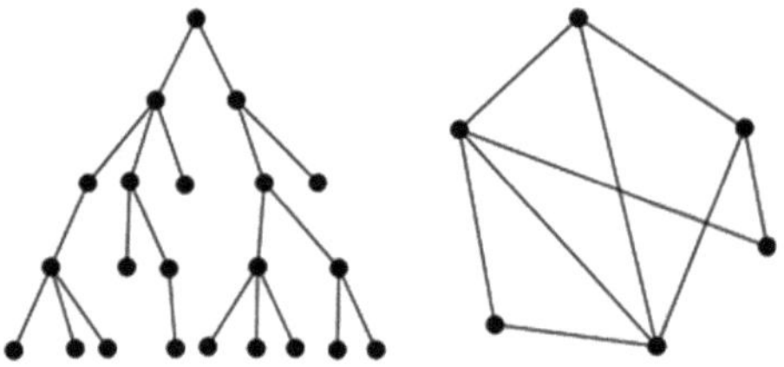

Fig. 6.4 .

As it turns out, it is quite straight-forward to express the solution of a sudoku as a graph coloring problem, and this idea connects the popular puzzle both to cutting edge research in abstract mathematics and to real-world mathematical applications. So, what do we mean when we say that solving a sudoku is equivalent to solving a graph coloring problem?

In mathematics, a graph is, of course, a structure composed of points (or vertices), that are joined by lines (or edges). These are commonly represented by pictures like the ones in Fig. 6.4:

We can consider a sudoku, composed of 81 cells in a 9×9 square grid, to be represented by a graph with 81 vertices. Each vertex is to be colored with one of 9 colors, corresponding to the numbers 1 through 9. Some of these colors are given, with the rest to be determined.

The nine cells in a common row (or a common column or a common 3×3 square) can be thought of as being joined pairwise by an edge. Solving the sudoku then amounts to finding a coloring of the graph with the nine colors, such that no two vertices with the same color are joined by a common edge.

When thought of in this way, it becomes clear that our daily newspaper sudoku is completely equivalent to a seemingly much more abstract problem. With this, we are already firmly in the middle of a practical research topic. The puzzle, considered purely for the sake of the enjoyment of finding its solution, has led us directly into the world of mathematical applications. Now that we understand this, we can strip away the camouflage and take a look at where graph coloring can lead in mathematical research.

First, let us consider a practical application of graph coloring, namely the problem of job scheduling.

Let us assume that we have a certain number of jobs that need to be done in some order. Certain of these jobs may be in conflict with each other, i.e. there may be some reason why they cannot be dealt with simultaneously. (For instance, the same person may be required to fullfil two tasks or the same machine may be needed for two distinct steps in production.) It is possible to represent the scheduling problem by drawing vertices of a graph corresponding to the jobs. Any two jobs that conflict with another can then be joined by an edge. The smallest number of colors with which it is possible to color the vertices of the graph without like-colored vertices ever being joined by a common edge then gives us information on the most efficient way for the jobs to be scheduled. This model can translate not only to concrete "jobs" that need to be done by people, but also to organizational problems ranging from the assignment of vehicles to individual trips for a delivery company to the assignment of frequencies to terrestrial television broadcasters in geographically conflicting areas.

Next, let us have a look at a more theoretical graph coloring problem that happens to be right at the cutting edge of modern mathematical research, namely the question of the chromatic number of the plane, also known as the Hadwiger-Nelson problem.

The problem can be stated in the following way. What is the smallest number of colors with which it is possible to color the points of the plane in such a way that no two points at unit distance have the same color?

Much has been written about this problem (see, for instance (Soifer 2008)), but despite more than half a century of intense research, the problem has not yet been solved. In fact, as easy as the problem is to state and understand, it is one of those intractable mathematical questions that are really devilishly difficult to grasp. It may well be that the problem cannot even be completely solved without making some non-standard assumptions, like the validity of the Axiom of Choice. It is relatively straight forward to show that the number in question must be larger than 3 and it can also be shown that it must be smaller than 8, but values of 4, 5, 6 or 7 are still possible.

A famous coloring problem of a related type, located somewhere on the spectrum between purely theoretical and practical, is the four-color map problem. For many years, there existed a conjecture, since famously proven with the help of computer-based methods, that any map in the plane can be colored by at most four colors in such a way that no two countries sharing a common border have the same color.

We see that the same sudokus that we know so well from purely recreational mathematics are related quite directly to problems both in concrete applications of mathematics and in high-level research in pure mathematics.

8c: Sudoku, Graphs and Coloring in School.

Neither sudokus nor graph theory are a standard school topic in most countries. Recently, many schools have taken to using something closely related to sudoku in order to give students an opportunity to practice mental calculation, namely kenken. (Note that KenKen is a registered trademark. Interested readers can find a large number of such problems at (http://www.kenkenpuzzle.com). The puzzles are sometimes also referred to under other names, such as Kendoku.)

For those not familiar with kenken, a brief introduction seems in order. As is the case for sudoku, a kenken puzzle is a square grid, and the goal of the puzzle is to place numbers in the grid in such a way that none of the numbers repeat in any column or row. If the size of the grid is nxn, the numbers from 1 through n are to be placed in the cells of the grid. Unlike sudokus, however, no digits are given in advance. Instead, certain areas are given, in which the numbers can be combined by addition, subtraction, multiplication or division with some given result. For instance, if two cells are joined to a 2 × 1 rectangle with the symbol "4 + ", this means that the two cells are to contain two different digits with the sum 4, and therefore one must contain the digit 3, and the other the digit 1. In some cases, there is more than one combination possible, as for instance for "2-". This could be the result of 3–1, 4–2, 5–3, and so on. Furthermore, if a single cell contains only one number without an operation, this number can be considered as given in that cell. An example of such a puzzle is shown in Fig. 6.5.

However, use of these puzzles in the classroom is not normally a path to understanding about graph coloring. The didactic idea behind the use of this in the classroom is for the students to get a better feel for number combinations in simple elementary calculations, and kenken gives an amusing context to such calculations.

Simple graph theoretical ideas are, however, also often championed for inclusion in the school curriculum (see, for instance Smithers 2005), especially in schools that are preparing students for computer programming. Most school systems, however, do not currently include this subject in their curricula. Students preparing for mathematical olympiads do, however, routinely deal with elementary graph theoretical ideas, as this is a common topic of olympiad problems in the so-called Combinatorics category. An example of such a problem will be given in 8e.

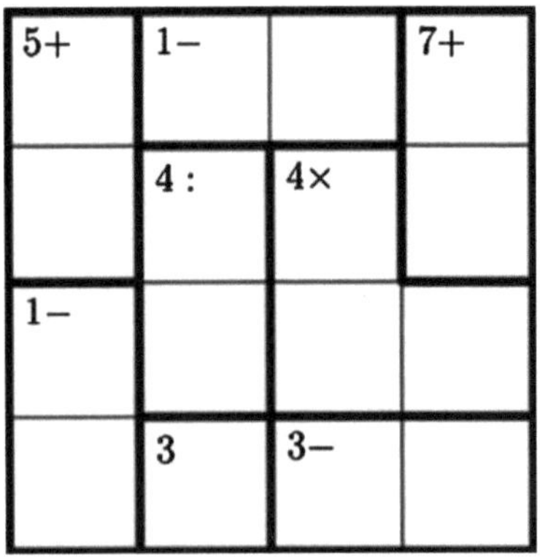

Fig. 6.5 .

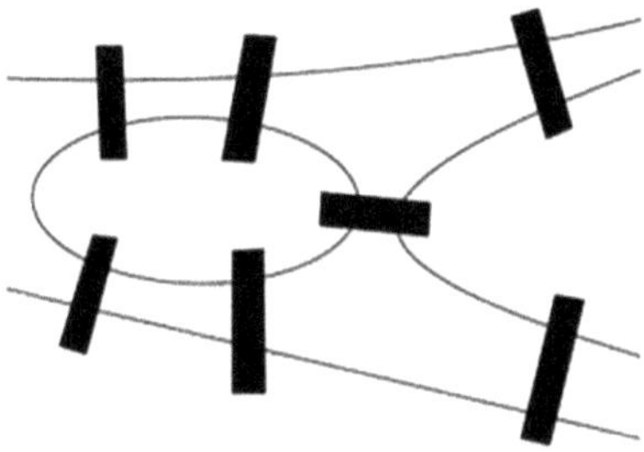

Fig. 6.6 .

A	*B*	*C*
B	*C*	*A*
C	*A*	*B*

Fig. 6.7 .

8d: History and Didactics: Graph Theory and more
If any graph theoretical ideas make it to the classroom at all, a likely candidate for inclusion is the classic Königsberg Bridge problem of Leonhard Euler (1707–1783). This problem, asking whether it is possible to cross each of the seven bridges in old-time Königsberg exactly once in one walking tour of the town, which straddles a river with islands as shown in Fig. 6.6 is the starting point of modern graph theory.

Students may not know anything about the history of the city of Königsberg (now the Russian city of Kaliningrad), but the question is a very practical one that can be readily understood. Also, its solution can be developed by simple logic, without resorting to any high-level mathematical tools. Giving some historical context can certainly make the topic more interesting for many students, and this is also a good excuse to name-drop Leonhard Euler in class.

Another interesting historical sidebar that might be mentioned in this context, is the Latin Square. A Latin Square is an nxn array, with n symbols written in the cells in such a way that each of the n symbols is represented once in every row and in every column of the array. This is also a subject studied by Euler, and the name is derived from his work, in which he used Latin letters as his symbols. A 3 $\times$ 3 example, such as could be found there, is shown in Fig. 6.7.

A Sudoku is, of course, a Latin square with some special restrictions, in which the symbols are digits. These mathematical objects have been studied to quite some extent since the 18th century. The idea behind them is closely related to (but not to be confused with) the idea behind the so-called Magic

Squares, in which the sums of numbers in all rows and columns (and often also diagonals) are equal.

Both these topics are typically seen as purely recreational, but as shown here, they are at the very foundation of an important section of mathematics that ranges through the whole rainbow, from recreational to applicable.

8e: An example of a graph coloring problem from an international competition.

An example of a nice competition problem concerning graph coloring is the following problem from the International Tournament of the Towns (Spring 1990, Senior O level):

(a) Some vertices of a dodecahedron are to be marked so that each face contains a marked vertex. What is the smallest number of marked vertices for which this possible?
(b) Answer the same question, but for an icosahedron.

(Recall that a dodecahedron has 12 pentagonal faces which meet in threes at each vertex, while an icosahedron has 20 triangular faces which meet in fives at each vertex).

In order for a student to solve this problem successfully, it is helpful to realize that it is indeed a graph coloring problem. The vertices of the polyhedron being considered can be thought of as the vertices of graphs, and the edges of the polyhedron as edges of these graphs. Of course, this is a three dimensional concept, but the graphs in 3-space can be projected onto a plane (for instance, from a point on the circumscribed sphere of the polyhedron onto the tangent plane diametrically opposite to this point), resulting in corresponding plane graphs with completely analogous properties. Since we then wish to "mark" vertices, we can think of this as coloring all the vertices of the graph with two colors, say black and white, with black corresponding to "marked" vertices and white to "non-marked" vertices.

The solution to part (a) is then quite simple. Since each vertex lies on three faces of the dodecahedron, marking any vertex gives three faces a marked vertex. Since there are 12 faces, we must certainly mark at least 12:3 = 4 vertices. We can see in Fig. 6.8 (a graph representing the dodecahedron's vertices and edges), that a marking of four vertices (represented by the full points) is indeed possible, such that each face has a marked vertex.

Part b is a bit more sophisticated. We can see in Fig. 6.9 that a marking of six vertices such that each face has a marked vertex is possible.

It remains to be shown that such a marking of five (or less) vertices is not possible. We can prove this by contradiction.

Let us assume that it is possible to mark five vertices in such a way that each face has a marked vertex. We consider the graph (as shown above) and delete all edges with the exception of those joining two marked vertices,

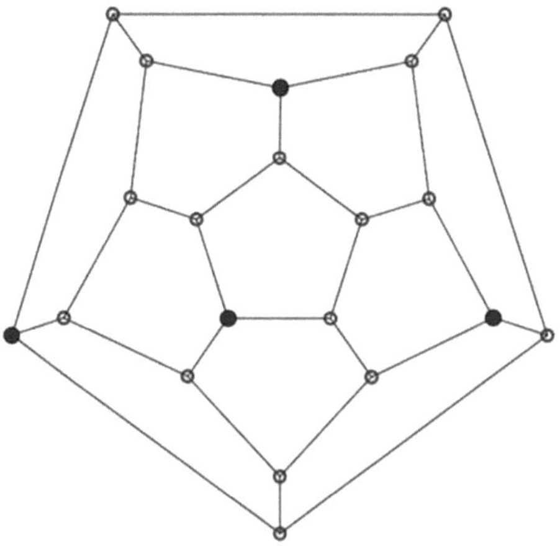

Fig. 6.8 .

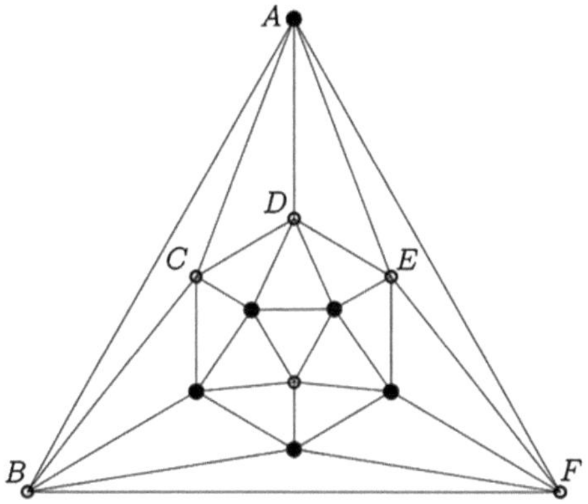

Fig. 6.9 .

and consider the number of components of the resulting graph. (Recall that a "component" of a graph consists of a subset of the vertices, connected by edges of the graph.) In any of these components, a first marked vertex contributes to 5 faces, but any succeeding vertex in this component can only contribute to at most 3 further faces that do not yet have a marked vertex. If there are at most 5 marked vertices and at most two components, the marked vertices can contribute to at most $5 + 5 + 3 + 3 + 3 = 19$ faces. We see that the graph must consist of at least three components. At least one of these components must then consist of only one marked vertex. Let us assume that this is vertex A in the figure above. This means that none of the vertices B, C, D, E and F is marked, and four of the remaining vertices must be marked. This is not possible, however, since these four would then certainly all be in the same component, in contradiction to the assumption that they contribute to faces in at least two separate components. We see that at least six vertices must be marked, as claimed, thus finishing the proof.

6.9 Conclusion

Mathematical instruction should include all aspects of the subject and engage students in whatever way they can be led to be interested in the subject. This is different for each person. Some will be excited by abstract math problems independent of any applications in the real world. This includes mathematical puzzles, mathematical games or individual pure math research. Others will be excited by the opportunity for applications, for instance in physics or other areas.

In this paper, I have attempted to argue that a complete treatment of any mathematical topic in school should include aspects from the complete Rainbow of Mathematics, in order to help every student of the subject find something suited to her or his tastes. A mathematical topic can be introduced starting from most any mathematical problem, be it a number puzzle (number theory, coding), a triangle problem (olympiad geometry, school trigonometry, land surveying) or a practical application. I have also tried to argue the fact that the world of mathematical competitions offers a strong tool, independent of where a student hops on board the math train.

Let us briefly return to the fundamental questions on the value of mathematics as a core subject in secondary education as posed in the introduction. Here are some answers I believe we could and should all agree on, considering all that I have presented here.

Question 1: Why do we think that mathematics is an important subject in school?
It seems clear to me that there are essentially three equally valid answers to this question.

First of all, mathematics is necessary for many things. Some elementary things, like basic number skills, are obvious prerequisites to life in a modern industrial-technological world. Other things are not of such import to everyone, but since school is meant to prepare students for their future professions and for their tertiary studies, a great deal of mathematical knowledge must be at their disposal when they leave high school, simply to prepare them for this. This is the practical argument.

Secondly, mathematics is interesting and enjoyable. This is true in many ways. Logical abstraction is a fundamental human thought process that has fascinated humanity for EONS. Individual mathematical problems are often interesting for their own sake, and finding their solutions is an enjoyable process. Students should certainly be offered the opportunity to experience this enjoyment for themselves. Mathematics competitions can play a large role in this, even if not every individual enjoys them in the same manner. Still, math is important because math is fun. This is the recreational argument.

Finally, studying mathematics schools abstract rational thought. Ideally, this should be true of most subjects in school, but the abstract world of pure mathematics is certainly the optimal ecosystem for such things to flourish. This is the abstract argument.

It is my firm belief that all three arguments are legitimate and strong, and that the various aspects of mathematics must therefore all be strongly represented in any complete curriculum.

Question 2: How do we best get students interested in the type of mathematics we want them to learn?
The answer to this question is, of course, dependent on the individual student's interests. Some students will be drawn in by the mathematical abstractions themselves. For some, the most interesting aspect will lie in potential applications. For yet others, it may be the historical context, the development of human thought through the generations. And for some, it may simply all be a game, and playing around with puzzles will prove the best path to the subject. All of these gateways are perfectly legitimate, and it seems clear to me that optimal teaching practices must offer at least a little bit of everything.

Question 3: What are we ultimately preparing students for in their mathematics classes?
Again, my answer to this question must necessarily be quite wide. We certainly want students to enjoy mathematics. Whether this is the most important aspect, or even important at all, will be up to individual teachers to decide. To my way of thinking, this is the base of all else, and students who do not have at least a semblance of enjoyment in their class work cannot be expected to fully appreciate the subject.

We are certainly teaching our students to prepare them for their future roles in society. This aspect cannot and should not be ignored. In this context, we must also prepare them for university. The tertiary institutions cannot be expected to start from scratch; human brains must have some developed mathematical competence by the time puberty is over, otherwise it is too late.

I would also argue that we should be training students' capacity for systematic logical thought and offering them as much general knowledge (here is that pesky concept of "Allgemeinbildung" again) as possible. If this is not to be imparted in the schools, then where?

References

Fischer, R. (2001). Höhere Allgemeinbildung. In A. Fischer et al. (Ed.), *Situation und Ursprung von Bildung* (Franz-Fischer-Jahrbuch 2001, pp. 151–161). Leipzig: Universitätsverlag.

Skovsmose, O. (1994). *Towards a philosophy of critical mathematics education*. Dordrecht: Kluwer Academic Publishers.

Smithers, D. B. (2005). Graph theory for the secondary school classroom. In *Electronic theses and dissertations* (Paper 1015). http://dc.etsu.edu/etd/1015.

Soifer, A. (2008). *The mathematical coloring book: Mathematics of coloring and the colorful life of its creators*. New York: Springer.

https://en.wikipedia.org/wiki/Mathematics_of_Sudoku.

http://www.kenkenpuzzle.com.

Chapter 7
Competition Aims to Develop Flexibility in the Classroom

Ingrid Semanišinová, Matúš Harminc and Martina Jesenská

Abstract We present a method for an implementation of Multiple Solution Tasks in the classroom in a way that should motivate students to solve problems in different ways. The method concerns a competition in problem solving for groups of students. Each group has to find and record such a solution of a given problem that, in their opinion, appears with the least frequency among the solutions of all groups in the class. For illustration, we present a few problems and corresponding different strategies which arose in the classroom and show how flexibility was demonstrated during a competition. Additionally, we discuss other benefits of including competitions in the classroom, namely creating connections among mathematical concepts and stimulating deeper understanding of concepts for students. For the teacher the method opens a possibility for developing flexibility and analysing the quality of students' knowledge and their level of understanding of mathematical concepts and relationships among them.

Keywords Flexibility · Competition · Multiple solution tasks · Problem solving strategies

I. Semanišinová (✉) · M. Harminc
Faculty of Science, Institute of Mathematics, Pavol Jozef Šafárik University, Jesenná 5, 04001 Košice, Slovakia
e-mail: ingrid.semanisinova@upjs.sk

M. Harminc
e-mail: matus.harminc@upjs.sk

M. Jesenská
Giles Academy, Boston, Lincolnshire, UK
e-mail: tina.jesenska@gmail.com

A. Soifer (ed.), *Competitions for Young Mathematicians*,
ICME-13 Monographs, DOI 10.1007/978-3-319-56585-9_7

7.1 Introduction and Motivation

The ability to solve diverse problems is considered as one of the most important cognitive activities in professional and personal life. An important component of the problem solving process is the ability that humans have of adjusting their response to changing circumstances and conditions and thereby exhibit flexibility. People who are characterized by more flexible thinking than others have a greater ability to adapt to special features of the environment and to produce more creative and appropriate solutions to problems. The success in solving mathematical problems is also closely related to the ability of students to use different problem solving strategies and to be flexible in using them. According to Silver (1997), flexibility in mathematical problem solving is the number of different ways that a student uses to solve, express or justify a problem. During the study of mathematics, students repeatedly face new situations that require the use of different strategies. It is natural that the strategy which a student will learn and successfully use in one situation may not be applicable in other situation, although it is similar. Outstanding results in mathematics depend on increasing the flexibility of a solver (Elia et al. 2009).

Including Multiple Solution Tasks (MSTs) in mathematics lessons is one of the possibilities of developing flexibility during problem solving (Levav-Waynberg and Leikin 2012). Flexibility, which is activated within a problem, leads a solver to observe each of the components of a problem so that the student's knowledge of the components becomes more complete and interconnected and she or he can more easily invent new strategies (Elia et al. 2009). Consequently, it develops the ability to determine the most efficient way to solve the problem and to make appropriate strategy choices and demonstrate adaptivity (Warner et al. 2009).

On the other hand, mathematics teachers usually prefer to solve problems in only one way in the classroom. They often argue that they do not have enough time to solve the problems in different ways and they must provide students with a secure tool to solve standard problems. They also often claim that students do not want to solve a problem in other way, if they have already solved the problem correctly. Therefore, it is difficult to motivate them to look for other solutions and to understand different approaches.

Teachers' arguments have motivated us to look for a method for an implementation of MSTs in the classroom in such a way that almost all students will be stimulated to solve a problem in different ways. We suggest the game—MSTs competition to challenge students to consider multiple solutions to a problem. In designing the competition our aim was to prepare such rules for competition that would show mathematics educators that creativity is not the domain of only a few exceptional individuals, but rather by using suitable mathematical activity, creativity can be fostered broadly in the classroom.

7.2 MSTs Competition Rules and Suggestions

The base for the initial rules for the competition were observations which were conducted during two school terms in four groups of pre-service teachers (in the 3rd and the 4th year of study, there were 10–13 members per group) and in one classroom of 16-year- old students with enhanced education in mathematics (25 students in the group). On the basis of our observations and corresponding forthcoming conclusions we designed the process of preparation and realisation of MSTs competition.

Selection of problems for the competition

We recommend using four or five MSTs problems for one standard lesson. Problems included in the competition could be both routine and non-routine. The selected problems should be solvable within 5–10 min for a chosen group of students, so that the students have enough time to look for other solution strategies. Before the competition, the teacher should prepare several possible solutions to the selected problems in order to show students possible approaches to problem solutions when competition ends.

The rules of the competition

- Before we start to work in the classroom, we have to arrange three- or four-member groups of students.
- Before the competition, students are asked to solve each problem in as many ways as they can. Then they have to choose one solution, which they consider to be the rarest in the class and pass it to the teacher.
- To avoid distraction, students receive only two problems at the beginning of the competition (out of four or five chosen MSTs problems). From 5 to 10 min after the beginning of the competition, students can get next problem, and after 15 min, students can get the last problem(s). If some group does not take all the problems from the teacher, he/she will give them to this group after 25 min from the beginning of the competition. Our experience shows that this organization for obtaining problems increases the chance that students in the group begin to communicate with each other on different solution strategies for the problems received.

If a group obtains all problems at once, the group has a tendency to assign each problem to one member and it leads to limited communication within group.

- Students may pass the solutions to the teacher at any time during the competition.
- Evaluation: If the group solves the problem correctly it gets 1 point + "number of groups in the class"–"number of groups in the class that had submitted the same solution". If the group solves the problem incorrectly it gets 0 points.
- The group that gets the most points wins.

Due to our empirical experience we recommend three member groups. We have realised there is good cooperation between three students during the solving process. Two member groups are very small to fulfil our goals. Four or more students in the group might cause the activity of some students to decrease.

If students have little experience with MSTs problems, we recommend for the competition to choose problems for which it is expected that students will easily find at least one solution strategy.

An example of a problem and its evaluation

Problem: Solve the quadratic equation $x^2 = 4x$.

1st group solution: We use the factorisation $x(x - 4) = 0$ *and we get two solutions* $x = 0$, $x = 4$.

2nd and 3rd group solution: Either $x = 0$ *or* $x \neq 0$ *and then we can divide equation by x and we get* $x = 4$.

4th, 5th and 6th group solution: We move all the terms to one side of equation, and we get $x^2 - 4x = 0$. *Then we use the formula for finding the roots of a quadratic equation.*

According to the rules, the assignment of points is following:

1st group gets $1 + 6 - 1 = 6$ *points.*

2nd and 3rd groups get $1 + 6 - 2 = 5$ *points.*

4th, 5th and 6th groups get $1 + 6 - 3 = 4$ *points.*

All students may use paper for their private notes and calculations, but the score of the group depends solely on the solution written on the paper which is given by the students to the teacher and explicitly marked as the one selected. It is expected that members of the group cooperate and discuss the solutions of each problem and choose the solution which they anticipate to be the rarest one.

The teacher decides whether the problem solutions of different groups are the same or not. Based on Leikin and Levav-Waynberg (2007), the difference between the solutions may be reflected in using:

(a) Different representations of a mathematical concept;
(b) Different properties (definitions or theorems) of mathematical concepts from a particular mathematical topic; or
(c) Different mathematical tools and theorems from different branches of mathematics.

7.3 Task Examples with Students' Solutions

We present five problems from a competition and the different corresponding strategies, which arose in two classrooms during this competition. The classrooms are from two different Slovak secondary schools with enhanced education in mathematics (52 seventeen- to eighteen-year-old students). Two problems are routine (involving an absolute value equation and a triangle), and the remaining three (involving cyclists, a hexagon and children grouping) are non-routine. Students have 45 min to solve these problems in groups (9 groups in each class).

Cyclists: *Two cyclists raced around the football field. The average speed of the first cyclist was 18 km/h and the average speed of the second one was 21 km/h. The second cyclist started to race when the first one already passed 300 m. The race ended after completing six laps. How long is one lap, if we know that both cyclists passed the finish line at the same time?*

Solution strategies:

1. Comparing distance:

   a. $18\text{t}+0.3=21\text{t}$, $\text{x}=\frac{21t}{6}$, where t is the time, when both cyclists are racing and x is the length of one lap.
   b. $18\text{t}=21\left(\text{t}-\frac{0.3}{18}\right)$, $\text{x}=\frac{18t}{6}$, where t is the time, when the first cyclist is racing and x is the length of one lap.

2. Comparing time: $\frac{6\text{X}-300}{18}=\frac{6\text{X}}{21}$, where x is the length of one lap.
3. Comparing speed: $\frac{0.3}{18}=\frac{1}{60}$ which means that the speed of the first cyclist is 300 m/min. Distance travelled by the second cyclist per minute is $\frac{21}{60}=0.35$ km. Speed of the second cyclist is 350 m/min. The group drew the picture (see Fig. 7.1).

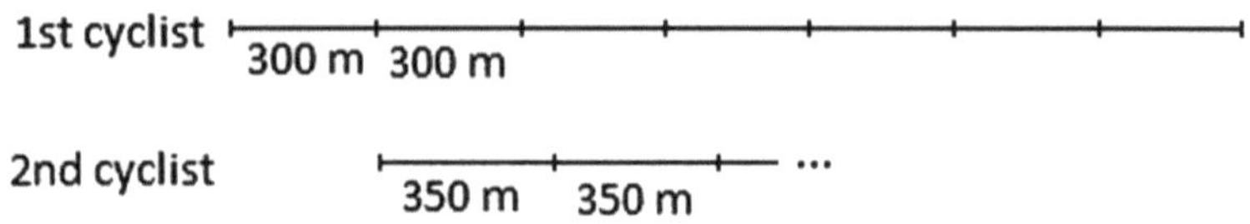

Fig. 7.1 A group of students models the race using this picture

lcm (300, 350) = 2100 m. $\frac{2100}{6} = 350$ m. A lap is 350 m long.

4. One group first used the second strategy but then, to the end of finding a rarer solution, the group handed into the teacher the following solution: $\frac{21}{18} \cdot 0.3 = \frac{7}{6} \cdot 0.3 = 0.35$. The lap is 350 m long. The method of solution was incorrect but accidently leads to the correct answer.

Because of the unusual formulation of the problem, some groups did not find any correct solution to the problem. Most groups solved the problem using strategy 1a and had difficulty in finding any other strategy. Comparing distance is the standard method to solve problems of this type. Strategy 3 was considered surprising by most of students because it uses number line representation and also different mathematical content. Students also found this strategy clear and easy to understand. This strategy was influenced and stimulated by group interactions. One member of the group started to draw a picture in order to understand the problem. Other members later found the mathematical content which was used to find the correct answer. In strategy 4, students tried to utilise the speed ratio. They interpreted the expression obtained, $\frac{7}{6} \cdot 0.3$, as a product of the speed ratio and 0.3 km as a head start. However, this interpretation is incorrect. If the group realised that the ratio of time is inverse to the speed ratio and used this fact to determine that the first cyclist finished the race in 6 min and the second in 7 min, they could also have realised that the product $7 \cdot 0.3$ is the length of whole race and $\frac{7}{6} \cdot 0.3$ is the length of 1 lap. Using these arguments, the solution strategy would be assessed as a correct one. Such a solution strategy occurred during pre-research among pre-service teachers.

Absolute value equation: Solve the absolute-value equation: $|x - 3| = |x + 5|$.

Strategies of solution:

1. Utilisation of algebraic properties of absolute-value and solving linear equations on the intervals

$$(-\infty, -5), \langle -5, 3) \text{ and } \langle 3, \infty).$$

2. Utilisation of geometric properties of absolute-value, searching for the number which is the same distance from the numbers 3 and −5 on the number line (see Fig. 7.2a).
3. Squaring the equation (both sides are positive) and solving the linear equation (see Fig. 7.2b).
4. Using graphs of functions $f\colon y = |x-3|$ and $g\colon y = |x+5|$.

Strategy 1 is a standard solution strategy to this type of task in our school. Four groups found only this solution strategy. Groups that found strategy 2, worked with a number line. But one of the groups used this strategy analytically (with no picture). They wrote that the distance between numbers −5 and 3 is 8, half of 8 is 4, so $x = -5 + 4 = -1$. Groups that found strategy 3 were very proud of it. They thought this strategy would be rare, but three groups in each class handed it into the teacher. As for strategy 4, a lot of groups had it among their solutions but only one handed in this solution to the teacher. This group achieved the maximum number of points using a standard solution strategy. Most students chose this problem as the easiest for finding more than one solution. Five groups found three different solution strategies.

Hexagon: *Let K, L, M, N, O, P be, respectively, the midpoints of the sides of regular hexagon ABCDEF. Join the points K, L, M, N, O, P to get hexagon KLMNOP. What is the ratio of the area of hexagon ABCDEF to the area of hexagon KLMNOP?*

(a) **(b)**

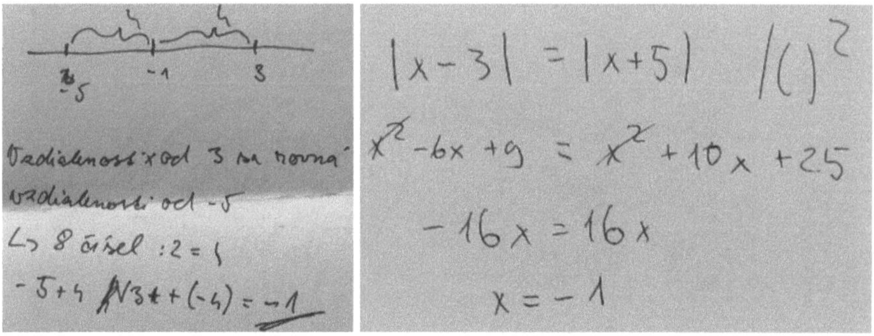

Fig. 7.2 **a** Absolute value—strategy 2 **b** Absolute value—strategy 3

Solution strategies:

1. Dividing both hexagons into six equilateral triangles and calculating the lengths of the triangles' sides and heights using the Pythagorean Theorem. Calculating the area of the equilateral triangles and the ratio of the areas of the hexagons.
2. Dividing both hexagons into six equilateral triangles and calculating the lengths of triangles' sides and heights using properties of trigonometric functions in right-angled triangles. Calculating the area of the equilateral triangles and the ratio of the areas of the hexagons.
3. Calculating the area of hexagon ABCDEF as in strategy 1. Subtracting the area of six isosceles triangles from the area of ABCDEF in order to calculate the area of the hexagon KLMNOP. Calculating the ratio of the areas of the hexagons.
4. Using the property that the ratio of the areas of the hexagons is the same as the ratio of the areas of the circumscribed circles.
5. Dividing the initial picture into congruent triangles using the properties of the centroid of a triangle; students found two different possibilities—18:24 and 36:48 (see Fig. 7.3).

Most of the groups solved the problem using strategy 1, 2 or 3. Some of them made a mistake during calculation. Only one group used strategy 4. They used intuitively the property that the ratio of the areas of the hexagons is the same as the ratio of the areas of the circumscribed circles. During the discussion about their solution they said they are not sure that this property really works. It was a good opportunity to start looking for the proof of this

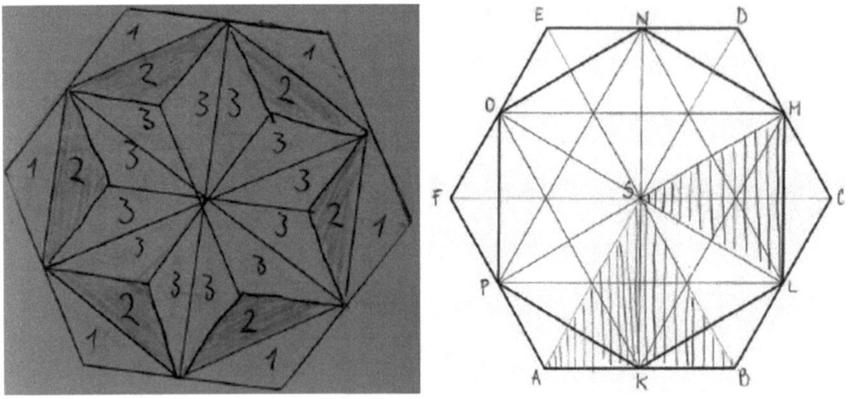

Fig. 7.3 Dividing the hexagon into the congruent triangles—two possibilities

property. Strategy 5 arose in two groups that first used strategy 1. One group divided the larger hexagon into 24 triangles and got the ratio 18:24 and other group divided it into 48 triangles and got the ratio 36:48. The fifth strategy was considered by most of the students as very surprising and beautiful. Only one group discovered three different solution strategies to this problem, namely strategies 1, 2 and 3.

Children grouping: *Four children: Anna, Barbara, Cyril and Daniel went to spend the night at their grandparents' house. Their grandparents have two separate bedrooms for them (one downstairs and another upstairs). In how many different ways can the grandparents assign children to bedrooms? For example: Anna, Barbara, Cyril and Daniel will sleep in the room upstairs and nobody will sleep downstairs (Batanero et al.* 1997).

Solution strategies:

1. The set of outcomes consists of the collection of all possibilities for children to be downstairs, perhaps encoded as all subsets of the set {A, B, C, D}.
2. In order to count the number of ways to arrange children in two rooms students count the number of possibilities of arranging 3 children, 2 children and 4 children in some room (see Fig. 7.4a).
3. Students arrange children in one room, divide the possibilities into the five subsets and then write the expression $\binom{4}{0} + \binom{4}{1} + \binom{4}{2} + \binom{4}{3} + \binom{4}{4}$ or a similar expression.

All groups, except one, organised the set of outcomes (solution strategy 1) into two columns. They considered it important to record the status in both rooms (see Fig. 7.5, on the left). One group realised that the second room is the complement of the first one and wrote possibilities for one room only

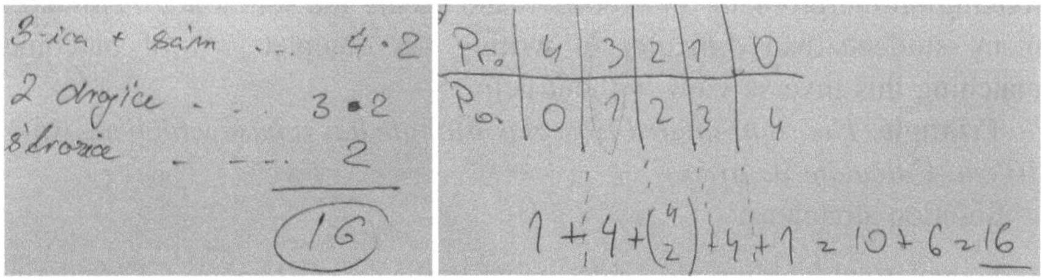

Fig. 7.4 **a** Children grouping—strategy 2 **b** Children grouping—strategy 3

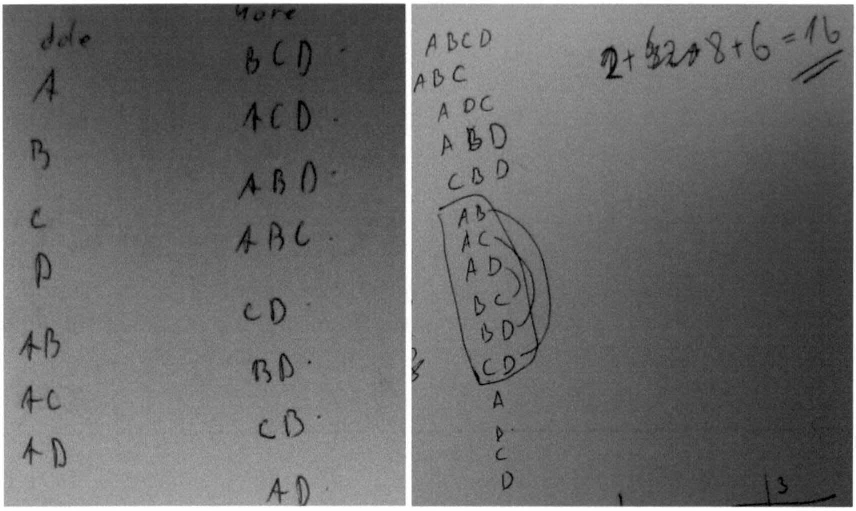

Fig. 7.5 Children grouping—strategy 1

(see Fig. 7.5, on the right). Strategy 2 together with strategy 1 were the most common strategies. Strategy 3 was used by five groups; one such solution can be found in Fig. 7.4b.

During the competition, no group found the method of solution that leads to the expression 2^4. They realised this method of solution during the discussion after the competition when the teacher asked them not to distribute children to rooms, but to distribute rooms to children (e.g. the grandparents give room keys to each child). The set of outcomes that was created after the reformulation of the problem is shown in Fig. 7.6.

After the presentation of this method of solution, some students discovered the connection between the result of the strategy 3 and the 5th row of Pascal triangle which they know is equal to 2^4. Students said that the discussion about this problem helped them to realise the possibility of looking on the problem "conversely" and seeing the solution 2^4. However, many students had difficulty in writing the complete set of outcomes matching this inverse view without help.

Triangle: *The right-angled isosceles triangle has a base which measures 10 cm. Calculate its area.*

Solution strategies:

1. Calculating the remaining sides using Pythagorean Theorem. Calculating the area of the triangle.

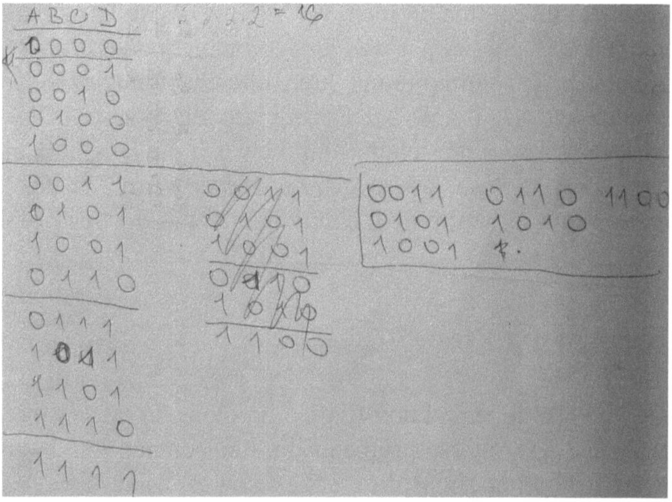

Fig. 7.6 Children grouping—set of outcomes after reformulation

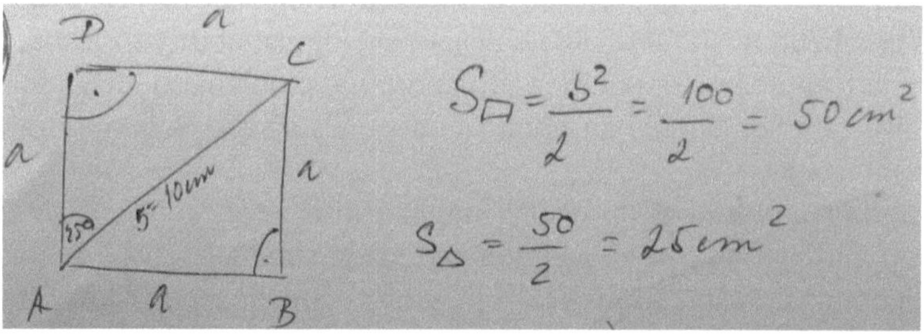

Fig. 7.7 Triangle—strategy 4

2. Calculating the remaining sides using trigonometric functions. Calculating the area of the triangle.
3. Dividing the triangle by its height, calculating the height using trigonometric functions and calculating the area of the triangle.
4. Utilising the fact that two congruent right-angled isosceles triangles make a square and using the "diagonal" formula to calculate the area of that square and consequently the area of the triangle (see Fig. 7.7).
5. Utilising the property that the height and median of an isosceles triangle are the same and that the median is equal to half of the base (from the Thales Theorem), we can directly identify necessary values and calculate the area of triangle.

This is the second routine problem we have used. Students usually found strategies 1, 2 and 3. Strategy 4 was less common and strategy 5 was used by only one group. This group found three different strategies for solving the problem (strategies 1, 4 and 5). During the discussion many students admitted that they forgot the relationship between the length of the median and the length of the base, which can be deduced from Thales Theorem. Three groups found three different solution strategies to this problem.

7.4 Discussion and Next Steps

We included both routine and non-routine problems in the competition. We think that including routine problems in the competition helps weaker groups to start thinking about different strategies. They have less trouble finding one and consequently two or more different strategies. Strong groups have enough time to look for an unexpected solution, e.g. squaring in the absolute value equation problem or utilisation of Thales Theorem in the triangle problem. Our experience shows that using routine problems in the MSTs competition should help students develop the ability to use the most appropriate approach for solution of the routine problem. Students may also recognize that sometimes an application of a standard solving strategy is disadvantageous.

Including non-routine problems might cause weaker groups to have a problem finding one correct solution and cause stronger groups to find only one solution strategy. However, the groups are likely to have different solution strategies and therefore more groups will get a maximum of points. For instance, using the hexagon problem in one class caused 4 groups to have wrong solutions, while the remaining 5 groups, which each had a correct solution, chose different strategies. That means that each of the 5 groups with a correct solution got 9 points. Moreover, solving non-routine problems requires creative thinking and the application of a certain heuristic strategy to understand the problem situation and find a way to solve the problem.

During the realisation of these experiences we come across a difficulty concerning the process of evaluation. For example, we think that the difference between strategies 1 and 5 is "bigger" than that between strategies 1 and 2 in the hexagon problem, although the groups got the same number of points. Nevertheless, if we try to distinguish such solutions (for instance classify solutions by their similarity, see e.g. Levav-Waynberg and Leikin 2012), the evaluation becomes unclear and difficult to understand for students. We also considered the possibility that, during the evaluation and discussion, students choose one solution for each problem which would get the maximum number of points, and that this solution be given one extra point, but we did not try it in the classroom. This suggestion is based on the experience that in the discussions students usually clearly expressed the opinion that some solution surprised them and was beautiful and/or unexpected. We also expect that the opportunity to participate in the evaluation motivates students to think more deeply about the solutions presented.

During the evaluation, the presentations of the different solutions were very fruitful because they show students various representations and properties, and create mathematical connections, which, as a result, helps to develop their flexibility. After presentations of different solutions to all of the problems we asked students the following questions:

1. Choose a problem for which you found it easy to find more than one solution.
2. In your opinion, which problem had a surprising solution?
3. Choose a problem which helped you to realise something new.

All students selected one of the routine problems as an answer to the first question. Students' responses to the second question helped us to identify which representations, solution strategies and connections they consider as unusual, exceptional, and original. Most students chose solutions that use pictures or manipulations with a given geometric shape as surprising (cyclists—strategy 3, absolute value equation—strategy 2, hexagon—strategy 5 and triangle—strategy 5). Analytic solution strategies using equations or calculation of areas or lengths and/or using standard formulas were not considered as surprising although some of these strategies were rare in the class. This confirms the claim formulated in Presmeg (1986) which states that most of the teachers preferred an analytic method of solution when solving mathematical problems.

For the mathematics teachers who would like to include MSTs competition in their lessons regularly, it could be useful to get a collection of problems with expert solution spaces including the most complete set of solutions to a problem. We think that such a collection could help teachers recognise the different possibilities of how one could represent the problem, to what mathematical content it can be assigned, and which mathematical connections could be created.

We assume that when students become familiar with the rules of competition, it is not necessary to spend the whole lesson on an MSTs competition; we can include one or two problems each week and evaluate the results of the groups, for example, every month.

7.5 Conclusion

Our experience showed us that the method presented can help mathematics teachers motivate students to look for new, non-standard solutions of the problems. Students are usually well-motivated to understand the "winning" solution strategy and to appreciate a nice one. During the process of repetition and systematisation of the knowledge of some topic, it may help students to organise and integrate the mathematical concepts, mathematical theorems and methods of solving mathematical problems which are pertinent to the topic. It can stimulate deeper understanding of concepts and principles and expand knowledge of the topic.

For the teacher, the method also opens a possibility of analysing the quality of students' knowledge and their level of understanding. When looking for another solution strategy, students are forced to leave the safe solution strategy of a problem preferred in the standard mathematics lesson. Mistakes that occurred among the students while looking for a new strategy may show the teacher that providing a secure tool for solving problems does not lead the student to conceptual understanding of the solving strategy presented.

Group work provides the opportunity to discuss ideas and listen to peers, to exchange ideas and hence to develop students' ability to communicate and reason. Students may discuss strategies and solutions, ask questions, and examine consequences and alternatives. The work in groups may also involve cooperative as well as independent work.

Mathematics educators accept that solving problems in different ways may help to develop students' creativity, especially two of its components—fluency and flexibility. In the current education system in Slovakia, gifted students do not receive enough attention and do not have enough opportunities to develop their giftedness during standard mathematics lessons. Teachers usually pay more attention to average students and to students who have problems with mathematics and usually provide them one safe and general solution strategy. The other reason for providing and training routine strategies lies in the fact that there is an external testing of mathematical knowledge in Slovakia (both in primary schools and secondary schools as well). Therefore, teachers pay attention to the development of such knowledge and algorithms that lead to a satisfactory test result.

We think that the method presented shows an approach that enables development of creativity of gifted and regular students together with stimulation of deeper understanding in students. The competition could be used as a tool for examining the mathematical flexibility of students, as a tool for analysis of how students discover, understand and use connections among mathematical ideas and, moreover, for nonstandard testing of students' knowledge. The exploitation of the method presented also allows students to communicate, analyse and evaluate their mathematical thinking and problem solving strategies.

Acknowledgements This work was partially supported by the Slovak Research and Development Agency under the contract No. APVV-0715-12.

References

Batanero, C., Navaro-Pelayo, V., & Godino, J. D. (1997). Effect of the implicit combinatorial model on combinatorial reasoning in secondary school pupils. *Educational Studies in Mathematics, 32,* 181–199.

Elia, I., van den Heuvel-Panhuizen, M., & Kolovou, A. (2009). Exploring strategy use and strategy flexibility in non-routine problem solving by primary school high achievers in mathematics. *ZDM Mathematics Education, 41*(5), 605–618.

Leikin, R., & Levav-Waynberg, A. (2007). Exploring mathematics teacher knowledge to explain the gap between theory-based recommendations and school practice in the use of connecting tasks. *Educational Studies in Mathematics, 66,* 349–371.

Levav-Waynberg, A., & Leikin, R. (2012). The role of multiple solution tasks in developing knowledge and creativity in geometry. *Journal of Mathematical Behavior, 31,* 73–90.

Presmeg, N. C. (1986). Visualisation and mathematical giftedness. *Educational Studies in Mathematics, 17,* 297–311.

Silver, E. A. (1997). Fostering creativity through instruction rich in mathematical problem solving and problem posing. *ZDM-Zentralblatt fuer Didaktik der Mathematik, 3,* 75–80.

Warner, L. B., Schorr, R. Y., & Davis, G. E. (2009). Flexible use of symbolic tools for problem solving, generalization, and explanation. *ZDM Mathematics Education, 41*(5), 663–679.

Chapter 8
Discovering, Development, and Manifestation of Mathematical Talent

Iliana Tsvetkova

Abstract Most parents want to think their child a gifted. That is generally considered a norm. But there is a difference between a high-achieving pupil in school and a child who is intellectually gifted. Discovering giftedness in the early school years is not always an easy task. However, gifted children have special traits that may help identify them among other pupils. For the last century thousands of papers and books have been written about gifted pupils. There is hardly something that can be added to them. The contribution of this paper is the Bulgarian experience in mathematics competitions for the discovery, the progress, the development and the manifestation of gifted pupils.

Keywords Mathematics competitions · Discovering giftedness · Mathematical talent · Development of talent

8.1 Introduction

There are gifted pupils everywhere around the world but children have different gifts. Some talents, like music for example, manifest in early childhood. Others require children to gain some experience and knowledge to help us first find and then develop talent. Mathematical talent is of the latter type. Children have to learn how to read and think before we can somehow find out in which little head a great mathematical talent is hidden. The time in which children are in the primary school is appropriate to provoke curiosity and to arouse interest in mathematics. The years between

I. Tsvetkova (✉)
Sofia Mathematical School, Iskar str. 61, Sofia, Bulgaria
e-mail: iliana_tzvetkova@yahoo.com

A. Soifer (ed.), *Competitions for Young Mathematicians*,
ICME-13 Monographs, DOI 10.1007/978-3-319-56585-9_8

grades 5 and 7 is the time in which the mathematical talent, to a large extent, is discovered and developed. In the upper secondary school this talent can be displayed and manifested.

8.2 Stimulate Interest

Children of the 21-st century face many challenges for their curiosity. The new technological inventions and the virtual world draw their attention like a magnet and take up their time. This is why it is not easy to attract them to do mathematics. From this point of view, mathematics competitions are a field that may awake pupils' interest. All kids love to compete, and win, of course. To enlist them for mathematics, the most important thing is to make them feel pleasure in solving problems.

The most enthusiastic to participate in mathematics competitions are the students in the beginning of the schooling, grades 1–2. There are several mathematics competitions for pupils in the early school years in Bulgaria. Even for grade 1 there are two national and (at least) two regional mathematics competitions. These events are the best places where the talented and gifted students can be discovered. Parents who find their child gifted, or at least exceptional, enter the child for these competitions. Certainly, parents may not be aware of the giftedness of their child. Teachers can judge better because they have broader view on the pupils' abilities and their opinion is much more objective. This is why teachers also help in this process.

Mathematics competitions for pupils in early school years last for about 120 min. Most of the problems are multiple choice items. These items are suitable for grade 1 pupils by the end of the school year when they can read. Nevertheless, there is a competition in the beginning of the school year (the

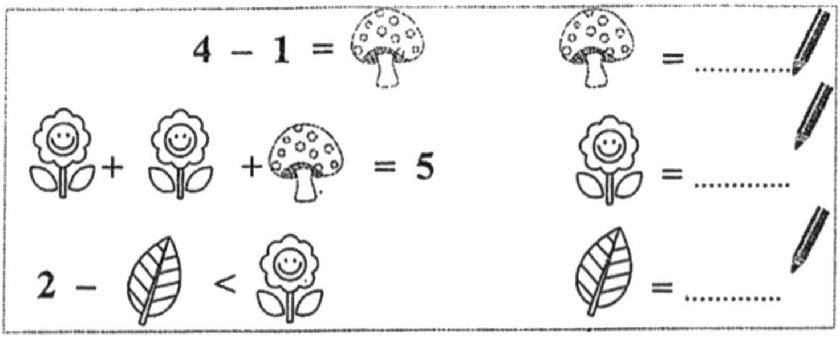

Fig. 8.1 First example from Sofia Mathematical Tournament, 2014

first days of November). At this time the pupils in grade 1 cannot read. This is why the grade 1 items are presented pictorially. Figures 8.1 and 8.2 present examples from the Sofia Mathematical Tournament, 2014.

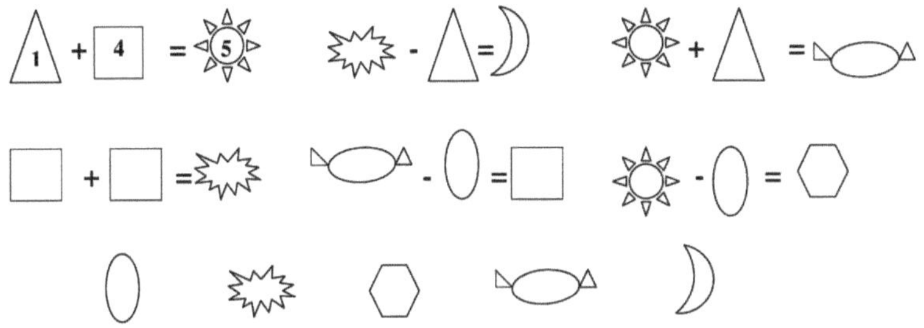

Fig. 8.2 Second example from Sofia Mathematical Tournament, 2014

Here are examples of two items from a mathematics competition, grade 1 (Easter Mathematics Competition in Bulgaria, 2011).

> On a playground there are balls, ropes, and hoops. Each child plays with one device only. Three children do not play with balls and ropes. Six children do not play with balls and hoops. Seven children do not play with ropes and hoops. How many children are there on the playground?
>
> A) 12 B) 13 C) 15 D) 16

> In each □ place either + or –, so that the number sentence is correct
>
> 2 □ 3 □ 4 □ 5 □ 6 = 12
>
> What are the symbols place between 3 and 4, and 4 and 5, respectively?
>
> A) +, + B) –, + C) –, – D) +, –

8.3 Discovery

The above mentioned competitions are the best stimulus for parents who find their child exceptional or gifted to consider applying for extracurricular activities offered by profile mathematical schools in Bulgaria. There are about 30 profile oriented schools in Bulgaria. They are secondary schools (grades from 8 to 12) situated in the provincial centers. Some of these schools offer extracurricular activities for pupils in grades from 2 to 4. These pupils study in different schools but attend the extracurricular activities on Saturdays. Also, some profile oriented schools accept children from grade 5. The entrance to these schools is based on an exam aiming to select the gifted and the most talented candidates.

As a case study, the experience of one of these schools, Sofia Mathematical School, is described below. It is a school specialized in training pupils from grade 5 to grade 12 (11–19 years old), who have a marked interest and talent in mathematics. The selection of the pupils is made on the basis of their results on several mathematics competitions. The most important of these is the one that is organized by the school itself. To be trained in Sofia Mathematical School, pupils should show interest in mathematics already in primary school and are directed to more specialized mathematics training, which is needed because every talent should be supported, developed and stimulated.

Sofia Mathematical School offers an opportunity to develop mathematical skills for pupils in grades from 2 to 4 (officially trained in other schools) by organizing extracurricular activities on Saturdays and Sundays. Highly qualified teachers train these pupils to solve non-standard problems helping them to develop their logical thinking. This training does not go beyond the arithmetic knowledge acquired during compulsory schooling but emphasizes their non-standard application. Teachers use charts, tables, pictures, etc. to activate pupils' intellectual abilities. Here is an example of a problem suitable for such training.

Problem A grandmother bought sweets to her three grandchildren. She divided the sweets among the grandchildren the following way: the oldest got half of all sweets and one more sweet; the second oldest got 2 sweets, then half of the remaining sweets and finally 3 more candies; the younger got 4 sweets, then half of the remaining sweets and one more sweet. The grandmother ate the last two sweets. How many sweets did the grandmother buy?

The solution is based on the principle of inversion (back–forward), and can be described by the following scheme:

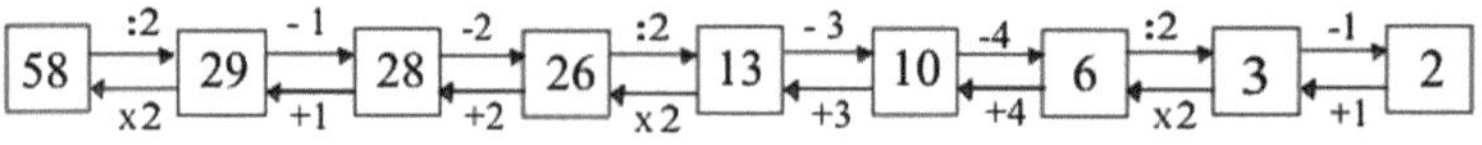

Most pupils attending these extracurricular activities preserve their interest in mathematics and the development of their mathematical abilities. After finishing grade 4 they apply for training in Sofia Mathematical School. The competition that is used for the final selection test consists of 15

multiple choice items, 5 open-ended short response items and 2 open problems whose solution should be explained and justified in detail.

Here is an example of such a problem.

> The snow was falling in huge flakes. Julia watched them through the window and counted the most beautiful snowflakes. To not forget how many she counted, she decided to record them in a notebook every five minutes. She drew ❄ to represent 12 snowflakes. Here is her drawing:
>
> 12:00 – 12:05 ❄, ❄
>
> 12:05 – 12:10 ❄, ❄, ❄, ❄, ❄
>
> 12:10 – 12:15 ❄, ❄, ❄
>
> 12:15 – 12:20 ❄
>
> (A) At what time interval did Julia count the most snowflakes?
> (B) How many snowflakes did Julia count between 12:10 and 12:15?
> (C) How many snowflakes in all were counted?
> (D) A few snowflakes "landed" on Julia's glove. Then ❄ of them melted. Then as many snowflakes "landed" as there were at the moment and ❄, ❄ more. The wind blew half of snowflakes from the glove and ❄, ❄, ❄ remained. How many snowflakes "landed" on the glove at the beginning?

8.4 Progress

Students who are accepted to study at the Sofia Mathematical School may attend additional training. For them the school organizes extracurricular work in mathematics, 3–4 academic hours weekly, mostly on Saturdays. Mathematics teachers from the school develop specially prepared programs for this training. These programs expand the topics studied in school and also contain some topics that are not studied in the compulsory curriculum. The participation in these "Saturday schools" is not compulsory and is free of charge for the pupils. In younger age groups (grades 5–7) there are many students eager to attend these activities. Later, when the difficulty of the studied material increases, the number of participants decreases. Usually those that drop the "Saturday schools" are pupils who achieve good results mainly thanks to their diligence and hard work, without having special mathematical talent. Some talented but lazy, sloppy and inconsistent pupils also leave these activities.

In order to get an idea of the work done in the "Saturday schools" a topic called "invariants" that is not connected to the school curriculum is presented below. The study of the topic can start from grade 5 and continue to the end of the secondary school. The problems that involve invariants are typical examples of tasks that do not need particular mathematical knowledge but whose solution requires a non-standard approach, logical thinking, and a lot of creativity.

What follows below is a part of the “Invariants” topic that is suitable for students in grades 5–7. Later in the chapter an extension of the topic that is suitable for upper grades (9–12) will be presented.

An invariant is a quantity or an indication that remains unchanged under certain transformation. The method of invariants for problem solving requires the solver to discover a feature that is unchanged under a given operation and to explain why a certain situation can be achieved or not. Sometimes it is necessary to make additional constructions or considerations.

8.4.1 Invariance and Parity (Kostadinova 2012)

Problem 4.1 Grasshopper jumps in a straight line. Every jump is 1 m. After a while, he returns to his original position. What is the parity of the number of jumps he made?

Problem 4.2 Not all alien beings have the same number of hands. A number of such aliens hold hands so that no hand is left free. Prove that the number of aliens with an odd number of hands is even.

Problem 4.3 Maya wrote 20 integers on the board, seven of which were odd. She erased two of the numbers and wrote the sum of their squares instead. Maya repeated this operation several times until only one number remained. What is the parity of this number?

Solution Let a and b be the numbers erased. Table 8.1 shows the change in the number of even and odd numbers after the execution of the operation.

Table 8.1 Change in the number of even and odd numbers

Erase a	Erase b	Write a^2+b^2	Number of even	Number of odd
Even	Even	Even	Decrease by 1	No change
Odd	Odd	Even	Increase by 1	Decrease by 2
Odd	Even	Odd	Decrease by 1	No change
Even	Odd	Odd	Decrease by 1	No change

Notice that the number of odd numbers either does not change or decreases by 2. In the beginning there are an odd number of odd numbers. Therefore, the last number is also an odd number

8.4.2 *Invariance and Coloring*

Problem 4.4 In a small rectangular garden flowers are blooming in 3 lines and there are 11 flowers per line. A bee starts from a flower adjacent to the flower in the upper left corner and flies from flower to flower going only to horizontally or vertically neighboring flowers. Is it possible for the bee to visit each flower only once?

Fig. 8.3 The garden represented as a chessboard

Solution Figure 8.3 represents the garden as a chessboard. The bee starts from a black square and alternately visits squares of different color. Since the number of white squares is one more than the number of black squares, it is not possible for the bee to visit each square only once.

8.4.3 *Invariance and Divisibility*

Problem 4.5 The teacher wrote the numbers 1, 2, 3, …, 2015, 2016 on the blackboard. Peter erases some of the numbers and writes the remainder of their sum modulo 11 instead. After the execution of this operation several times, only two numbers remain on the blackboard. One of them is 1001. What is the other number?

Hint. The invariant is the remainder of the sum $1 + 2 + 3 + \cdots + 2015 + 2016$ modulo 11. The answer is 6.

8.5 Development

After grade 8, only the truly gifted, talented and hardworking pupils attend "Saturday schools". As in all areas, talented students who make systematic efforts to develop and enrich their talent may perform well. And when it comes to children, the teacher (or the trainer) plays an important role. He/she must not only have excellent training (in mathematics, in this case), but also teach pupils to continue the development of their abilities by themselves, and find a way to motivate and encourage them when they have difficulties.

At this stage of the development of talent, motivation is the most important trait, because the achievement of excellence requires a lot of work and effort, denial of leisure and the usual teenage activities. In order to keep up their interest in the increasingly difficult mathematics, the competitions play an important role. They allow the pupils to prove themselves as "good mathematicians" and also to increase their prestige. Participation in national and a number of regional mathematics competitions is only a step in the development and the preparation of the talented students. It maintains their competitive shape and helps them to assess their level of preparation.

Every school year the Bulgarian Ministry of Education publishes a calendar of the national and regional school students' competitions in Bulgaria. The timetable of mathematics competitions is quite overloaded. The list starts with the National Mathematics Olympiad—one of the oldest Olympiads of this type in the world. It has more than 60 years of history. The Bulgarian Mathematics Olympiad consists of three rounds (for students in grades 7–12) and two rounds (for students in grades from 4 to 6). The results of the final round are important for the selection of the Bulgarian team for the International Mathematics Olympiad. All Bulgarian mathematicians and many other people whose profession is connected to mathematics participated in the Bulgarian Mathematics Olympiad.

There are two types of national mathematics competitions in Bulgaria: popular and elite. The former focus on a broad group of students of different ages (grades from 1 or 2 to 8 or 12), who do not necessarily have to have additional mathematics preparation but only "good logical thinking". These competitions are held at the same time in different cities. Usually the contest papers consist of 15–30 multiple choice items. Based on the results of these competitions some students receive a diploma for participation and/or a prize. This is a good way to stimulate students from regions that have not developed a system for extracurricular work in mathematics or for which the mathematics achievement of their students is not on a high level. The most popular competitions of this type are: Chernorizets Hrabar (http://www.math.bas.bg/ch/), Mathematical Tournament Ivan Salabashev (http://www.

math.bas.bg/salabashev/), European Kangaroo (http://www.aksf.org/). The last one is a type of international mathematics competitions held on the same day in many European countries as well as in the USA and Canada.

Elite mathematical competitions are for students of grades from 5 to 12 (or from 8 to 12). The contest papers for the elite competitions consists of 3 or 4 mathematical problems. The participants should present complete and well-grounded solutions. To enter such competition one needs not only talent but should also be systematically well prepared in mathematics. This is why most students that participate study in profile mathematical schools. The well-known elite national competitions in Bulgaria are: Autumn Mathematical Tournament, Winter Mathematical Competitions, Spring Mathematical Tournament "Atanas Radev", and Mathematics Competition for Linguistic Profile Schools.

The larger goal, however, is the participation in international competitions. Being selected for the team for such a competition is not an easy task. Usually the selection of the team goes through several rounds of tests and takes into account the results of the Bulgarian National Mathematical Olympiad. Motivation is very important in this stage as well. Participation in international contests is not only proof of abilities and talent, but can also be an opportunity to visit other countries and network with young people with similar interests.

The most prestigious of these opportunities is participation in the International Mathematics Olympiad (IMO) (https://www.imo-official.org/). It is very difficult to be selected for the IMO, because the team consists of 6 students from grades 11 or 12, but there are many rivals. The situation is similar for the Balkan Mathematical Olympiad (BMO) (http://en.wikipedia.org/wiki/Balkan_Mathematical_Olympiad). BMO has a junior version for students up to 15.5 years of age. For the last decade Bulgarian students have taken part in the International Zhautykov Olympiad in Kazakhstan (http://www.artofproblemsolving.com/community/c3241_international_zhautykov_olympiad). These competitions are very demanding. Usually the contestants have 3 or 4 mathematical problems for each of two consecutive days. To solve these problems one needs deep mathematical knowledge that goes far beyond the mathematics learned at school. The winners of these competitions are usually accepted to study mathematics (or studies related to mathematics) at many famous universities around the world.

International mathematics competitions for "younger students" (i.e. up to age 15) are much more attractive for pupils. Many of them are organized in Asia. The selection of the participants is made at the school level. This makes it possible for more pupils to enter the competitions. Each country may send up to 4 teams for a competition. It is not only the exotic destinations that attract pupils, but also the organization of the contests that is different from Bulgarian traditions and therefore is interesting for the participants. Usually there are two rounds (individual round and team round) in these competitions. During the individual round students work individually on a test consisting of short-answer and/or multiple-choice items. The team round presents several open-ended mathematical problems for each of the teams to solve. The students first distribute the problems among the team members, so that each of them thinks about the solution of at least one of the problems. Some of the solutions are developed by the whole team. The team round is usually very attractive and emotional for the contestants. Participation in cultural events that require students to present songs, dances, etc. of their country is a part of these competitions. The closing ceremony where the winners receive their prices are also magnificent and exciting. This way participation in these competitions play an important role in motivating students to develop their mathematical talent.

Mathematics competitions, depending on their type and organization, develop different aspects of talent. Competitions that present many problems of multiple-choice or short-answer type develop promptness of thought, intuitiveness, quick assessment of the situation, helpfulness of the memory and even correct judgment when to take a calculated risk. Certainly, some of these features are universal, not only mathematical.

Competitions that presents few but demanding mathematical problems requiring complete, extended, well-grounded solutions develop deep logical thinking, a high level of analysis and synthesis, a lot of mathematical knowledge, skills to reformulate the task to reduce it to something familiar, the finding of different ways to attack a completely unknown problem, etc. and all these things happen within a few hours. The students that have developed their talent in this direction usually become professional mathematicians.

The team competitions equally develop both mathematical and social aspects of talent, for example, to know not only your own strengths but also those of the other team members. These competitions require skills for teamwork, ability to critically appreciate the work of the team members and at the same time trust in their ability to think. The team competitions also develop ability to quickly understand and perceive others' solutions, to compare them with your own or alternative solutions, and to make a choice, to judge when to agree with the opinion of the other team members and when to assert your own. All these things need to happen for the best performance of the team.

The preparation of students in this ages (grades 8–12) in Sofia Mathematical School continue in the "Saturday schools". Some of the students also take part in the specialized preparation conducted by the leaders of the

Bulgarian national mathematics team. A variety of topics are learned in the "Saturday schools" and they are demanding. Some of the topics are continuation of the topics learned earlier, in the lower secondary school. Such a topic is "Invariants", part of which is presented in part 4 of this chapter. Here are some more problems from this topic suitable for the upper secondary school students.

8.5.1 Invariance and Divisibility

Problem 5.1 The teacher wrote the numbers $1, 2, 3, \ldots, n$ on the blackboard. Peter erases some of the numbers and writes the remainder of their sum modulo k instead, where k is a natural number. After carrying out this operation several times, only two numbers remain on the blackboard. One of them is mk where m is any natural number. What is the other number that remains? (This is a generalization of Problem 4.5.)

8.5.2 Invariance and Operations

Problem 5.2 A series of natural numbers $x_1, x_2, x_3, \ldots, x_n$ is written on the blackboard. The following operation on the series is admissible: randomly choose three of the numbers *a, b, c* and substitute *a* with $(a + 1)$ or $(a - 1)$, *b* with $(b + 2)$ or $(b - 2)$, *c* with $(c + 3)$ or $(c - 3)$, and do not change the rest of the numbers. Is it possible to obtain the following series $x_n + 1, x_n, x_{n-1}, \ldots, x_1$ in a finite number of admissible steps?

Solution Denote by *S* the sum of the numbers in the series. After carrying out the admissible operation, the value of *S* changes by one of the numbers $1+2+3=6$, $1+2-3=0$, $1-2+3=2$, $1-2-3=-4$, $-1+2+3=4$, $-1+2-3=-2$, $-1-2+3=0$, $-1-2-3=-6$. Since all these numbers are even, the parity of *S* does not change. The parity of the sums of the numbers of the initial and the final series are different, therefore the answer is "no".

Problem 5.3 (LIII Bulgarian National Olympiad in Mathematics, Final round, 2004) Consider all possible strings consisting of the letters *a* and *b*. In such string, the following substitutions are admissible: $aba \to b$, $b \to aba$, $bba \to a$, $a \to bba$. The initial string is $\underbrace{aa\ldots a}_{2003}b$. Is it possible obtain $b\underbrace{aa\ldots a}_{2003}$ applying admissible substitutions?

Solution We will show that the execution of any of the admissible substitution does not change the parity of the number of the letter a in even (odd) positions in the string. Indeed, let the substitution $aba \to b$ be applied over w_1abaw_2, where w_1 and w_2 represent the strings of letters surrounding the place where the substitution takes place. The new string is w_1bw_2. All a's belonging to w_1 stay in the same position, and all a's from w_2 move two positions to the left. Similarly, for any of the other substitutions—the a's before the substitution remain in the same position, and the ones after it move either two position to the left or two positions to the right. The number of a's in even positions in $b\underbrace{aa\ldots a}_{2003}$ is 1002, while in $\underbrace{aa\ldots a}_{2003}b$ there are 1001. Therefore, it is not possible to obtain the final string from the initial one.

8.5.3 *Invariance and a Change in the Parity of the Number of Elements*

Problem 5.4 There are several zeroes, ones and twos written on the blackboard. The following operation is admissible: delete two different digits and write the third one instead (e.g. if 1 and 2 are deleted, 0 is written instead; if 0 and 1 are deleted, 2 is written instead). Prove that if, after a finite number of applications of the admissible operation, only one digit remains on the blackboard, this digit does not depend on the order in which these operations are executed.

Solution Let p be the number of 0's, q the number of 1's, and r the number of 2's in the initial arrangement. After the execution of the admissible operation all three numbers p, q and r, change by 1, so they all change their parity simultaneously. If only one digit remains on the blackboard, then one of the numbers p, q or r is equal to 1, and the other two are equal to 0. Therefore, the parity of one of these numbers is different from the parity of the other two. The respective digit is the one that remains on the blackboard.

8.5.4 *Other Invariants*

Problem 5.5 Each of 20 given cards contains one of the digits 0, 1, 2, …, 9, so that each digit is written on exactly 2 cards. Is it possible to arrange these cards in such a way that the two 0 s are next to each other, there is one card between the cards containing 1's, there are two cards between the cards containing 2's, and so on, and finally there are nine cards between the cards containing 9's?

Solution Consider all possible arrangements of the cards with the numbers a and b written on them. If between the a-cards, there is exactly one b-card, then between the b-cards there is exactly one a-card. If between the a-cards there are two b-cards, then $b < a$, and between the b-cards there are no a-cards. If between the a-cards there are no b-cards, then between the b-cards there are either two a-cards or none. This way we conclude that there are even number of cards between every two cards with equal numbers written on them. On the other hand, the total number of cards between every two cards is $1 + 2 + 3 + \cdots + 9 = 45$, which is an odd number. This contradiction proves that it is impossible to arrange all the cards in the desired way.

Problem 5.6 A natural number is written on the blackboard. Every minute this number is divided or multiplied by two or three, so that the result is a natural number as well. The initial number is 12. Is it possible after an hour to obtain the number 54?

Solution Let's represent 12 as $2^2 \cdot 3$. Each number written on the blackboard is a product of powers of two and three. For the number $A = 2^{a_1}.3^{a_2}$ denote by $S(A) = a_1 + a_2$ and let $f(A)$ be the remainder of $S(A)$ modulo 2, $f(A) \in \{0, 1\}$. Every minute, either a_1 or a_2 changes its parity, so the value of $f(A)$ also changes. The value of $f(A)$ is the same as the initial value at every even minute. We have $S\ (12) = 2 + 1 = 3$ and $f(12)\ \equiv 1 (\text{mod } 2)$, while $S\ (54) = 3 + 1 = 4$, $f(54)\ \equiv\ 0 (\text{mod } 2)$. Since $f(12)\ \neq\ f(54)$ and an even number of minutes (1 h = 60 min) has passed, it is not possible to obtain 54 starting from 12.

8.6 Manifestation

Most of the students in Sofia Mathematical School spend 8 years studying in the school (from grade 5 to grade 12). These are eight years of very intensive preparation in mathematics both in the classroom and in the "Saturday schools". The best way to keep students motivated is to give them the opportunity to manifest their talent. Mathematics competitions provide an excellent opportunity in this respect. This is not only the International Mathematical Olympiad where only 6 students per year from Bulgaria participate. The good thing is that there are numerous national and international competitions that allow children of different ages and different levels of training to express their talent.

The existence of some mass and attractive competitions accessible for many students is needed because they increase interest in mathematics and keep this interest alive. The best students can easily win these competitions. They need to express and develop their talent in more demanding and elite contests. The awards from participation in mathematics competitions, the public attention from the school, the town, or the region, media attention, and a sense of satisfaction from the achievements support the development of talent. The number of contestants in the upper grades is diminishing because the competition problems are more demanding and require more effort. But the students that remain are really talented because the requirements for participation in the hardest competitions cannot be achieved only through diligence and perseverance.

Those that reach a level of development of talent required to participate in international competitions, usually continue their education in prestigious universities in mathematics or in fields related to mathematics and its applications, i.e. economics, finance, IT.

Below three of the very many examples of successful realization of graduates of the Sofia Mathematical School are listed. They entered the school from grade 5 and continued their education in the same school until the end of grade 12, participated in many competitions, won awards and have followed a career as professional mathematicians.

(i) Ljudmila Kamenova (http://www.math.sunysb.edu/~kamenova/): Gold Medal in the 13th Balkan Mathematical Olympiad, 1996; Silver Medal in the 37th International Mathematical Olympiad, 1996; National Diploma for excellence in the fields of natural and mathematical sciences, 1996; Over 20 first prizes in national and regional mathematical competitions from 1990 to 1996; Second Prize in the National Mathematical Olympiad for University Students, 1997; First Prize in the 5th International Mathematics Competition for University Students, 1998; Norman Levinson award for MIT graduate student, September 2001–May 2002; Mentor recognition award for the Siemens-Westinghouse Competition, 2003; Research Assistant Fellowships with professor Gang Tian, MIT 2003, 2005, 2006; Elected an honorary member in the Golden Key International Honour Society in 2014.

(ii) Tzvetelina Tzeneva (https://www.linkedin.com/in/tzvetelina-tzeneva-34450a25): Silver Medal at the International Mathematical Olympiad 2005; Bronze Medal at the International Mathematical Olympiad 2006; Silver Medal at the International Mathematical Olympiad 2007;

Shapiro Prize for Academic Excellence Princeton University 2009; The Peter A. Greenberg'77 Prize Princeton University 2011.

(iii) Bozhidar Velichkov (http://www.velichkov.it/): Silver medal winner (2003) and a gold medal winner (2004) Zhautykov Mathematical Olympiad Kazakhstan.

The discovery, the progress, and the development of mathematical talent takes a long time. It needs enormous effort from both the student and the teacher. Mathematics competitions play a very important role in this process. Their influence on the development of some famous Bulgarian mathematicians is described in Bankov (2013).

References

Bankov, K. (2013). Mathematics competitions and their influence on the professional development of the Bulgarian mathematicians. *Newsletter of the International Group for Mathematical Creativity and Giftedness* (4).

Kostadinova, M. (2012). Invariants. In S. Grozdev, V. Nenkov & S. Doychev (Eds.), *Hand-out for excellence in mathematics.* Foundation "M. Balkanski" (In Bulgarian).

Part V
Original Competitions

Chapter 9
International Mathematical Tournament of Towns

Nikolay Konstantinov and Sergey Dorichenko

> Why do we conduct the Tournament of Towns? Because we want everything to be well in our house. And our house is the whole world.
>
> *N.N. Konstantinov*

> When a new problem is invented, what should one do? Propose it at an olympiad or throw is out? I think that the same question faces a composer who has created a new melody: he offers it to people if he believes that it will be a gift to them.
>
> *N.B. Vasyliev*

Abstract This article is devoted to the International Mathematical Tournament of Towns, a high-level contest for high school students. We will tell you how this contest appeared, what are its aims, features and distinctions from other olympiads, what towns and countries participate in it and how one can take part in this contest. A significant part of the article is devoted to examples of problems from the Tournament of Towns, and to solutions of some of these problems

Keywords Mathematics · Competition · Olympiad · Tournament · International · Mathematical · Tournament of towns · Mathematical problems · Problem solving

Nikolay Borisovich Vasyliev was one of the key figures in the jury of the Soviet Union Mathematical Olympiad from 1966 to 1979. A.N. Kolmogorov,

---

N. Konstantinov · S. Dorichenko (✉)
Moscow School #179, Moscow, Russia
e-mail: sdorichenko@gmail.com

A. Soifer (ed.), *Competitions for Young Mathematicians*,
ICME-13 Monographs, DOI 10.1007/978-3-319-56585-9_9

who was the jury chairman, gathered a team of bright young mathematicians, who defined the spirit and style of this competition. But in 1979 government officials disbanded the jury. Then a natural but ambitious idea appeared: to create a new olympiad whose organization would be maximally independent of official Soviet institutions, so that it could be controlled by the mathematical community. That is how the Tournament of Towns was created, and N.B. Vasyliev took an active part in this work. He served as the chairman of the Central Jury of the Tournament of Towns till his death in 1998: he selected problems, and was the author of many of them. He also used problems sent by the readers of "Kvant" (Quantum) magazine for its problem section. Thanks to N.B. Vasyliev, the Tournament of Towns retained the scientific style of the Soviet Union Mathematical Olympiad while avoiding some of its organizational shortcomings.

We dedicate this article to Nikolay Vasyliev.

Today the Tournament of Towns is conducted by a large group of strong mathematicians and organizers, so the high quality of this olympiad is a teamwork result.

9.1 What Is the Tournament of Towns?

The Tournament of Towns is a worldwide problem solving competition in mathematics for high school students. Its scale is illustrated by the list of participating towns, see the end of this text. Each participating town has a place where students come to solve the problems and write down their solutions; then solutions are graded locally and the best of them are sent for central grading.

So what is special about the Tournament of Towns? How is it different from the many other mathematical competitions and why is it being held?

9.2 Goals of the Tournament

The Tournament organizers try to select interesting and beautiful problems that require a nonstandard approach. Solving these problem during the competition (or after the competition), they see mathematics from a different angle. In Russia, as in other countries, the standard math classes do not inspire creativity and often give the (wrong) impression that this science is a set of boring recipes for solving standard problems. In general, the school curriculum cannot keep up with the rapidly evolving modern world, which needs more and more creative people.

Unlike the International Mathematical Olympiad (where a student must pass a series of other competitions in order to participate), anyone can participate in the Tournament of Towns. Every year four rounds are held (a ordinary and advanced level, in autumn and spring), students may participate in all of them, and the final result depends only on the participant's best performance. Out of the many thousands participants of the multilevel International Mathematical Olympiad, only a few reach the final round. We believe that such a system discourages some students from doing mathematics, rather than attracting them.

9.3 We Are Looking for Talent!

The organizers of the Tournament of Towns, in Moscow and in other participating towns, are interested in discovering talented young people, who will later attend math schools and universities, and eventually work in scientific institutions. We are not interested in checking what a person has learned, but guessing what he can accomplish. This is why we use very liberal criteria for checking the student's work. The student's work should be written so that the jury can understand it, but in questionable situations, when it is uncertain whether the student skipped a proof because he was unable to find it or because it was obvious to him, the question is decided in the student's favor. This can lead to mistakes, when a solution is marked correct, while the student doesn't actually fully understand it. It is possible that we are doing a disservice to the students, by encouraging them to write down their solutions in a manner that will not be accepted in formal exams that they eventually will have to take. This is why we warn the students that most exams have stricter requirements for written solutions.

9.4 The Tournament of Towns and the Moscow Mathematical Olympiad

The advanced level of the spring round of the Tournament of Towns takes place on the same day as the Moscow Mathematical Olympiad, so the spring round (A-level) isn't held in Moscow. This system was implemented for several reasons: first, the spring semester is overloaded with various olympiads, secondly, for a long time there existed the idea of running the Moscow Mathematical Olympiad in other cities, and the Tournament of Towns partially solves this. Holding these two olympiads on the same day allows them to unite their bank of problems, and a lovely problem, suggested for one of these olympiads can be used for both. These two events are closely connected, and the list of successful participants of both competitions is considered as a whole.

9.5 The Tournament of Towns as a Sport

What attracts students to the Tournament? Not only the interesting problems, but that the Tournament is a sports competition. The need to test one's skills and to compete with others is in the nature of many people, especially youngsters. An athlete cannot expect his results to be kept a secret. One of the rules of the Tournament is that all results and materials are open to the public.

The athletic side of the Tournament is in conflict with its scientific side. For those who have just begun to be interested in mathematics, this conflict will be unnoticeable. During the Tournament the student is given 5 h. For beginners this is more than enough time to demonstrate their skills, but for a more advanced student, who is capable of solving the hardest problems of the Tournament, 5 h is not enough, and for them the Tournament becomes a timed competition. This is in conflict with the spirit of science. To compensate this flaw, the papers are graded by looking at three of the student's best solutions. The challenge of the Tournament is that solving three of the hardest problems is comparable to winning the International Mathematical Olympiad, so striving for an even higher results in olympiads is no longer necessary.

9.6 The Tournament of Towns and Professional Science

What should a student who has reached such a high level aspire to? He no longer needs olympiads, he needs unsolved mathematical problems. As one of the founders of mathematical olympiads in Russia, Boris Nikolaevich Delone, said: "An olympiad lasts only 5 h, but you need 5000 h to solve a serious mathematical problem".

The Tournament of Towns is followed by a summer school, where the students work in a format close to that of a professional mathematician. This school is called the Summer Conference of the Tournament of Towns. Unfortunately, it is held only for a small number of students (70–80 people of the 10 000 participants). Students spend a week there solving problems in a free format. Some of these problems are unsolved problems.

9.7 Summer Conference

The Tournament's Summer conference are unlike scientific conferences in the usual sense of the word. They do not have plenary lectures, sectional work-groups or even official programs. These conferences are more like informal gatherings to which students are invited along with accompanying teachers. One of the purposes is to give gifted students the chance to work on research problems. That is why the organizers propose very interesting projects—difficult problems or cycles of problems frequently connected with real mathematical research. Even the presentation of the statements of such a project can take up a whole lecture, and the presentation of all the projects takes a whole day. Each participant chooses one or two projects, which he will research as deeply as possible.

Solving such problems takes a long time and requires considerable intellectual efforts. So the solving process is rather informal. Usually, several days are given for its attempt, which can either be individual or collaborative.

The participant's achievements in the Tournament is the main criterion for the invitation to the Conference: those who achieve the highest results are invited. Invitations are also sent to the winners of other prestigious competitions such as All-Russian Olympiad and the IMO. Thus some students from cities and towns where the Tournament is not held can also come to the Conference.

Groups of students from different cities are usually headed by the teachers who organise the Tournament in their city. Many of them take part in work of the jury of the Conference. The composition of the jury is not predetermined.

All the participants of the Conference can enjoy sufficient rest, intensive creative work and interesting contacts.

9.8 The Jury Does Not Assign Places

There are no formal competitions during the conference. The Jury simply takes note of what problems are solved, and in the diplomas handed out to the participants only contain the list of their achievements, but no comparison to the other participants. The overall list of achievements is published, and students can judge by themselves whose achievement is higher.

This is close to real life: there is no jury which can say who is better—Galileo or Newton, Bohr or Einstein, Gauss or Euler. Such decisions would be, firstly, of no use to anyone, and secondly, anyone who wants to know can decide for himself.

During the Tournament of Towns no ranking is performed. In the records published at the end of the Tournament, each of winner's diplomas only contains the best results of the participant.

There is a level (12 points for the 2015/2016 school year) starting from which students receive diplomas, but the maximal results are much higher.

9.9 Diplomas and Awards

In addition to the diplomas for the winners of the Tournament handed out by the Central Jury, the local jury in each participating town can give out their own awards based on their own criteria. In Moscow, students who got at least 5 points but less than 12 are awarded by the Moscow Jury.

Five points approximately corresponds to solving one problem (not the easiest one). This has the following meaning: the difference between a student who solved one problem of medium difficulty and a student who solved nothing is a lot greater than the difference between the student who solved one problem and the student who solved five problems. In the first case the difference is qualitative, in the second it is quantitative. Plus you can add to the students who solved one problem during the Tournament, those who solved it after the competition (for example, while riding the Moscow underground, where, as one often hears, problem solving goes especially well).

9.10 Our Wish for the Participants

In conclusion we have the following request to our participants. All around the world, including Moscow, educators are unjustly shifting their focus from teaching to competitions. Olympiads, tournaments, math battles and other events, which were intended as a means to check the mathematical abilities of students, have gone way beyond the learning of mathematics needed to form these abilities. We advise students to pay more attention to studying and less to competitions. The main things in studying is to work systematically and not to rush. The outstanding Russian mathematician Igor Rostislavovich Shafarevich once wrote that a wonderful trait of Moscow mathematical circles is that every question is discussed for as long as it is needed, with no rush. Such style of work is in conflict with the busy rhythm of our lives. To achieve this style you only need to select all the interesting possibilities—the necessary ones, and out of the necessary ones—the most interesting. Don't spend your time on nonsense, although this is not easy. Don't go around with your eyes closed, try to look at the world with your own eyes, do not blindly follow the authors of books and concepts. Good luck!

9.11 Regulations of the Tournament

The Tournament is held each year in two rounds—spring and fall. Students and their cities can take part in either round or both, taking local conditions into consideration. If a certain city participates in both rounds, a student in this city has the right of choosing to take part in only one of them. This does not prevent the student from achieving a good result, because the student's score for the Tournament is the maximum (and not the sum) of the scores in the two rounds.

Each round has two levels—O-level (ordinary) and A-level (advanced). They are scheduled approximately two weeks apart. Here students have the right of choice as well. They may attempt either level or both. The score for the round is the maximal (not the sum) of the scores in the two levels. The questions in the O-level are less complicated and are accessible to beginners. However, students are awarded less points for solving these questions. Nevertheless, students can get enough points to win Diplomas if they solve the hardest three O-level problems. Questions in the A-level are more complicated. The most difficult ones are often solved only by a few participants. A beginner probably has no chance of obtaining any points from these questions. On the other hand, an exceptional student can be awarded two or three times as many points for them as for O-level questions.

Students who exceed a certain minimum score are awarded a Diploma from the Central Jury. Also, each participating town obtains a certain score. A town's score is the average score of the town's best N students' score, where N is the town's population divided by one hundred thousand. If a town's population is less than 500,000, N is then taken to be 5; but the town's score is then multiplied by a handicap factor.

9.12 Towns that Participate in the Tournament

One of the key traits of the Tournament of Towns is the diversity of participating regions. Because of different cultures, school programs, quality of life and many other factors, the local organizers of the Tournament do things in their own way. For example, in many schools across Argentina the O-level round serves as a qualifier for the A-level round, which is held in two cities—

Buenos Aires and Bahia Blanca. In Iran, the Tournament is held as a team competition, in Toronto the O-level is preceded by a set of problems from math circles in Moscow. Taiwan is not a city, but a whole country (a politically independent part of China), but in the list below Taiwan is mentioned as a city because the Taiwanese organizers requested to count their region as one big city with a population of 22 million. Several cities in Bulgaria and Israel participate in the Tournament, but they do not send the participant's work to the Central Jury.

The total population of the participating cities is around 100 million. Every year around 1000 people are given a winner's diploma of the Tournament, so for the past 37 Tournaments around 35 thousand people were awarded.

Only a small part of all the students on Earth can participate in the Tournament. If one takes into account that there are two million illiterate adults, millions of children who cannot attend school, then it can be seen that our event is still too small to make a difference in the overall mathematical culture of mankind.

Nevertheless some students are lucky enough to discover that mathematics is an endless world of the most refined intellectual creation, capable of satisfying the need of a thinking person. It does not matter if they made this discovery at some olympiad, the Tournament of Towns, at a math circle, or any other way. Even if mathematics did not become their profession, it entered their life forever, leaving a mark on their future work, no matter in which area it might be.

9.13 List of Towns and Rating

Notations:

Pop/1000—Population of the town, divided by 1000
NoP—Number of Participants
Dipl—Number of Diplomas
Max—Maximum of points, received by participants of the town
Av—Mean score of the best student's results
Coeff—Coefficient for towns with population less than 500 thousands people
Rat—Rating of the town (For towns that sent to the Central Examination papers of 5 or more students)

36 tournament of towns

Town	Location	Pop/1000	NoP	Dipl	Max	Av	Coeff	Rat	Rank
Kurgan	Russian Fed.	325	34	26	37.5	29.32	1.22	35.77	1
Troy	Michigan, USA	82	12	8	24	21.6	1.52	32.83	2
Dolgoprudny	Moscow reg, Russian Fed.	94	30	29	26.67	21.13	1.51	31.91	3
Ulyanovsk	Russian Fed.	615	19	18	34.67	29.39	1	29.39	4
Jeju	Rep. of Korea	435	29	29	30	27.02	1.08	29.18	5
Irvine	California, USA	223	12	7	30	21.25	1.35	28.69	6
Gwangmyeong	Rep. of Korea	341	14	13	26.25	23.32	1.2	27.98	7
Yaroslavl	Russian Fed.	559	23	18	35	25.95	1	25.95	8
Mokpo	Rep. of Korea	247	14	13	25.33	19.6	1.32	25.87	9
Kirov	Russian Fed.	483	48	46	25.33	24.87	1.02	25.36	10
Maikop	Adygea, Russian Fed.	144	27	16	22.5	17.2	1.44	24.77	11
Belgrade	Serbia	1232	31	26	37.33	24.44	1	24.44	12
Novorossiysk-BC	Krasnodar reg., Russian Fed.	262	7	7	22	18.7	1.3	24.31	13
Naberezhnye Chelny	Tatarstan, Russian Fed.	524	33	26	26	24.03	1	24.03	14
Windsor	Canada	5	5	3	32.5	14.2	1.62	23	15
Novorossiysk-2	Krasnodar reg., Russian Fed.	262	7	4	25.33	17.6	1.3	22.88	16
Nizhny Tagil	Sverdlovsk reg., Russian Fed.	358	20	13	23	19.13	1.18	22.58	17
Ulsan	Rep. of Korea	1163	35	35	31.5	22.21	1	22.21	18
Zaporizhia	Ukraine	768	43	14	29.33	22.14	1	22.14	19
Novosibirsk	Russian Fed.	1547	27	23	32	22.03	1	22.03	20
Zagreb	Croatia	792	5	5	25	21.9	1	21.9	21
Saransk	Mordovija, Russian Fed.	299	17	5	21.25	17.32	1.25	21.64	22
Vologda	Russian Fed.	306	33	15	18.75	17.35	1.24	21.51	23
Helm	Poland	72	22	11	15	14	1.53	21.42	24
Minsk	Belarus	1912	68	54	29.33	21.25	1	21.25	25
Krasnodar-BC	Russian Fed.	805	22	21	26	21.12	1	21.12	26
Kragujevac	Serbia	150	23	5	17	14.33	1.44	20.64	27
Moscow	Russian Fed.	12184	1382	368	33.33	20.24	1	20.24	28
Omsk	Russian Fed.	1160	32	27	26.25	20.17	1	20.17	29
Petropavlovsk-Kamchatsky	Kamchatka, Russian Fed.	181	10	3	18	14.37	1.4	20.11	30
Chelyabinsk	Russian Fed.	1182	9	9	29.33	19.95	1	19.95	31
Seongnam	Rep. of Korea	994	9	7	42	19.78	1	19.78	32
Seosan	Rep. of Korea	163	6	3	18.75	13.68	1.42	19.43	33
Ufa	Bashkortostan, Russian Fed.	1077	35	16	30.67	18.33	1	18.33	34
Almaty	Kazahstan	1485	32	32	22.5	17.91	1	17.91	35
Ekaterinburg	Russian Fed.	1396	35	17	22.67	17.79	1	17.79	36
Pereslavl-Zalessky	Yaroslavl reg., Russian Fed.	40	8	2	22.67	11.33	1.57	17.79	37

Town	Location	Pop/1000	NoP	Dipl	Max	Av	Coeff	Rat	Rank
Kuala Lumpur	Malaysia	1809	11	10	28.75	17.73	1	17.73	38
Changwon	Rep. of Korea	1089	11	11	29.33	17.47	1	17.47	39
Kostroma	Russian Fed.	271	7	6	15	13.53	1.29	17.46	40
Daejeon	Rep. of Korea	1535	20	16	24	17.23	1	17.23	41
Graz	Austria	276	7	4	20	13.4	1.28	17.15	42
Gwangju	Rep. of Korea	1477	14	12	29.33	17.11	1	17.11	43
Zhukovsky	Moscow reg., Russian Fed.	107	8	2	20	11.27	1.49	16.79	44
Petrozavodsk	Karelia, Russian Fed.	268	7	3	16	12.98	1.29	16.75	45
Novi Sad	Serbia	341	16	7	16	13.95	1.2	16.74	46
Sochi-BC	Krasnodar reg., Russian Fed	473	6	6	17.33	16.17	1.03	16.65	47
Toronto	Canada	2615	29	20	34.5	16.58	1	16.58	48
Erevan	Armenia	1068	12	10	23	16.57	1	16.57	49
Kazan	Tatarstan, Russian Fed.	1176	10	7	28	16.31	1	16.31	50
Elizovo	Kamchatka, Russian Fed.	38	5	2	14.67	10.28	1.58	16.25	51
Buenos Aires	Argentina	2890	44	23	28.75	15.8	1	15.8	52
Vitebsk	Belarus	373	15	4	16.25	13.52	1.16	15.68	53
Saint Petersburg	Russian Fed.	5028	74	34	30.67	15.68	1	15.68	54
Calgary	Canada	1096	13	6	28	15.66	1	15.66	55
Perm	Russian Fed.	1013	75	15	18.75	15.49	1	15.49	56
Charlotte	North Carolina, USA	792	5	2	34.67	15.27	1	15.27	57
Jeonju	Rep. of Korea	654	6	5	22	15.14	1	15.14	58
Astana	Kazahstan	828	31	14	16	15.09	1	15.09	59
Melbourne	Australia	4250	40	27	30	14.75	1	14.75	60
Perth	Australia	1832	13	11	22.67	14.68	1	14.68	61
Samara	Russian Fed.	1171	24	6	20	14.65	1	14.65	62
Suwon	Rep. of Korea	1170	10	8	22.5	14.47	1	14.47	63
Goyang	Rep. of Korea	1073	8	5	28.75	14.41	1	14.41	64
Ivanovo	Russian Fed.	409	5	3	23	12.95	1.11	14.37	65
Kiev	Ukraine	2849	26	21	34.5	14.27	1	14.27	66
Chita	Russian Fed.	335	23	2	20	11.35	1.21	13.73	67
Tomsk	Russian Fed.	557	8	3	21	13.6	1	13.6	68
Malmö	Sweden	309	5	2	15	10.92	1.24	13.54	69
Seoul	Rep. of Korea	10117	45	26	33	13.24	1	13.24	70
Saratov	Russian Fed.	840	7	5	17.33	13.08	1	13.08	71
Cheboksary	Chuvashya, Russian Fed.	464	6	4	13.33	12.27	1.04	12.76	72
Anyang	Rep. of Korea	609	5	2	18.75	12.4	1	12.4	73
Isfahan	Iran	1583	16	8	18	12	1	12	74
Pohang	Rep. of Korea	520	5	2	23	11.85	1	11.85	75
Busan	Rep. of Korea	3525	25	11	28	11.73	1	11.73	76
Hamburg	Germany	1718	22	7	20	11.71	1	11.71	77
Bremen	Germany	544	8	2	14	11.35	1	11.35	78

Town	Location	Pop/1000	NoP	Dipl	Max	Av	Coeff	Rat	Rank
Gumi	Rep. of Korea	374	6	1	16.25	9.7	1.16	11.25	79
Daegu	Rep. of Korea	2492	15	8	18	11.08	1	11.08	80
Cheongju	Rep. of Korea	838	7	2	18	10.71	1	10.71	81
Yazd	Iran	486	5	1	13	10.4	1.02	10.61	82
Voronezh	Russian Fed.	1023	62	2	15	10.5	1	10.5	83
Rostov-on-Don	Russian Fed.	1103	12	4	17.33	9.7	1	9.7	84
Kharkiv	Ukraine	1449	10	3	20	9.43	1	9.43	85
Penza	Russian Fed.	521	20	1	17	8.8	1	8.8	86
Volzhsky	Volgograd reg., Russian Fed.	326	9	0	11.25	7	1.22	8.54	87
Kropotkin	Krasnodar reg., Russian Fed.	80	9	0	7.5	5.55	1.52	8.44	88
Panama	Panama	600	6	1	19	8.04	1	8.04	89
Incheon	Rep. of Korea	2899	15	5	15	8.04	1	8.04	90
Tabriz	Iran	2383	6	2	12	8	1	8	91
Tyumen	Russian Fed.	679	12	0	10.5	7.92	1	7.92	92
Surgut	Khanty-Mansiy AO, Russian Fed.	332	80	0	5	4.7	1.21	5.69	93
Sovetsk	Kaliningrad reg., Russian Fed	41	54	0	7.5	3.33	1.57	5.23	94
Protvino	Moscow reg., Russian Fed.	37	5	0	1.25	0.85	1.58	1.34	95
Ejsk-BC	Krasnodar reg., Russian Fed.	85	3	3	16	14.67	1.52	22.29	–
Vancouver	Canada	603	4	3	13	12.25	1	12.25	–
Yongin	Rep. of Korea	909	4	4	24	19.73	1	19.73	–
Makhachkala	Dagestan, Russian Fed.	578	4	4	17.5	15.38	1	15.38	–
Volgograd	Russian Fed.	1018	3	3	16	14	1	14	–
Lund	Sweden	107	2	1	16	8.63	1.49	12.85	–
Essentuki	Stavropol reg., Russian Fed.	103	1	0	0	0	1.5	0	–
Luga	Leningrad reg., Russian Fed.	36	3	3	18	15.67	1.58	24.75	–
Kaliningrad	Russian Fed.	448	1	0	8	8	1.06	8.48	–
Nizhny Novgorod	Russian Fed.	1259	2	2	17.33	16.66	1	16.66	–
Boroujen	Iran	49	3	2	12	10.33	1.56	16.12	–
Khanty-Mansiysk	Russian Fed.	90	3	3	15	13.67	1.51	20.64	–
Berkley	California, USA	112	1	1	12	12	1.48	17.76	–
Kerman	Iran	573	3	2	15	11.33	1	11.33	–

Notes:

1. This rating is counted only for towns that sent to the Central examination 5 or more papers.

2. In towns of the Krasnodar land with the note (BC) the Tournament was held by the Bernoulli Centre (Krasnodar).

3. The results of Bremen only for the spring round. The results of Bremen in the fall round are included into the results of Hamburg.

4. Every Iranian team is counted in this rating as one team-participant.

9.14 How to Enter the Tournament

The Tournament is open to all towns and cities anywhere in the world. If in the city there is either an education organization (university, institute, school, etc.) which would like to organize the Tournament there, or a group of leaders or even a single teacher who can act as a town committee, this is quite sufficient for the city's participation in the Tournament.

CONTACTS to get the problems and perhaps to join the Tournament: prof. Nikolay Konstantinov (President), Sergey Dorichenko (Chairman of the Jury),
turnir.gorodov@gmail.com, turgor@mccme.ru.

9.15 Selected Problems from Different Years of the Tournament

During its existence the Tournament of Towns has presented its participants over a thousand problems. We present here a select few of these problems for a more meaningful introduction to the Tournament. There is no deep meaning in the choice of problems or their ordering. We wanted to demonstrate the variety of themes—what a participant sees, when he begins solving the problems of yet another round. We tried to order them by increasing order of difficulty (next to the problem is the number of points which was awarded for its complete solution) The format of the present publication prevents us from discussing the solutions. We warn you that:

1. In many of them the answer is surprising or even paradoxical.

2. Some problems are very difficult and were solved by very few participants.

The English version of the problems and solutions included into this article mostly were written by Andy Liu; they are taken from website https://www.math.toronto.edu/oz/turgor/ and from a series devoted to the Tournament of Towns, published by Peter Taylor in the Australian Mathematics Trust. We are very grateful to Andy Liu, Peter Taylor, Olga Zaitseva-Ivrii and Victor Ivrii.

TOURNAMENT 33, Fall 2011
Junior questions, O Level

1. [3 points] P and Q are points on the longest side AB of triangle ABC such that $AQ = AC$ and $BP = BC$. Prove that the circumcentre of triangle PQC coincides with the incentre of triangle ABC.

V.V. Proizvolov

TOURNAMENT 17, Fall 1995
Senior questions, O Level

2. [3 points] A square is placed in the plane and a point P is marked in this plane with invisible ink. A certain person can see this point through special glasses. One can draw a straight line and this person will say on which side of the line the point P lies. If P lies on the line, the person says so. What is the minimal number of questions one needs to find out if P lies inside the squares or not?

A.Ya. Kanel-Belov

TOURNAMENT 33, Spring 2012
Junior questions, O Level

3. [3 points] A treasure is buried under a square of an 8×8 board. Under each other square there is a message which indicates the minimum number of steps needed to reach the square with the treasure. Each step takes one from a square to another square sharing a common side. What is the minimum number of squares we must dig up in order to bring up the treasure for sure?

N. Strelkova

TOURNAMENT 14, Spring 1993
Junior questions, O Level

4. [3 points] Each of two houses A and B is divided into two flats. Several cats and dogs live there. It is known that the fraction of cats in the first flat of A (the ratio between the number of cats and the total number of animals in the flat) is greater than the fraction of cats in the first flat of B, and the fraction of cats in the second flat of A is greater than the fraction of cats in the second flat of B. Is it true that the fraction of cats in house A is greater than the fractions of cats in house B?

A. K. Kovaldji

TOURNAMENT 9, Spring 1988
Senior questions, O Level

5. [Variant for Moscow participants.] A point has been chosen in a 3-dimensional space. Is it possible to arrange four balls in the space so that they do not touch

either the point or each other, but "hide" the point in that any ray emanating from the point meets one of the balls?

Problem from Leningrad

TOURNAMENT 4, Spring 1983
Junior questions, O Level

6. [8 points] A pedestrian walked for 3.5 h. In every period of 1 h duration he walked 5 km. Is it true that his average speed was 5 km per hour?

N.N. Konstantinov

TOURNAMENT 10, Fall 1988
Junior questions, O Level

7. [3 points] It is known that the proportion of people with fair hair among people with blue eyes is more than the proportion of people with fair hair among all people. Which is greater, the proportion of people with blue eyes among people with fair hair, or the proportion of people with blue eyes among all people?

Folklore

TOURNAMENT 12, Fall 1990
Junior questions, O Level

8. [4 points] Suppose two positive real numbers are given. Prove that if their sum is less than their product then their sum is greater than four.

N.B. Vasiliev

TOURNAMENT 17, Spring 1996
Junior questions, O Level

9. [4 points] The two tangents to the incircle of a right-angled triangle ABC the are perpendicular to the hypotenuse AB intersect it at points P and Q. Find $\angle PCQ$.

M.A. Evdokimov

TOURNAMENT 28, Fall 2006
Senior questions, O Level

10. [4 points] Three positive integers x, y and z are written on the blackboard. Mary records in her notebook the product of any two of them and reduces the third number on the blackboard by 1. With the new trio of numbers, she repeats the process, and continues until one of the numbers on the blackboard becomes zero. What will be the sum of the numbers in Mary's notebook at that point?

Ye. Gorsky, S.A. Dorichenko

TOURNAMENT 30, Spring 2009
Junior questions, O Level

11. [4 points] Let a^b denote the number a^b. The order of operations in the expression 7^7^7^7^7^7^7 must be determined by parentheses (5 pairs of parentheses are needed). Is it possible to put parentheses in two distinct ways so that the value of the expression be the same?

A.K. Tolpygo

TOURNAMENT 31, Spring 2010
Junior questions, O Level

12. An angle is given in a plane. Using only a compass, one must find out
(a) [2 points] if this angle is acute. Find the minimal number of circles one must draw to be sure.
(b) [2 points] if this angle equals $31°$. (One may draw as many circles as one needs.)

G. Feldman, D.V. Baranov

TOURNAMENT 32, Fall 2010
Senior questions, O Level

13. [5 points] From a police station situated on a straight road infinite in both directions, a thief has stolen a police car. Its maximal speed equals 90% of the maximal speed of a police cruiser. When the theft is discovered some time later, a policeman starts to pursue the thief on a cruiser. However, he does not know in which direction along the road the thief has gone, nor does he know how long ago the car has been stolen. Is it possible for the policeman to catch the thief?

G.A. Galperin

TOURNAMENT 26, Spring 2005
Junior questions, O Level

14. [5 points] M and N are the midpoints of sides BC and AD, respectively, of a square $ABCD$. K is an arbitrary point on the extension of the diagonal AC beyond A. The segment KM intersects the side AB at some point L. Prove that $\angle KNA = \angle LNA$.

A.V. Akopyan

TOURNAMENT 35, Fall 2013
Junior questions, O Level

15. [6 points] Eight rooks are placed on a 8×8 chessboard so that no two rooks attack each other. Prove that one can always move all rooks, each by a move of a knight so that in the final position no two rooks attack each other

as well. (In intermediate positions several rooks can share the same square).

E.V. Bakaev

TOURNAMENT 11, Fall 1989
Senior questions, O Level

16. [3 points] The numbers 2^{1989} and 5^{1989} are written out one after the other (in decimal notation). How many digits are written altogether?

G.A. Galperin

TOURNAMENT 7, Spring 1986
Senior questions

17. [4 points] Vectors coincide with the edges of an arbitrary tetrahedron (possibly non-regular). Is it possible for the sum of these six vectors to equal the zero vector?

Problem from Leningrad

TOURNAMENT 8, Spring 1987
Junior questions, O Level

18. [3 points] We are given two three-litre bottles, one containing 1 litre of water and the other containing 1 litre of 2% salt solution. One can pour liquids from one bottle to the other and then mix them to obtain solutions of different concentration. Can one obtain a 1.5% solution of salt in the bottle which originally contained water?

S.V. Fomin

TOURNAMENT 4, Fall 1982
Junior questions, O Level

19. [12 points] There are 36 cards in a deck arranged in the sequence spades, clubs, hearts, diamonds, spades, clubs, hearts, diamonds, etc. Somebody took part of this deck off the top, turned it upside down, and cut this part into the remaining part of the deck (i.e. inserted it between two consecutive cards). Then four cards were taken off the top, then another four, etc. Prove that in any of these sets of four cards, all the cards are of different suits

A. Merkov

TOURNAMENT 34, Spring 2013
Junior questions, O Level

20. [5 points] Eight rooks are placed on a 8×8 chessboard, so that no two rooks attack one another. All squares of the board are divided between the rooks as follows. A square where a rook is placed belongs to it. If a square is attacked by two rooks then it belongs to the nearest rook; in case these two rooks are equidistant from this square then each of them possesses a half of

the square. Prove that every rook possesses the equal area.

E.V. Bakaev

TOURNAMENT 9, Spring 1988
Senior questions, O Level

21. Pawns are placed on an infinite chess board so that they form an infinite square net (along any row or column containing pawns there is a pawn, three free squares, pawn, three squares, and so on, with only every fourth row and every fourth column containing pawns). Prove that it is not possible for a knight to tour every free square once and only once.

An old problem of A.K .Toplygo

TOURNAMENT 38, Fall 2016
Senior questions, O Level

22. [5 points] Of the triangles determined by 100 points on a line plus an extra point not on the line, at most how many of them can be isosceles?

E.V. Bakaev

TOURNAMENT 31, Spring 2010
Senior questions, O Level

23. [5 points] Assume that $P(x)$ is a polynomial with integer nonnegative coefficients, different from constant. Baron Munchausen claims that he can restore $P(x)$ provided he knows the values of $P(2)$ and $P(P(2))$ only. Is the baron's claim valid?

S.V. Markelov

TOURNAMENT 35, Fall 2013
Senior questions, O Level

24. [6 points] A spacecraft landed on an asteroid. It is known that the asteroid is either a ball or a cube. The rover started its route at the landing site and finished it at the point symmetric to the landing site with respect to the center of the asteroid. On its way, the rover transmitted its spatial coordinates to the spacecraft on the landing site so that the trajectory of the rover movement was known. Can it happen that this information is not sufficient to determine whether the asteroid is a ball or a cube?

E.V. Bakaev

TOURNAMENT 11, Fall 1989
Senior questions, A Level

25. [3 points] Is it possible to choose a sphere, a triangular pyramid and a plane so that every plane, parallel to the chosen one, intersects the sphere and the pyramid in sections of equal area?

Problem from Latvia

TOURNAMENT 38, Fall 2016
Junior questions, A Level

26. [5 points] The diagram shows an arbitrary triangle dissected into congruent triangles by lines parallel to its sides. Prove that the orthocentres of the six shaded triangles are concyclic.

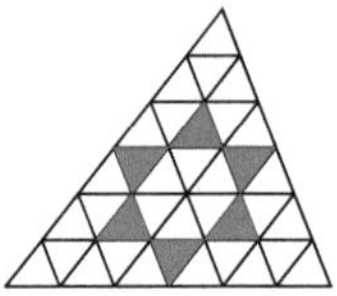

E.V. Bakaev

TOURNAMENT 24, Fall 2002
Senior questions, A Level

27. [6 points] A cube is cut by a plane so that the cross-section is a pentagon. Prove that the length of one of the sides of the pentagon differs from 1 m by at least 20 cm.

G.A. Galperin

TOURNAMENT 5, Spring 1984
Junior questions, A Level

28. [12 points] The two pairs of consecutive natural numbers $(8, 9)$ and $(288, 289)$ have the following property: in each pair, each number contains each of its prime factors to a power not less than 2. Prove that there are infinitely many such pairs.

A.V. Andjans

TOURNAMENT 17, Fall 1995
Junior questions, A Level

29. A journalist is looking for a person Z at a meeting of n persons. He has been told that Z knows all the other people at the meeting but none of them knows Z. The journalist may ask any person about any other person: "Do you know that person?" One person can be questioned many times. All answers are truthful.
(a) [3 points] Can the journalist always find Z by asking less than n questions?
(b) [3 points] What is the minimal number of questions which are needed to find Z?

G.A. Galperin

TOURNAMENT 6, Spring 1985
Senior questions, A Level

30. [8 points] The convex set F does not cover a semi-circle of radius R. Is it possible that two sets, congruent to F, cover the circle of radius R? What if F is not convex?

N.B. Vasiliev, A.G. Samosvat

TOURNAMENT 6, Spring 1985
Junior questions, A Level

31. [4 points] A square is divided into 5 rectangles in such way that its 4 vertices belong to 4 of the rectangles, whose areas are equal, and the fifth rectangle has no points in common with the sides of the square (see diagram).

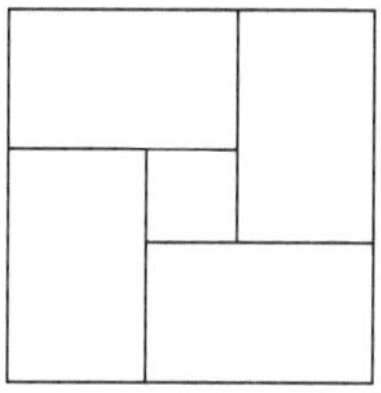

Prove that the fifth rectangle is a square.

V.V. Proizvolov

TOURNAMENT 4, Fall 1982
Senior questions

32. [15 points] Prove that for all natural numbers greater than 1

$$[\sqrt{n}] + [\sqrt[3]{n}] + \cdots + [\sqrt[n]{n}] = [\log_2 n] + [\log_3 n] + \cdots + [\log_n n].$$

V.V. Kisil

TOURNAMENT 4, Fall 1982
Senior questions

33. [8 points] Does there exist a polyhedron (not necessary convex) which could have the following complete list of edges?

$$AB, AC, BC, BD, CD, DE, EF, EG, FG, FH, GH, AH$$

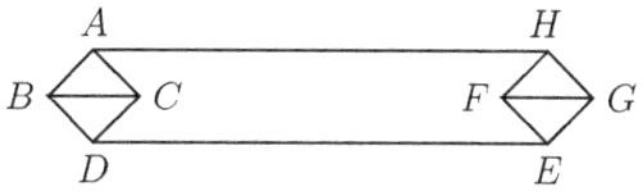

Folklore

TOURNAMENT 8, Spring 1987
Senior questions, A Level

34. [5 points] In a certain city only simple (pairwise) exchanges of apartments are allowed (if two families exchange flats, they are not allowed to participate in another exchange on the same day). Prove that any compound exchange may be effected in two days. It is assumed that under any exchange (simple or compound) each family occupies one flat before and after the exchange and the family cannot split up.

A. Shnirelman, N.N. Konstantinov

TOURNAMENT 16, Fall 1994
Senior questions, A Level

35. [4 points] The median AD of triangle ABC intersects its inscribed circle (with center O) at the points X and Y. Find the angle XOY if $AC = AB + AD$.

A. Fedotov

TOURNAMENT 28, Fall 2006
Senior questions, A Level

36. [6 points] Is it possible to split a prism into disjoint set of pyramids so that each pyramid has its base on one base of the prism, while its vertex on another base of the prism?

S. Slobodnik

TOURNAMENT 6, Spring 1985
Senior questions, A Level

37. A square is divided into rectangles. A "chain" is a subset K of the set of these rectangles such that there exists a side of the square which is covered by projections of rectangles of K and such that no point of this side is a projection of two inner points of two different rectangles of K.
(a) [12 points] Prove that every two rectangles in such a division are members of a certain "chain".
(b) [12 points] Solve the similar problem for a cube, divided into rectangular parallelepipeds (in the definition of chain, replace "side" by "edge").

A.I. Golberg, V.A. Gurevich

TOURNAMENT 9, Fall 1987
Junior questions, A Level

38. [8 points] There are 2000 apples, contained in several baskets. One can remove baskets and/or remove apples from the baskets. Prove that it is possible to then have an equal number of apples in each of the remaining baskets, with the total number of apples being not less than 100.

A.A. Razborov

TOURNAMENT 11, Spring 1990
Senior questions, A Level

39. [10 points] A cake is prepared for a dinner party to which only p or q persons will come (p and q are given co-prime integers). Find the minimum number of pieces (not necessary equal) into which the cake must be cut in advance so that the cake may be equally shared between the persons in either case.

D. Fomin, Leningrad

TOURNAMENT 10, Spring 1989
Senior questions, A Level

40. [6 points] A club of 11 people has a committee. At every meeting of the committee a new committee is formed which differs by 1 person from its predecessor (either one new member is included or one member is removed). The committee must always have at least three members and, according to the club rules, the committee membership at any stage must differ from its membership at every previous stage. Is it possible that after some time all possible compositions of the committee will have already occurred?

S.V. Fomin

TOURNAMENT 29, Fall 2007
Junior questions, A Level

41. [6 points] Michael is at the centre of a circle of radius 100 metres. Each minute, he will announce the direction in which he will be moving. Catherine can leave it as is, or change it to the opposite direction. Then Michael moves exactly 1 metre in the direction determined by Catherine. Does Michael have a strategy which guarantees that he can get out of the circle, even though Catherine will try to stop him?

A.I. Bufetov

TOURNAMENT 17, Fall 1995
Senior questions, A Level

42. [7 points]
Version for Nordic Countries
Six pine trees grow on the shore of a circular lake. It is known that a treasure is submerged at the mid-point T between the intersection points of the altitudes of two triangles, the vertices of one being at three of the six pines, and the vertices of the second one at the other three pines. At how many points T must one dive to find the treasure?
Version for Tropical Countries
A captain finds his way to Treasure Island, which is circular in shape. He knows that there is treasure buried at the midpoint of the segment joining the

orthocentres of triangles ABC and DEF, where A, B, C, D, E and F are six palm trees on the shore of the island, not necessarily in cyclic order. He finds the trees all right, but does not know which tree is denoted by which letter. What is the maximum number of points at which the captain has to dig in order to recover the treasure?

S.V. Markelov

TOURNAMENT 7, Spring 1986
Junior questions, A Level

43. [8 points] ("Sisyphian labour") There are 1001 steps going up a hill, with rocks on some of them (no more than 1 rock on each step). Sisyphus may pick up any rock and raise it one or more steps up to the nearest empty step. Then his opponent Aid rolls a rock (with an empty step directly below it) down one step. There are 500 rocks, originally located on the first 500 steps. Sisyphus and Aid move rocks in turn, Sisyphus making the first move. His goal is to place a rock on the top step. Can Aid stop him?

S. Yeliseyev

TOURNAMENT 26, Fall 2004
Junior questions, A Level

44. [7 points] Point K belongs to side BC of triangle ABC. Incircles of triangles ABK and ACK touch BC at points M and N respectively. Prove that $BM \cdot CN > KM \cdot KN$.

S.V. Markelov

TOURNAMENT 28, Spring 2007
Senior questions, A Level

45. [5 points] A convex figure F is such that any equilateral triangle with side 1 has a parallel translation that takes all its vertices to the boundary of F. Is F necessarily a circle?

S.V. Markelov

TOURNAMENT 28, Spring 2007
Junior questions, A Level

46. Nancy shuffles a deck of 52 cards and spreads the cards out in a circle face up, leaving one spot empty. Andy, who is in another room and does not see the cards, names a card. If this card is adjacent to the empty spot, Nancy moves the card to the empty spot, without telling Andy; otherwise nothing happens. Then Andy names another card and so on, as many times as he likes, until he says "stop."
(a) [5 points] Can Andy guarantee that after he says "stop," no card is in its initial spot?

(b) [5 points] Can Andy guarantee that after he says “stop,” the Queen of Spades is not adjacent to the empty spot?

A.V. Shapovalov, L.E. Mednikov

TOURNAMENT 26, Fall 2004
Senior questions, A Level

47. [6 points] A circle with the center I is entirely inside of a circle with center O. Consider all possible chords AB of the larger circle which are tangent to the smaller one. Find the locus of the centers of the circles circumscribed about the triangle AIB.

A.A. Zaslavsky

TOURNAMENT 29, Fall 2007
Junior questions, A Level

48. [7 points] Two players take turns entering a symbol in an empty cell of a $1 \times n$ chessboard, where n is an integer greater than 1. Aaron always enters the symbol X and Betty always enters the symbol O. Two identical symbols may not occupy adjacent cells. A player without a move loses the game. If Aaron goes first, which player has a winning strategy?

B.R. Frenkin

TOURNAMENT 30, Spring 2009
Junior questions, A Level

49. (a) [2 points] Find a polygon which can be cut by a straight line into two congruent parts so that one side of the polygon is divided in half while another side at a ratio of $1 : 2$.
(b) [3 points] Does there exist a convex polygon with this property?

S.V. Markelov

TOURNAMENT 19, Spring 1998
Senior questions, A Level

50. (a) [6 points] Two people perform a card trick. The first performer takes 5 cards from a 52-card deck (previously shuffled by a member of the audience), looks at them, and arranges them in a row from left to right: one face down (not necessarily the first one), the others face up. The second performer guesses correctly the card which is face down. Prove that the performers can agree on a system which always makes this possible.

A problem of M. Gardner

(b) [6 points] For their second trick, the first performer arranges four cards in a row, face up; the fifth card is kept hidden. Can they still agree on a system which enables the second performer to correctly guess the hidden card?

G.A. Galperin

TOURNAMENT 33, Fall 2011
Junior questions, A Level

51. [7 points] A car goes along a straight highway at the speed of 60 Km per hour. A 100 m long fence is standing parallel to the highway. Every second, the passenger of the car measures the angle of vision of the fence. Prove that the sum of all angles measured by him is less than 1100 degrees.

A. Shen

TOURNAMENT 30, Spring 2009
Senior questions, A Level

52. [6 points] Three planes dissect a parallelepiped into eight hexahedrons such that all of their faces are quadrilaterals (each plane intersects two corresponding pairs of opposite faces of the parallelepiped and does not intersect the remaining two faces). One of the hexahedrons has a circumscribed sphere. Prove that each of these hexahedrons has a circumscribed sphere.

V.V. Proizvolov

TOURNAMENT 32, Fall 2010
Senior questions, A Level

53. Two dueling wizards are at an altitude of 100 above the sea. They cast spells in turn, and each spell is of the form "decrease the altitude by a for me and by b for my rival" where a and b are real numbers such that $0 < a < b$. Different spells have different values for a and b. The set of spells is the same for both wizards, the spells may be cast in any order, and the same spell may be cast many times. A wizard wins if after some spell, he is still above water but his rival is not. Does there exist a set of spells such that the second wizard has a guaranteed win, if the number of spells is
(a) [2 points] finite;
(b) [5 points] infinite?

I.V. Mitrofanov

TOURNAMENT 31, Fall 2009
Senior questions, A Level

54. [8 points] Denote by $[n]!$ the product $1 \cdot 11 \cdot 111 \cdot \dots \cdot \underbrace{11...11}_{n \text{ ones}}$ (n factors in total). Prove that $[n+m]!$ is divisible by $[n]! \cdot [m]!$.

M.A. Bershtein

TOURNAMENT 31, Spring 2010
Junior questions, A Level

55. N horsemen are riding in the same direction along a circular road. Their speeds are constant and pairwise distinct. There is a single point on the road

where the horsemen can surpass one another. Can they ride in this fashion for arbitrarily long time? Consider the cases:
(a) [3 points] $N = 3$;
(b) [5 points] $N = 10$.

A.Klyachko, E. Frenkel

TOURNAMENT 25, Fall 2003
Senior questions, A Level

56. [7 points] An ant crawls on the outer surface of the box in a shape of rectangular parallelepiped. From ant's point of view, the distance between two points on a surface is defined by the length of the shortest path ant need to crawl to reach one point from the other. Is it true that if ant is at vertex then from ant's point of view the opposite vertex be the most distant point on the surface?

S.V. Markelov

TOURNAMENT 38, Fall 2016
Senior questions, A Level

57. [9 points] Is it possible to cut a 1×1 square into two pieces which can cover a disk of diameter greater than 1?

A.V. Shapovalov

TOURNAMENT 31, Spring 2010
Senior questions, A Level

58. [8 points] Quadrilateral $ABCD$ is circumscribed around the circle with centre I. Let points M and N be the midpoints of sides AB and CD respectively and let $\frac{IM}{AB} = \frac{IN}{CD}$. Prove that $ABCD$ is either a trapezoid or a parallelogram.

N. Beluhov, A. Zaslavsky

TOURNAMENT 26, Fall 2004
Senior questions, A Level

59. [8 points] Let $\angle AOB$ be obtained from $\angle COD$ by rotation (ray AO transforms into ray CO). Let E and F be the points of intersection of the circles inscribed into these angles. Prove that $\angle AOE = \angle DOF$.

I.I. Bogdanov, P.A. Kozhevnikov

TOURNAMENT 25, Spring 2004
Senior questions, A Level

60. [7 points] The parabola $y = x^2$ intersects a circle at exactly two points A and B. If their tangents at A coincide, must their tangents at B also coincide?

S.V. Markelov

TOURNAMENT 27, Fall 2005
Senior questions, A Level

61. [7 points] In triangle ABC bisectors AA_1, BB_1 and CC_1 are drawn. Given

$$\angle A : \angle B : \angle C = 4 : 2 : 1,$$

prove that $A_1B_1 = A_1C_1$.

S.I. Tokarev

TOURNAMENT 28, Fall 2006
Senior questions, A Level

62. [7 points] Let

$$1 + \frac{1}{2} + \frac{1}{3} + \cdots + \frac{1}{n} = \frac{a_n}{b_n},$$

where a_n and b_n are relatively prime. Show that there exist infinitely many positive integers n, such that $b_{n+1} < b_n$.

S.V. Markelov

TOURNAMENT 9, Fall 1987
Senior questions, A Level

63. [8 points] A certain town is represented as an infinite plane, which is divided by straight lines into squares. The lines are streets, while the squares are blocks. Along a certain street there stands a policeman on each 100th intersection. Somewhere in the town there is a bandit, whose position and speed are unknown, but he can move only along the streets. The aim of the police is to see the bandit. Does there exist an algorithm available to the police to enable them to achieve their aim?

A.V. Andjans

TOURNAMENT 2, Spring 1981
Senior questions

64. On an infinite "squared" sheet six squares are shaded as in the diagram. On some squares there are pieces. It is possible to transform the positions on the pieces according to the following rule: if the neighbour squares to the right and above a given piece are free, it is possible to remove this piece and put pieces on these free squares.

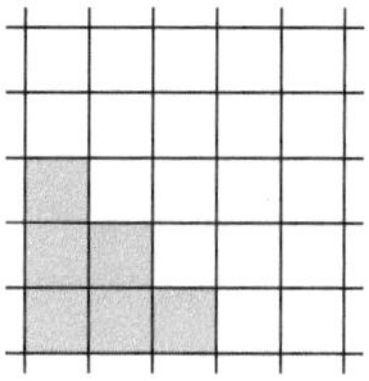

The goal is to have all the shaded squares free of pieces. Is it possible to reach this goal if
(a) [8 points] In the initial positions there are 6 pieces and they are placed on the 6 shaded squares?
(a) [8 points] In the initial positions there is only one piece, located in the bottom left shaded square?

M. Kontsevich

TOURNAMENT 30, Spring 2009
Senior questions, A Level

65. [9 points] Initially the number 6 is written on a blackboard. At nth step an integer k on the blackboard is replaced by $k + gcd(k, n)$. Prove that at each step the number on the blackboard increases either by 1 or by a prime number.

M. Frank

TOURNAMENT 29, Fall 2007
Junior questions, A Level

66. The audience arranges n coins in a row. The sequence of heads and tails is chosen arbitrarily. The audience also chooses a number between 1 and n inclusive. Then the assistant turns one of the coins over, and the magician is brought in to examine the resulting sequence. By an agreement with the assistant beforehand, the magician tries to determine the number chosen by the audience.
(a) [4 points] Prove that if this is possible for some n, then it is also possible for $2n$.
(b) [5 points] Determine all n for which this is possible.

S. Gribok

TOURNAMENT 32, Fall 2010
Senior questions, A Level

67. [14 points] A square is divided into congruent rectangles with sides of integer lengths. A rectangle is *important* if it has at least one point in common with a given diagonal of the square. Prove that this diagonal bisects the total area of the important rectangles.

V.V. Proizvolov

TOURNAMENT 4, Spring 1983
Senior questions, A Level

68. [30 points] k vertices of a regular n-gon P are coloured. A colouring is called *almost uniform* if for every positive integer m the following condition is satisfied:

If M_1 is a set of m consecutive vertices of P and M_2 is another such set then the number of coloured vertices of M_1 differs from the number of coloured vertices of M_2 at most by 1.

Prove that for all positive integers k and n ($k \leq n$) an almost uniform colouring exists and that it is unique within a rotation.

M. Kontsevich

TOURNAMENT 38, Fall 2016
Senior questions, A Level

69. [12 points] A finite number of frogs are placed on distinct integer points on the real line. At each move, a single frog jumps by 1 to the right provided that the new location is unoccupied. Altogether, the frogs make n moves, and this can be done in m ways. Prove that if they jump by 1 to the left instead of to right, they can still make n moves in m ways.

F. Petrov

TOURNAMENT 35, Fall 2013
Senior questions, A Level

70. [14 points] A closed broken self-intersecting line is drawn in the plane. Each of the links of this line is intersected exactly once and no three links intersect at the same point. Further, there are no self-intersections at the vertices and no two links have a common segment. Can it happen that every point of self-intersection divides both links in halves?

A.V. Shapovalov, A. Lebedev

Links and Books
About Tournament of Towns

https://www.turgor.ru/en

https://www.math.toronto.edu/oz/turgor

International Mathematics Tournament of Towns. Books 1–6. Australian Mathematics Trust. Books 1–3 by P. Taylor, book 4 by A. Storozhev and P. Taylor, book 5 by A. Storozhev, book 6 by A. Liu and P. Taylor.

L. Mednikov, A. Shapovalov. Tournament of Towns: world of mathematics in problems. MCCME, 2012. (in Russian)

A. Tolpygo. One thousand problems from the Tournament of Towns. MCCME, 2010. (in Russian)

Some Answers, Hints and Solutions

2. The answer is 3.

3. The answer is 3.
Hint: Let the first two squares we dig up be at the lower left corner and the lower right corner.

4. The answer is no.

5. The answer is yes.

6. The answer is no.

7. The proportion of people with blue eyes among people with fair hair is more than the proportion of people with blue eyes among all people.

9. The answer is 45°.

10. The answer is xyz.

11. The answer is yes.
Hint: (7^(7^7))^7 = (7^7)^(7^7).

12. (a) *(Solution by Olga Ivrii.)* Let O be the vertex of the given angle. Let P be any point on one arm of the angle other than O. Draw a circle with centre P and radius OP. If the other arm is tangent to the circle, then the given angle is a right angle. If the other arm intersects the circle in two points, then the given angle is acute. If the other arm misses the circle, then the given angle is obtuse. Hence the task can be accomplished using the compass only once.

(b) *(Solution by Wen-Hsien Sun.)*

Let O be the vertex of the given angle. Draw a circle ω with centre O and arbitrary radius, cutting the two arms of the angle at A_0 and A_1 respectively. Using A_1A_2 as radius, mark off on ω successive points A_2, A_3, …so that $A_0A_1 = A_1A_2 = A_2A_3 = \ldots$. Then $\angle A_0OA_1 = 31°$ if and only if $A_{360} = A_0$ but $A_k \neq A_0$ for $1 \leq k \leq 359$, and we have gone around ω exactly 31 times.

13. The answer is yes.

14. Let AC cut MN at O, and extend BA to cut KN at P. Since PL is parallel to NM and O is the midpoint of NM, A is the midpoint of PL. Hence triangles PAN and LAN are congruent to each other, so that $\angle KNA = \angle LNA$.

15. Observe that condition "no two rooks attack one another" means exactly that

(a) Each horizontal has 1 rook,

(b) Each vertical has 1 rook.

We break movement into two steps:

Step 1: Rooks from verticals 1, 2, 5, 6 move 2 squares right—to verticals 3, 4, 7, 8 respectively; rooks from verticals 3, 4, 7, 8 move 2 squares left—to verticals 1, 2, 5, 6 respectively. Obviously both conditions (a), (b) remains fulfilled.

Step 2: Rooks from horizontals 1, 3, 5, 7 move 1 square up—to horizontals 2, 4, 7, 8; rooks from horizontals 2, 4, 7, 8 move 1 square down—to horizontals 1, 3, 5, 7 respectively. Obviously both conditions (a), (b) remains fulfilled.

As a result each rook made a knight's move.

16. The answer is 1990.

Suppose that 2^{1989} has m digits and 5^{1989} has n digits. Then $10^{m-1} < 2^{1989} < 10^m$ and $10^{n-1} < 5^{1989} < 10^n$. Therefore

$$10^{m+n-2} < 2^{1989} \cdot 5^{1989} = 10^{1989} < 10^{m+n}.$$

Hence $1989 = m + n - 1$ and $m + n = 1990$.

17. The answer is no.

18. The answer is no.

22. The answer is 150.

23. The answer is yes.

Let $P(x) = a_0x^n + a_1x^{n-1} + \cdots + a_n$, where the coefficients are non-negative integers. Suppose $P(2) = b$. Then $b = a_0 2^n + a_1 2^{n-1} + \cdots + a_n > a_0 + a_1 + \cdots + a_n$. It follows that we have $b^n > b^{n-1}(a_0 + a_1 + \cdots + a_n) \geq a_1 b^{n-1} + \cdots + a_{n-1}b + a_n$. Now $\frac{P(b)}{b^n} = a_0 + \frac{a_1 b^{n-1} + \cdots + a_{n-1}b + a_n}{b^n}$. Then $a_0 = \lfloor \frac{P(b)}{b^n} \rfloor$, where n is the largest integer for which $P(b) \geq b^n$. In an analogous manner, $a_1 = \lfloor \frac{P(b) - a_0 b^n}{b^{n-1}} \rfloor$, and so on. It follows that $P(x)$ is uniquely determined, and the Baron is right!

24. The answer is yes.

Consider a sphere of radius r and a surface of cube with the side a with the same center. Observe that if $a = \sqrt{2}r$ the sphere touches each edge at its midpoint and therefore it intersects each face of the cube along circle of radius $r/\sqrt{2}$ in its center like on the figure below (we draw only three visible faces):

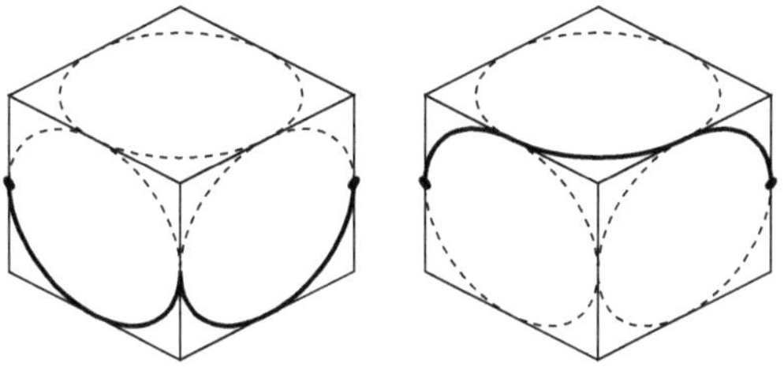

Then any path consisting of arcs of these circles belongs to both sphere and the surface of the cube and one can connect two symmetric points marked on the figure by such path. Therefore *it can happen that such information is not sufficient to determine whether the asteroid is a ball or a cube.*

25. The answer is yes.

27. Proof by a contradiction. Assume that pentagon has sides ranging from 0.8 to 1.2. To get a pentagon in cross-section of a cube, a plane has to cross five faces, two pairs of which are parallel. Therefore the pentagon has two pairs of parallel sides. Let us consider pentagon $BCDKL$ with $BC \parallel DK$ and $CD \parallel LB$. Then A be a point of intersection of BL and KD (extended). Note that $ABCD$ is a parallelogram. Due to triangle inequality $AL + AK > LK$, then $AB + AD > BL + LK + KD$. So, $BC + CD > BL + LK + KD$. Then even if BC and CD are two longest sides, $BC + CD \leq 2 \cdot 1.2 = 2.4$ while $BL + LK + KD \geq 3 \cdot 0.8 = 2.4$ which is contradiction.

28. A number is said to be *desirable* if it contains each of its prime factors to a power not less than 2. We construct an infinite sequence (a_n) by defining $a_1 = 8$ and $a_{n+1} = 4a_n(a_n + 1)$. We use induction on n to prove that both a_n and $a_n + 1$ are desirable for all n. For $n = 1$, both $a_1 = 8$ and $a_1 + 1 = 9$ are desirable. Suppose the result holds for all $n \leq k$. Then $a_{k+1} = 4a_k(a_k + 1)$ is desirable since each of 4, a_k and a_{k+1} is. Also, $a_{k+1} + 1 = (2a_k + 1)^2$ is desirable. Hence the result holds for all n.

29. (a) The answer is yes.
(b) The answer is $n - 1$.

30. The answers is yes.

33. The answers is yes.

35. The answers is 120°.

36. The answer is no.

39. The answer is $p + q - 1$.

40. The answer is no.

41. The answer is yes.

42. The answer is one point.

Label the trees A, B, C, D, E and F arbitrarily. Consider triangles ABC and DEF. Denote the centroids of these triangles by T_1 and T_2, and their orthocentres by H_1 and H_2. Then the homothety with centre T_1 and coefficient -2 maps the centre O of the island to H_1 (because it maps ABC to the triangle $A'B'C'$ such that A, B, C are the midpoints of $B'C'$, $C'A'$, $A'B'$ and H_1 is the circumcentre of $A'B'C'$), the homothety with centre T_2 and coefficient -2 maps O to H_2. So the midpoint T between T_1 and T_2 lies on the segment OH where H is the midpoint between H_1 and H_2 and $OH = 3OT$. Since T is the centre of gravity for all six given points and does not depend on the way we label the six trees, the Captain needs to dig only one hole.

43. The answer is yes.

45. (*Solution by Olga Zaitseva-Ivrii.*) No, the convex figure does not have to be a circle. Let AB be a horizontal segment of length 2. Draw a semicircle with diameter AB above AB. For any equilateral triangle of side 1, place its lowest vertex at the midpoint O of AB. If there are two choices, place either one at O. The other two vertices of the equilateral triangle always lie on the semicircle. Hence the convex figure bounded by AB and the semicircle has the desired property

46. (a) The answer is yes. Andy can call the cards out in order starting with the Ace of Spades, two of Spades down to the King of Spades, followed by the Hearts, the Diamonds and the Clubs. We refer to this as one cycle. In each cycle, each card can move at most once since it is called exactly once, and at least one card must move. Andy then makes another 51 cycles of calls. We claim that all moves are in the same direction, either all clockwise or all counter-clockwise. This is clear within each cycle. Consider the card X which is the last to move in a cycle, and let Y be the other card adjacent to the empty spot. Since Y does not move after X in this cycle, it must have been called before X. So in the next cycle, Y will be called before X, and follows X in the same direction. This justifies our claim. To go once around and return to its initial spot, a card must have moved 53 times, and this is not possible since Andy makes only 52 cycles of calls. If it is to be in its initial spot, it must not have moved at all. However, this is also impossible as otherwise at most 1 move could have been made, but in 52 cycles, at least 52 moves have been made. Therefore, after 52 cycles of calls, every card is in a spot different from its initial one.

(b) The answer is no. Construct a graph where each of the vertices represents one of the 52! permutations of the cards, with the first and the last adjacent to the empty spot. Two vertices are joined by an edge if and only if a call by Andy changes the two permutations to each other. Label the edge with the card called by Andy. In this graph, each vertex has degree 2, and the graph is a union of disjoint cycles. Consider the cycle containing the vertex representing the initial permutation. For each vertex, let a person starts there. Whenever Andy makes a call, the person moves along an edge labelled with that card to an adjacent vertex if possible, and stays put otherwise. We call a vertex safe if and only if in the permutation it represents, the Queen of Spades is not adjacent to the empty spot. By shifting each card clockwise into the empty spot in turns, we will arrive at permutations represented by safe vertices as well as permutations represented by unsafe vertices. Note that after each call, there is still one person on each vertex. Thus no matter what sequence of calls Andy may employ, he cannot get everyone to a safe vertex. It follows that there is an initial permutation for which Andy's sequence will leave the Queen of Spades adjacent to the empty spot.

48. Betty can guarantee a win.

49. (*Solution by Daniel Spivak.*) Divide the sides of a square in counterclockwise order in the ratio $1:2$. If we connect both pairs of points of division on opposite sides, the square is dissected into four congruent parts. If we connect only one pair, we have two congruent convex quadrilaterals. Disregard one of them, and the line connecting the other pair of points of division will dissect the remaining convex quadrilateral into two congruent part.

50. (b) The answer is yes.

51. Divide the points of observation into six groups cyclically, so that the points in each group are 100 m apart, the same as the length of the billboard.The diagram below shows the angles of visions from the points of a group.

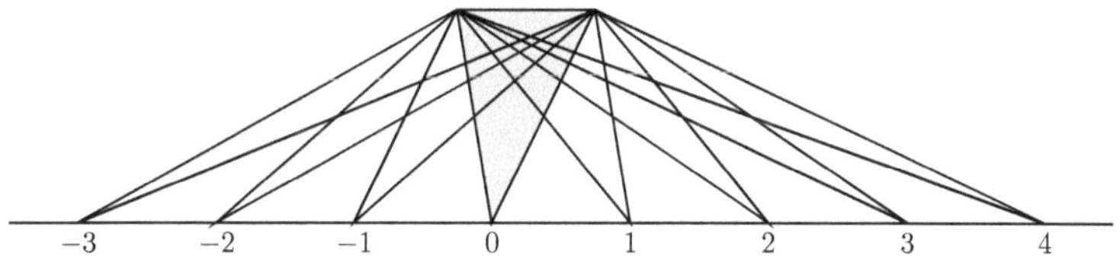

We now parallel translate all these points to a single point, along with their billboards and angles of vision, as shown in the diagram below.

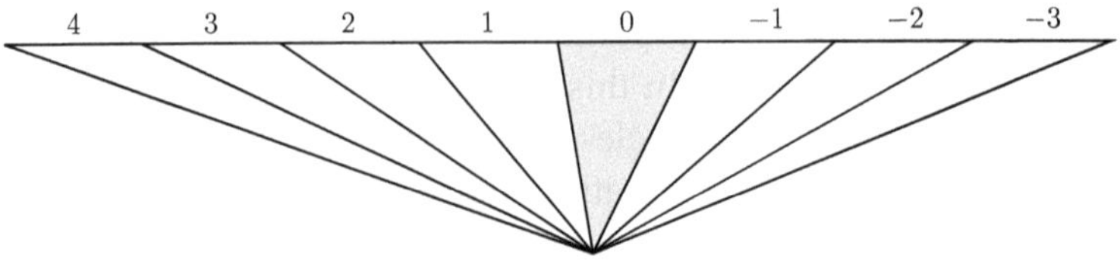

The sum of all these angles is clearly at most 180°. Since there are six groups of points of observations, the sum of all angles of vision is at most $6 \cdot 180° < 1100°$.

53. (a) The answer is no. With a finite number of spells, there is one for which $b - a$ is maximum. If the first wizard keeps casting this spell, the best that the second wizard can do is to maintain status quo by casting the same spell. Hence the second wizard will hit the water first, giving the first wizard a win.

(b) The answer is yes. In the nth spell, let $a = \frac{1}{n}$ and $b = 100 - \frac{1}{n}$. By symmetry, we may assume that the first wizard casts the nth spell. He is then $100 - \frac{1}{n}$ above water while the second wizard is $\frac{1}{n}$ above water. However, the second wizard wins immediately by casting the $(n+1)$-st spell. He will still be $\frac{1}{n} - \frac{1}{n+1} = \frac{1}{n(n+1)}$ above water while the first wizard is submerged in water since $(100 - \frac{1}{n}) - (100 - \frac{1}{n+1}) = -\frac{1}{n(n+1)}$.

55. (*Solution by Jonathan Zung.*) We use induction on the number n of runners. For $n = 1$, there is nothing to prove. Suppose the result holds for some $n \geq 1$, each with a distinct integer speed. Let M be the least common multiple of these speeds. If we add an $(n+1)$-st runner with speed 0 at the passing point, the result still holds. Now increase the speed of each of the $n+1$ runners by M. Since their relative speeds remain the same, the result continues to hold. In particular, it holds for $n = 3$ and $n = 10$.

56. The answer is no.

57. The answer is yes.

58. (*Solution by Jonathan Zung.*) Let P, Q, R and S be the points of tangency of the circle with AB, BC, CD and DA respectively. Let $\angle AIS = \angle AIP = \alpha$, $\angle BIP = \angle BIQ = \beta$, $\angle CIQ = \angle CIR = \gamma$ and $\angle DIR = \angle DIS = \delta$. Then $\angle AIB + \angle CID = \alpha + \beta + \gamma + \delta = 180°$. If $\angle AIB > 90°$, then $\angle CID < 90°$. The point I will be inside the circle with AB as diameter but outside the circle with CD as diameter. Hence $\frac{IM}{AB} < \frac{1}{2} < \frac{IN}{CD}$. Similarly, if $\angle AIB < 90°$, then $\frac{IM}{AB} > \frac{1}{2} > \frac{IN}{CD}$. Both contradict the hypothesis that $\frac{IM}{AB} = \frac{IN}{CD}$. Hence $\alpha + \beta = \angle AIB = 90°$ so that Q, I and S are collinear. Since both BC and DA are perpendicular to QS, they are parallel to each other.

60. The answer is no.

63. The answer is yes.

64. (a) The answer is no. (b) The answer is no.

65. Let us write down few first terms in the sequence:

Step #	1	2	3	4	**5**	**6**	7	8	9	10	**11**	**12**
Number in the cell	6 7	8	9	10	**15**	**18**	19	20	21	22	**33**	**36**
Increment	1	1	1	1	**5**	**3**	1	1	1	1	**11**	**3** ...

Let us denote by n the number of the step, $A(n)$ the number in the cell, $I(n) = A(n) - A(n-1)$ its increment.

One can notice the following pattern: If on some step n, $I(n) \neq 1$ then $A(n) = 3n$. (In the table corresponding columns are in bold).

Let $A(n) = 3n$ for some n. On the next step the number increases by $I(n+1) = gcd(n+1, 3n)$ and since n and $n+1$ are coprimes then $I(n+1) = gcd(n+1, 3)$. Thus, increment is either $I(n+1) = 1$ or $I(n+1) = 3$. In the latter case we have that $n+1$ is divisible by 3 so on the next step $I(n+2) = 1$ for certain.

This observation leads us to the following

Conjecture. *Let $A(n) = 3n$ for some n, and the next increment be $I(n+1) = 1$. Consider the nearest step $n+k$ when increment will be different from 1: $I(n+k) \neq 1$. Then $I(n+k)$ is a prime number and $A(n+k) = 3(n+k)$.*

To prove conjecture we use induction. We already checked the base for small numbers n. Let $A(n) = 3n$ for some n and $n+k$ be the nearest number with $I(n+k) \neq 1$:

Step #	n	$n+1$	$n+2$	$\dots$	$n+k-1$	$n+k$
Number in the cell	$3n$	$3n+1$	$3n+2$	$\dots$	$3n+k-1$	?

For increment $I(n+k)$ we have (using here and below $gcd(a, b) = gcd(a, a-b)$):

$$I(n+k) = gcd(n+k, 3n+k-1) = gcd(n+k, 3(n+k) - (3n+k-1)) = gcd(n+k, 2k+1).$$

Hence, $I(n+k)$ is divisor of $2k+1$.

Assume that $2k+1$ is not a prime and p is a prime divisor of $gcd(n+k, 2k+1)$. Since $2k+1$ is odd then $p \leq (2k+1)/3$. Therefore $p < k$. Let us look at step $n+k-p$. At this step an increment is

$$I(n+k-p) = gcd(n+k-p, 3n+k-p-1) = gcd(n+k-p, 3(n+k-p)-(3n+k-p-1))$$
$$= gcd(n+k-p, 2k+1-2p).$$

But since both $n+k-p$ and $2k+1-2p$ are divisible by p we see that on step $n+k-p$ increment differs from 1. This contradicts to the assumption that $n+k$ is the nearest step.

Therefore, $2k+1$ is a prime number and $I(n+k) = 2k+1$ and then $A(n+k) = A(n+k-1) + I(n+k) = (3n+k-1) + (2k+1) = 3(n+k)$.

Our conjecture is proven by induction and the problem solved.

66. (a) Given a row of n coins arbitrarily arranged heads and tails, and a number between 1 and n inclusive, the assistant can flip exactly one coin so that the magician can tell which number has been chosen. With a row of $2n$ coins and a number m between 1 and $2n$, the magician and the assistant place the numbers 1 to n in order in the first row of a $2 \times n$ array, and the numbers from $n+1$ to $2n$ in order in the second row. If the row number h and the column number k of the location of m are determined, then $m = (h-1)n + k$. The magician and the assistant also consider the $2n$ coins as in a $2 \times n$ array. Code each coin with heads up as 0 and each coin with tails up as 1. Compute the sum of the codes of the two coins in each column modulo 2 and regard the result as a linear array of n coded coins. By the hypothesis, the assistant can flip the qth coded coin to signal the number k to the magician. This can be achieved by flipping either of the two coins in the qth column. To signal the number h to the magician, the assistant will just use the bottom coin of the qth column, code 0 meaning $h = 1$ and code 1 meaning $h = 2$. If the bottom coin is not correct, flip it. Otherwise, flip the top coin.

(b) For $n = 1$, the assistant must flip the only coin. However, the chosen number can only be 1, and the magician does not require any assistance. Hence the task is possible. For $n = 2$, let the coins be coded as in (a). The assistant will just use the second coin, code 0 meaning $h = 1$ and code 1 meaning $h = 2$. If the second coin is not correct, flip it. Otherwise, flip the first coin. Hence the task is also possible. By (a), the task is possible whenever n is a power of 2. We now show that the converse also holds. Each of the 2^n arrangement of the coins codes a specific number between 1 and n. If n is not a power of 2, then $2^n = qn + r$ where q and r are the quotient and the remainder obtained from the Division Algorithm, with $r > 0$. By the Pigeonhole Principle, some number is coded by at most q arrangements. Each may be obtained by the flip of a single coin from exactly n other arrangements. The yields a total count of $qn < 2^n$. On the other hand, from each of the 2^n arrangements, we must be able to obtain one of these q arrangements by the flip of a single coin. This contradiction shows that the task is impossible unless n is a power of 2.

67. Divide the square into unit cells. Enumerate the (checkered) diagonals parallel to AC, starting from the lower left corner B. The rectangles from the condition of the problem, such that their lower left corner belongs to the kth diagonal will be called *bricks of the kth sort.*

Lemma 1 *The number of bricks of each sort does not depend of dissection.*

Proof Observe that the number of cells on the kth diagonal occupied by a brick of an nth sort depends only on n and k and not on the location of the brick.

There is a single brick of the 1st sort. Suppose the assertion is true for bricks of the first $k-1$ sorts. Then the number of cells occupied by them on the kth diagonal is independent of dissection. But the number of brick of the kth sort equals the number of the remaining cells of this diagonal. The lemma is proven.

Now we return to the problem. The bricks *under* diagonal AC are just bricks of several first sorts. Thus their number is independent of dissection. Reflect the square in its center to get a new dissection with this number of bricks *over* AC. Hence in the original dissection, the number of bricks *under* and *over* diagonal AC is the same. This clearly implies the assertion of the problem.

69. Let $S = \{a_1, a_2, \ldots, a_n\}$, where each a_k, $1 \leq k \leq n$, is either ℓ or r. It denotes a sequence of jumps where the kth jump is to the left if $a_n = \ell$ and to the right if $a_k = r$. Let $f(S)$ be the number of possible ways of carrying out S from the initial configuration of frogs. We claim that $f(\ell\ell \ldots \ell) = f(rr \ldots r)$. Note that $f(S\ell) = f(Sr)$ for any S. Clearly, we have equality up to the last jump, and the number of possible last jumps depends only on the number of groups of adjacent frogs at that point. We also have $f(S\ell rS') = f(Sr\ell S')$ for any S and S'. Again, we have equality up to the completion of S. Consider $S\ell rS'$. If the two switched jumps are made by different frogs, we can just switch their order and continues. If they are made by the same frog but not the leftmost one, we can replace them with jumps r and ℓ by the frog immediately to its left, with at least one space in between. Finally, if both jumps are made by the leftmost frog, we can replace them with jumps r and ℓ by the rightmost frog. These transformations allow us to conclude that $f(S)$ depends only on the length of S.

70. The answer is no.

Chapter 10
Madhava Mathematics Competition—A Recent Initiative in India

V.M. Sholapurkar

Abstract This article presents a survey of the Madhava Mathematics Competition, a recently started Mathematics competition for undergraduate students in India. The competition was started in the academic year 2009–10 and has received a tremendous response in last eight years. It can be seen as an extension of the Olympiad competitions to undergraduate classes. We therefore take a review of Olympiads in India and then discuss the various aspects of this new competition.

Keywords Madhava Mathematics Competition · Mathematical Olympiads

10.1 Introduction

In this article, we propose to present a recent initiative in India in the field of Mathematical Competitions. We shall present an account of Madhava Mathematics Competition—a mathematics competition for the students of undergraduate classes. The crux of the presentation lies in the discussion of the problems used for the competition and responses of the students in terms of some beautiful solutions to these problems. In the process, we shall also discuss the overall Indian scenario in the context of mathematics education at the school and college levels, the decisive role of this competition in enhancing the mathematical abilities of students, and some statistics, depicting the trends in the competition. A feedback of the competition from the student fraternity, teachers and senior mathematicians is an integral part of the article.

V.M. Sholapurkar (✉)
Department of Mathematics, S. P. College, Pune, India
e-mail: vmshola@gmail.com

A. Soifer (ed.), *Competitions for Young Mathematicians*,
ICME-13 Monographs, DOI 10.1007/978-3-319-56585-9_10

In Sect. 10.2, a survey of the impact of Mathematical Olympiad in India has been taken. The Mathematical Olympiad has provided a necessary impetus and motivation for launching the Madhava Mathematics Competition. In Sect. 10.3, we shall focus on the evolution of Madhava Competition, its growth and the nitty-gritties of organising the competition at the national level. This section also includes a brief Bio-Mathography of the 14th century Indian mathematician Madhava. In Sect. 10.4, some sample problems that appeared in the competition along with their solutions will be discussed. Some elegant solutions devised by the students while writing the competition have also been included. We shall see that a comparative analysis of the understanding of students in variety of topics at the undergraduate level is extremely illuminating. Section 10.5 is devoted to description of Nurture Camps—a distinctive feature of thc competition. In Sect. 10.6, a feedback of participants and mathematics teachers involved in the competition as well as comments of senior mathematician of the country have been included. In Sect. 10.7, we conclude with an epilogue and the future plans of expansion.

10.2 Mathematical Olympiads—A Precursor to Madhava Competition

India started participating in the International Mathematical Olympiad (IMO) in 1989. Following a few years of its being based in Bangalore, the nodal center of the activity was shifted to the Homi Bhabha Center for Science Education (HBCSE), Mumbai which is now in-charge of the competitions at all levels. The selection of the Indian team for IMO takes place in three stages. A regional level competition, called Regional Mathematical Olympiad (RMO) is conducted in about twenty five regions and then thirty students from each region are selected to participate in the Indian National Mathematical Olympiad (INMO). Only those students who are selected in RMO and those who have received an INMO certificate of merit are eligible to appear for the INMO provided they are in class XI or below. On the basis of the INMO, the top 30 students in merit from all over the country are chosen as INMO awardees. In addition to INMO awardees, the next 45 students who are in grade XI or lower and have done well in INMO, but have not qualified as INMO awardee are awarded INMO certificate of merit. These students are eligible to appear for INMO of the next year directly without qualifying through RMO, provided they are not in grade XII.

The INMO awardees are invited for a month long training camp in April–May each year at HBCSE, Mumbai. The INMO awardees of the previous years who are eligible for IMO 2016 and, in addition, who have satisfactorily gone through postal tuition throughout the year, are invited to the training camp as senior students. The junior students receive INMO certificate and a prize in the form of books. The senior students receive a prize in the form of books and cash. On the basis of a number of selection tests during the Camp, a team of the best six students is selected from the combined pool of junior and senior batch participants.

We now present sample problems, one from each of the tests—RMO, INMO and Selection Test.

Problem 10.2.1 RMO Problem[1] Find all integers k such that all the roots of the following polynomial are also integers:

$$f(x) = x^3 - (k-3)x^2 - 11x + (4k-8)$$

Solution 1. Suppose that for some value of k, all the roots of $f(x)$ are integers. We observe that the coefficient of k in the expression of the polynomial is $(-x^2+4)$; meaning that for $x = 2$ and $x = -2$, the value of the polynomial does not depend on k.
We get: $f(-2) = 18$ which is positive; and $f(2) = -10$ which is negative. So at least one root lies between -2 and 2.
Case 1: One of the roots is -1. This implies $f(-1) = 3k + 5 = 0$; so $k = -\frac{5}{3}$, which is not an integer.
Case 2: One of the roots is 0. This implies $f(0) = 4k - 8 = 0$; implying $k = 2$. In this case, the polynomial is: $f(x) = x^3 + x^2 - 11x = x(x^2 + x - 11)$. But the quadratic expression inside the bracket does not have integer roots.
Case 3: One of the roots is 1. This implies $f(1) = 3k - 15 = 0$; implying $k = 5$. In this case, the polynomial is $f(x) = x^3 - 2x^2 - 11x + 12 = (x-1)(x^2 - x - 12) = (x-1)(x-4)(x+3)$. So the roots of the polynomial are $1, 4, -3$ which are all integers, as required.
Hence, the only solution is $k = 5$; giving $f(x) = x^3 - 2x^2 - 11x + 12$ with roots $1, 4$ and -3.

Solution 2. Consider the polynomial $g(x) = f(x+2)$.
If the roots of $f(x)$ are p, q, r, then the roots of $g(x)$ are $p-2, q-2, r-2$.
Also, we note that the constant term of $g(x)$ is equal to $g(0) = f(2) = -18$; and its leading coefficient is still 1.
Hence the product of the roots of $g(x)$ is

$$(p-2)(q-2)(r-2) = -18 \tag{10.1}$$

[1]The problem is designed by Prashant Sohani, Regional Coordinator, Math Olympiad and Bronze Medalist in IMO in the year 2008.

Since p, q, r are all integers, so are $(p-2), (q-2), (r-2)$. For each possible factorization of (10.1), we will only check if p, q, r satisfy the correct relationship with the coefficient of x in $f(x)$, or in other words, whether $pq + qr + rp = -11$.
Accordingly, we get the following cases:

$(p+2, q+2, r+2)$	$pq + qr + rp$
$(1, 1, -18)$	41
$(1, -1, 18)$	−61
$(1, 2, -9)$	11
$(1, -2, 9)$	−31
$(-1, 2, -9)$	33
$(1, 3, -6)$	−1
$(1, -3, 6)$	−19
$(-1, 3, 6)$	−11
$(2, -3, 3)$	−5
$(-2, 3, 3)$	−7

We see that only the case of $(-1, 3, 6)$ satisfies the requirement of $pq + qr + rp = -11$.
It corresponds to the values of p, q, r as $-3, 1, 4$, and $f(x) = x^3 - 2x^2 - 11x + 12$.
Importantly, there exists a value of k, namely $k = 5$, that yields this polynomial.
Thus $k = 5$ is the only solution.

The reader must have observed that the first solution is more elegant for it makes a crucial observation about the coefficient of k. Once that is done, the problem becomes considerably easy! On the contrary, the second solution is somewhat routine.

Problem 10.2.2 INMO Problem Let ABC be a right angled triangle with $\angle B = 90°$. Let AD be the bisector of $\angle A$ with D on BC. Let the circumcircle of triangle ACD intersect AB again in E and let the circumcircle of triangle ABD intersect AC again in F. Let K be the reflection E in the line BC. Prove that $FK = BC$.

The reader is urged to solve the problem. (Hint: Show that $ACDE$ and $AFDB$ are cyclic quadrilaterals. Alternatively, one may also observe that K, D, F are collinear and further triangles AKF and ABC are similar.)

Problem 10.2.3 Shortlisted for IMO[2]

(1) There are n circles drawn on a piece of paper in such a way that any two circles intersect in two points and no three circles pass through the same point. Turbo the snail slides along the circles in the following fashion: Initially he move on one of the circles in clockwise direction. Turbo always keep sliding along the current circle until he reaches intersection with another circle. Then he continues his journey along this new circle and also changes the direction of moving i.e. from clockwise to anti-clockwise or vice versa.
Suppose that Turbo's path entirely covers all circles. Prove that n must be odd.

(2) Let r be a positive integer and $a_0, a_1, \ldots$ be an infinite sequence of real numbers. Assume that for all no-negative integers m and s, there exists a positive integer n in $[m+1, m+r]$ such that

$$a_m + a_{m+1} + \cdots + a_{m+s} = a_n + a_{n+1} + \cdots + a_{n+s}.$$

Prove that the sequence is periodic i.e. there exists some $p \geq 1$ such that $a_{n+p} = a_n$ for all $n \geq 0$.

The solutions of the shortlisted problems are not included here and interested reader may refer to the official website of IMO.

In the last 28 years, the Olympiad competitions have become very prestigious and competitive in India. The impact of Olympiads in the country has been multifold. The competitions provide a rigorous exposure to students with a very high mathematical aptitude and help in generating more interest in the subject. The difficulty level of problems in the Olympaid contests, as compared to that of problems appearing in their school examinations, is considerably high. As a result, bright students are motivated by this challenge and end up learning the mathematics with more zeal and enthusiasm. On the other hand, a large number of students who appear for RMO but do not necessarily make it to further levels, still learn more mathematics than what their school curriculum demands. As a consequence, the general mathematical know-how at the school level in the country has consistently increased in the post-Olympiad period. The Olympiad movement has evolved as one of the most treasured educational instrument in mathematical circles at school level. Yet another decisive contribution of Olympiad contests has been the strengthening of mathematical abilities of teachers who were involved in

[2]Both the problems are constructed by Dr. Tejaswi, ex-member of Math Olympiad Cell in India.

training the Olympiad students. In India, the knowledge of the mathematics teachers in school is generally restricted to the school curriculum which essentially deals with basic algebra, geometry and routine arithmetic involving applications in everyday life. Thus, these teachers are generally not in position to tackle problems in Number Theory and Combinatorics and also more advanced Olympiad topics in algebra and geometry. The pool of teachers working for the cause of Olympiad therefore, mainly consists of teachers teaching in undergraduate colleges.

10.3 Madhava Mathematics Competition: Concept and Scope

Before we proceed to discuss about Madhava Mathematics Competition in detail, the reader, we hope, would be delighted to know about the 14th Century Indian Mathematician Madhava and his work in nutshell.

10.3.1 Madhava—The Inventor of Calculus

Madhava is regarded as the founder of the most influential mathematical tradition in India that began in the middle of fourteenth century and continued for about next 250 years. Madhava and his disciples were from Kerala (on the south–west coast of India) and their writings are mainly in Sanskrit and local vernacular language, Malayalam. The Madhava school is a typical example of Indian 'Guru-Shishya Parampara' characterized by flow of knowledge through a chain of teachers and their disciples, from one generation to the next. The only known pupil of Madhava was Parameswara, a very prolific mathematician and an authority in Astronomy. He wrote about twenty-five texts on astronomy and was known for his contributions in eclipse observations. The tradition continued with Parameswara's son Damodara and his disciples Nilakantha and Jyesthadeva. Nilakantha is known for his work 'Tantra-sangraha' and Jyesthadeva for his masterly text 'Yuktibhasa'. Their pupil Sankara wrote a commentary based on Yuktibhasa. All these works mention the contributions of Madhava. **The major achievements of Madhava are astonishing. In particular, one can trace back the invention of Calculus and its applications to trigonometric functions in Madhava's work.**

Some of his major accomplishments have been listed below:

(1) **Madhava-Leibniz Series** For π

$$\frac{\pi}{4} = 1 - \frac{1}{3} + \frac{1}{5} - \frac{1}{7} + \cdots$$

The Madhava-Leibnitz series being slow in converging, and hence not useful in computing π, an ingenious sequence of correction terms for its partial sums was introduced, using which the computation could be effected efficiently.

(2) **Arc-tangent Series** For $0 \leq \theta \leq \frac{\pi}{4}$,

$$\theta = \tan\theta - \frac{\tan^3\theta}{3} + \frac{\tan^5\theta}{5} - \cdots$$

(3) **Madhava's Numerical Value** For π: A numerical value of π as found by Madhava is 3.14159265359 which is correct to eleven decimal places. In fact they has a great fascination for determining values of π to great accuracy.

(4) **Series expansion for Sine and Cosine**

$$\sin\theta = \theta - \frac{\theta^3}{3!} + \frac{\theta^5}{5!} - \cdots$$

$$\cos\theta = 1 - \frac{\theta^2}{2!} + \frac{\theta^4}{4!} + \cdots$$

It is extremely illuminating and interesting to understand the recursive methods employed by the mathematicians of the Kerala School to arrive at these and other such results, especially in the light of the fact that the work was carried out in the Pre-Newtonian period. For a comprehensive account on the mathematics of Kerala School and in particular the contributions of Madhava, the reader is refereed to Plofker (2009), Diwakaran (2007), Joseph (2009).

We now present a detailed account of the establishment of Madhava Mathematics Competition and its emergence as a national level competition at the undergraduate level.

10.3.2 Mathematics Competitions: A Prologue to Madhava Competition

In the cultural history of any society, it is observed that several artists, authors, musicians, athletes, etc. have started their careers with a competition at an early age. In the realm of Science, mathematics included, competitions held at early age help in generating curiosity and interest in the subject and triggers the mind for pursuing intellectual quest. The competitions allow students to stretch their capacities and go beyond the regular curriculum. We sincerely feel that rather than the end result of a competition, the preparation for the competition has far greater value in developing a permanent interest in the subject. Competitions, taken in right spirit, teaches one to face competitive situations in life and inculcates the healthy spirit of accepting defeats with an open mind as well as enjoying the success in a dignified manner. An ultimate objective of a mathematics competition is to provide a platform for students, outside the regular structure of teaching-learning processes that would allow them to become good at mathematics. Thus a well designed Math Competition is certainly an effective educational tool and assumes a very high potential in creating a society with a better perspective for mathematics. Several mathematicians and educationists have therefore taken a keen interest in organizing a good math competition at all stages of learning mathematics. There are literally thousands of math competitions that take place across the globe. The nature of these competitions display a huge variety in terms of level of mathematics, objective type or writing solutions of problems, online or paper based, regional, national or international, prize money involved, supported by government or not, etc. In this labyrinth of mathematics competitions, we shall now localize to Indian scenario and set up a context for introducing Madhava Mathematics Competition for undergraduate students in India.

India stands for pluralism in terms languages, food, dress codes, cultural ethos, etc. and educational system is not an exception. The country is divided into 29 states with each state having their own School Boards that governs school education in the state. There are a few central boards that cater to schools across the country and implement a uniform curriculum in all states. Students have a choice to take their school education either in English or in a regional language as also to chose a state board or a central board. This multi-parametric educational system poses several challenges in designing a competition at a national level. Many states in the country have association of mathematics teachers or mathematical forums that conduct local level mathematics competitions. For example, in the state of Maharashtra in India, Mathex competitions are being conducted across the state for last 50 years. These competitions have gained a very good reputation and many students have been motivation to study more mathematics through these competitions. These local level competitions are very crucial in preparing the mind-set of students (and parents) for reaching out to competitive examinations beyond school horizons. Many of these competitions are held at early stages of schooling. In last 25 years, these local competitions have began to emerge as a precursor to Math Olympiad competitions. As mentioned earlier, Olympiad competitions, at both regional and national levels, have attracted the attention of school students and their parents in a substantial way. A success in Olympiads is considered to be a remarkable achievement in the career of a student and are justly proud of their accomplishment! The name, fame and charisma of Olympiads is phenomenal and the obvious reason for this growing impact is the high quality of intriguing mathematical problems offering students an opportunity to scratch their brain and bring in their innovation to arrive at beautiful solutions. This excitement of disentangling the knot (or knots!) in a problem is the crux of the Olympiad mathematics. One of the main motives for extending the Olympiad competition from school to undergrad level is to retain the continuity of experiencing the charm of solving challenging problems and therefore generate enthusiasm and love for the subject the among students at a right level, from where they would probably take up mathematics as their life-time intellectual pursuit.

10.3.3 The Madhava Mathematics Competition: Operational Aspects

The Madhava Mathematics Competition was started in the academic year 2009–10, 20 years after India first participated in IMO. In India, XIIth grade is a crucial year for students in the sense that after XIIth grade they have variety of options to go for professional courses such as Engineering, Architecture, Medicine, Law, Management etc. These courses are job oriented and generally students aspire to get admission to one of these professional courses. Most of the students who shine in Olympiad competitions prefer to join elite institutes such as Indian Institute of Technology (IIT). As a result, students entering into undergraduate stream and not getting into professional courses are generally thought of as academically poor. Of course, in recent years the trend is encouraging because many Olympiad toppers are opting for career in pure sciences through the opportunities provided by the national level elite institutions such as Indian Institute of Science (IISc), Chennai Mathematical Institute (CMI), etc. All these undergrads pursuing pure sciences, especially those who are interested in mathematics were in a way deprived of having math competition like Olympiads. The Madhava Competition has attempted to remedy the situation in a modest way.

The commitment, enthusiasm and drive of teachers of Department of Mathematics, S. P. College, Pune, led to the enterprise of initiating a mathematics competition at the undergraduate level. At this point the reader should note that S. P. College, Pune is a hundred year old, highly reputed educational institute in India catering to the needs and aspirations of more than 7000 students. These teachers submitted a proposal for financial assistance to the National Board for Higher Mathematics (NBHM), a board of mathematicians monitoring the funding of several activities across the country. The NBHM readily agreed to support the activity. Professor S.G. Dani, a renowned mathematician from TIFR and then Chairman of NBHM took keen interest in promoting such a competition. In fact, he named the competition as Madhava Mathematics Competition with a view to spread the name and work of Madhava.

As a pilot project in the first year, the competition was launched in two cities Pune and Mumbai, both located in the state of Maharashtra in the western part of the country. Right in the first year, the competition received very good response, This prompted the organizers to expand it outside the state of Maharashtra and turn it into a National level competition. Any expansion plan certainly demands a lot of active manpower and fortunately, many college teachers came forward and coordinated the event in their regions. The following table indicates the growth of the competition in diverse parts of the country:

Madhava Competition - No of appeared students

Year	2009 - 2010	2010 - 2011	2011 - 2012	2012 - 2013	2013-2014	2014-2015	13 th Dec. 2015
Date of Exam.	3/1/2010	9/1/2011	8/1/2012	6/1/2013	5/1/2014	4/1/2015	13/12/2015
Total No. of students appeared for exam.	820	2644	3880	5151	7672	8327	9041
No. of Regions	2	6	12	17	18	17	19
Regions	Pune	Pune	Pune	Nanded	Nanded	Ahmedabad	Ahmedabad
	Mumbai	Mumbai	Mumbai	Hyderabad	Hyderabad	Ahmednagar	Ahmednagar
		Ahmednagar	Ahmednagar	Kolkata	Kolkata	Allahabad	Allahabad
		Nasik	Nasik	Allahabad	Allahabad	Almora	Almora
		Ahmedabad	Ahmedabad	Ahmedabad	Ahmedabad	Bhuvaneswar	Bhuvaneswar
		Baroda	Baroda	Ahmednagar	Ahmednagar	Darjiling (North Bengal)	Darjiling (North Bengal)
			Nanded	Mumbai	Mumbai	Goa	Goa
			Kolkata	Baroda	Baroda	Hyderabad	Hyderabad
			Hyderabad	Nashik	Nashik	Indore	Indore
			Ernakulum	Jharkhand	Jharkhand	Jarkhand (Chattisgad)	Jarkhand (Chattisgad)
			Allahabad	Pune	Pune	Kerla	Kerla
			Goa	Almora	Almora	Kolkata	Kolkata
				Kerala	Kerala	Mumbai	Mumbai
				Indore	Indore	Nanded	Nanded
				Goa	Goa	Nashik	Nashik
				Varanasi	Varanasi	Pune	Pune
				Bhuvaneswar	Bhuvaneswar	Varanasi	Varanasi
					Darjiling	-	Delhi
							Banglore

Though the competition is yet to reach all states of the country, the overwhelming response from students and teachers is an encouraging indicator that in the near future it will reach all parts of the country.

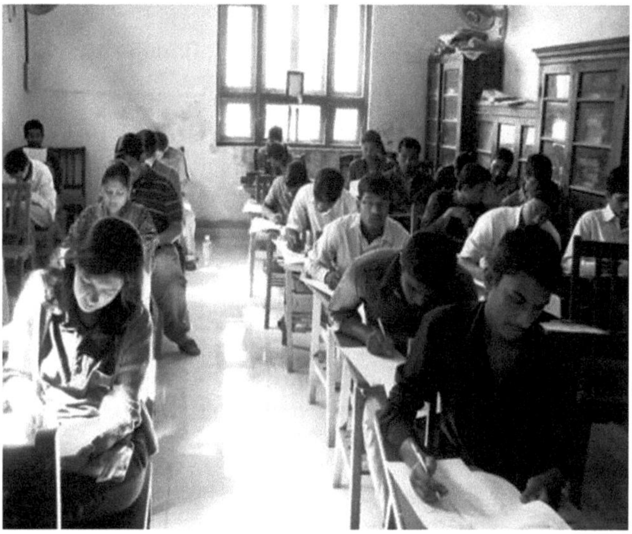

We now turn our attention to three major aspects of the competition viz. Rules for the Competition, Curriculum for the competition and Sample problems of the competition.

10.3.4 Rules for the Competition

The following set of rules have been laid down for the competition:

(1) A three hour competition with maximum score 100
(2) Questions of three types: Objective (Multiple Choice), Short Answer Problems (Less Difficult) and Long Answer Problems (More Difficult)
(3) Meant for second year undergrad students, but interested students of lower standards may also appear
(4) First, Second and Third Prizes and several Cheer Prizes
(5) All participants would get certificates
(6) Prize winners to be invited for a Nurture Camp
(7) Spot entries are allowed

10.3.5 Topics for the Competition

As mentioned earlier, in India, there is no common curriculum for all undergrad students. Every University has the autonomy to frame their own curriculum. Therefore it is difficult to set up a common set of topics for the competition. However, we decided include the following topics, because they typically characterize undergraduate mathematics:

(1) Calculus of one variable: Continuity and differentiability of a function of one real variable, integration (as anti derivative), elementary differential equations
(2) Matrices: Rank and determinant of a matrix, system of linear equations
(3) Coordinate geometry of two and three dimensions
(4) Elementary number theory: divisibility, modular arithmetic, Fermat's little theorem
(5) Elementary Combinatorics: Permutations and combinations
(6) Algebra: Polynomials-relation between roots and coefficients, sets, functions etc.
(7) General logical puzzles

Note that the topics for the competition include Calculus of one variable. In India, Calculus enters the curriculum in XIth grade. The secondary school education ends in Xth grade and the XIth and XIIth gardes constitute Higher Secondary Education. After XIIth grade, as mentioned earlier, students have several choices for pursuing their career. Students willing to go for career in pure sciences have option of entering into a 3 year Bachelor of Science (B.Sc.) course. The students of these classes are termed as Undergraduate Students. The Madhava competition is typically meant for these undergrads. Thus the undergrads have already learned Calculus for a couple of years and they also undergo a more regiourous course on Calculus of one variable in the first year of their undergraduation. In fact, the emphasis of Madhava Competition is more on typical undergrad topics such as Calculus, Matrices, etc.

10.4 Sample Problems and Analysis of Results

10.4.1 Sample Problems

We now present the sample problems, a few of each types for understanding the level of the questions that have appeared in the competition. We place on record that at this early stage of the competition, we have used various sources (generally not available to undergrad students in India) like Larson (1984), Zeitz (1999), for setting up the questions in the competition. Thus the problems posed in the competition are not necessarily new ones and here we have tried to cite the references of the original sources wherever possible. But the absence of a reference does not mean the originality on the part of author.

Problem 10.4.1 Objective Type

(1) The value of $\lim_{n\to\infty} \dfrac{1\cdot 1! + 2\cdot 2! + \cdots + n\cdot n!}{(n+1)!}$

(a) 1 (b) 2 (c) $\frac{1}{2}$ (d) does not exist. [2 Marks]

(2) Let $A = \begin{pmatrix} 1 & 2 & \dots & n \\ n+1 & n+2 & \dots & 2n \\ \vdots & \ddots & & \vdots \\ (n-1)n+1 & (n-1)n+2 & \dots & n^2 \end{pmatrix}$. Select any entry and call it x_1. Delete row and column containing x_1 to get an $(n-1)\times(n-1)$ matrix. Then select any entry from the remaining entries and call it x_2. Delete row and column containing x_2 to get $(n-2)\times(n-2)$ matrix. Perform n such steps. Then $x_1 + x_2 + \cdots + x_n$ is

(A) n (B) $\frac{n(n+1)}{2}$ (C) $\frac{n(n^2+1)}{2}$ (D) None of these. [2 Marks]

Problem 10.4.2 Short Answer Problem

(1) Let H be a finite set of distinct positive integers none of which has a prime factor greater than 3. Show that the sum of the reciprocals of the elements of H is smaller than 3. Find two different such sets with sum of the reciprocals equal to 2.5. [6 Marks]

(2) Let $f : [0,1] \to [0,1]$ be a function defined as follows: $f(1) = 1$ and if $a = 0.a_1a_2a_3\cdots$ is the decimal representation of a (which does not end with a chain of 9's), then $f(a) = 0.0a_10a_20a_4\cdots$. Discuss the continuity of f at 0.392. [6 Marks]

Problem 10.4.3 Long Answer Problem

(1) Let $p(x)$ be a polynomial with positive integer coefficients. You can ask the question: What is $p(n)$ for any positive integer n? What is the minimum number of questions to be asked to determine $p(x)$ completely? Justify. [13 Marks]

(2) Give an example of a function which is continuous at exactly two points and differentiable at exactly one of them. Justify your answer. [13]

(3) In an $m\times n$ matrix over N the only operations allowed are multiplying a row by 2 or subtract 1 from every member of a column. Can you reach a zero matrix in finitely many steps? Justify your answer. [12]

Readers are encouraged to tackle these problems and not to refer to the Sect. 10.4.3 where we discuss the solutions of the problems. We shall not discuss the solutions of the objective type problems here. For some of the problems, we shall also get a chance to see a solution given by a students which is different from the official solution.

We shall now take a look at the overall performance of the students in the competition.

10.4.2 *Analysis of the Results*

The following table indicates the performance of students in last three years.

Students' performance		
Year	Total no. of students	Marks above 30
December 2015	9041	37
January 2015	8327	61
January 2014	7672	59

The performance chart clearly reveals that the competition is very challenging for most of the students. There are several reasons for the low scores of students in the competition. A few main points causing the undesirable performances have been listed below:

(1) Students of undergraduate classes in India are generally not exposed to a competitive problem solving situation either in classrooms or in examinations.
(2) The emphasis of the teaching-Learning processes at the undergrad level is unfortunately not inclined towards problem solving.
(3) Not more than 5% of the students participating in Madhava competition have gone through Olympiad competitions. As mentioned earlier, most of the Olympiad students enter into professional courses and thus move outside the net of this competition.
(4) The University examinations are memory based, predictable and not oriented to test the problem solving capacities. For example, most of the universities ask to state and prove Lagrange's mean Value Theorem and many students end up solving it correctly. However, in Madhava competition we observed that not even 10% of the students are able to solve a problem based on mean value theorem.
(5) Students have fear for topics in Calculus, mainly due to $\epsilon - \delta$ definitions.

In the year 2011, we took a review of the performance of the students. It was revealed that even the better students have not done so well in Calculus problems. The following diagram clearly indicates this fact.

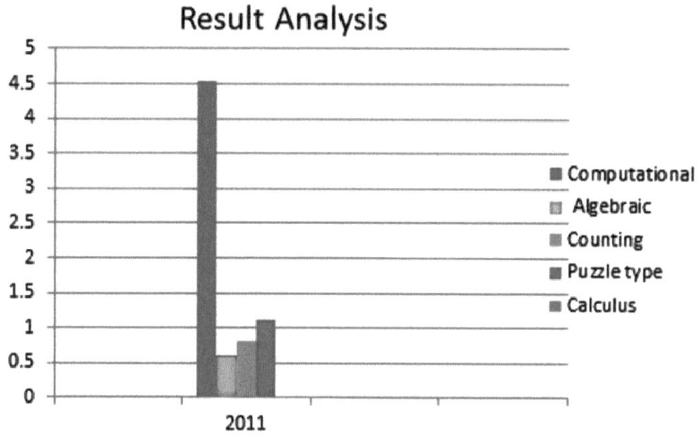

In fact, these students from higher academic bracket are typically past Olympiad students and they tend to capitalize on their Olympiad mathematics and solve problems on Number Theory and Combinatorics. With a view to encourage students to solve Calculus problems, we then decided to diminish the level of Calculus problems. The performance on Calculus problems then got enhanced as can be seen from the table given below indicating the topic-wise distribution of top 25 students.

Year	Algebraic problems	Puzzles	Calculus
December 2015	16	15	10
January 2015	18	0	08

The number of prizes offered every year varies according to the frequency of marks. The number of Cheer Prizes vary accordingly. The following table gives the information about the number of prizes awarded every year, the highest score in the competition and the cut-off for winning a prize.

Year	No. of prize winners	Highest score	Cut-off for a prize
2010	16	69	25
2011	19	64	43
2012	25	74	40
2013	32	80	38
2014	10	74	47
January 2015	14	85	44
December 2015	8	92	47

The overall topper in all these years has scored 92 marks. We wish to mention that the student scoring 92 marks Mr. Pranav Nuti, has a strong Olympiad background and in fact has won a Bronze Medal in the IMO, 2013.

We now discuss the solutions of the problems stated earlier. We shall present official solution as well as elegant solutions given by students wherever possible.

10.4.3 Solutions

(1) Let H be a finite set of distinct positive integers none of which has a prime factor greater than 3. Show that the sum of the reciprocals of the elements of H is smaller than 3. Find two different such sets with sum of the reciprocals equal to 2.5.

Solution: The given condition implies that every $n \in H,\ n$ is of the form $n = 2^{\alpha}3^{\beta},\ \alpha, \beta \geq 0$. Since H is finite, $\exists k \in \mathbb{N}$ such that $\alpha \leq k,\ \beta \leq k$ for each $n \in H$. This implies [6 Marks]

$$\sum_{n\in H}\frac{1}{n} \leq 1 + \sum_{i=1}^{k}\frac{1}{2^i} + \sum_{j=1}^{k}\frac{1}{3^j} + \sum_{i=1}^{k}\sum_{j=1}^{k}\frac{1}{2^i 3^j}$$

$$= 1 + \sum_{i=1}^{k}\frac{1}{2^i} + \sum_{j=1}^{k}\frac{1}{3^j} + \left(\sum_{i=1}^{k}\frac{1}{2^i}\right)\left(\sum_{j=1}^{k}\frac{1}{3^j}\right)$$

$$= \left(1 + \frac{1}{2} + \cdots + \frac{1}{2^k}\right)\left(1 + \frac{1}{3} + \cdots + \frac{1}{3^k}\right)$$

$$= \left(\frac{1 - \frac{1}{2^{k+1}}}{1 - 1/2}\right)\left(\frac{1 - \frac{1}{3^{k+1}}}{1 - 1/3}\right) < \left(\frac{1}{1/2}\right)\left(\frac{1}{2/3}\right) = 2\left(\frac{3}{2}\right) = 3.$$

Let $H = \{1, 2, 3, 4, 6, 8, 12, 24\}$. Then $\sum_{n\in H}\frac{1}{n} = 2.5$.

Let $H = \{1, 2, 3, 4, 6, 8, 12, 36, 72\}$. Then $\sum_{n\in H}\frac{1}{n} = 2.5$.

(2) Let $p(x)$ be a polynomial with positive integer coefficients. You can ask the question: What is $p(n)$ for any positive integer n? What is the minimum number of questions to be asked to determine $p(x)$ completely? Justify. [13]

Solution: Let $p(x)$ be a polynomial with positive integer coefficients say, $p(x) = a_0 + a_1x + a_2x^2 + \cdots + a_kx^k$. We can ask the question: what is $p(1)$?

Let $p(1) = N$.

Then $N = a_0 + a_1 + a_2 + \cdots + a_k > a_i, \forall i$.
Then $p(N) = a_0 + a_1 N + a_2 N^2 + \cdots + a_k N^k$.
Now express $p(N)$ to base N, then ith digit gives a_i, $\forall i$ which determines $p(x)$.
Alternatively, one may ask questions (i) what is value of $p(10)$? and (ii) what is value of $p(10^n)$?, where n is the number of digits of $p(10)$. These two questions also determine the polynomial completely. The reader can check that the coefficients of $p(x)$ can be determined from the answers to these questions. Also, it is easy to prove that one question is not enough.

(3) Give an example of a function which is continuous at exactly two points and differentiable at exactly one of them. Justify your answer. [13]

Solution:
Define the function $f : \mathbb{R} \to \mathbb{R}$ thus:

$$f(x) = \begin{cases} x^2 & \text{if } x \text{ is rational} \\ x^3 & \text{if } x \text{ is irrational} \end{cases}$$

We show that f is continuous only at 0 and 1, and differentiable only at 0. For this, consider a real number a. Then as $x \to a$ through rational values, $f(x) = x^2 \to a^2$, and as $x \to a$ through irrational values, $f(x) = x^3 \to a^3$. So the limit $\lim_{x\to a} f(x)$ will exist if and only if the above two limits are equal i.e. if and only if $a^2 = a^3$ i.e. $a^2(a-1) = 0$ i.e. $a = 0$ or $a = 1$. Thus f is continuous at 0 since $\lim f(x) = \lim x^2 = 0 = f(0)$. Similarly, f is continuous at 1. But when $a \neq 0, 1$, $\lim_{x\to a} f(x)$ does not exist; so f is discontinuous at a.

Next, let $g(x) = [f(x) - f(a)]/(x-a)$. Let a be rational. As $x \to a$ through irrational values, $\lim g(x) = \lim\{[x^3 - a^2]/(x-a)\}$ is not finite if $\lim [x^3 - a^2] \neq 0$ i.e. if $a^3 \neq a^2$ i.e. if $a \notin \{0, 1\}$. Hence $f'(a)$ does not exist (finitely) if $a \notin \{0, 1\}$. Let $a = 0$. Then $\lim_{x\to 0} g(x) = \lim_{x\to 0}[f(x)/x] = 0$. So $f'(0)$ exists and is 0.

But as $x \to 1$ through rational values, $\lim g(x) = \dfrac{x^2-1}{x-1} = 2$, while as $x \to 1$ through irrational values, $\lim g(x) = \lim \dfrac{x^3-1}{x-1} = 3$. Hence $f'(1)$ does not exist.

Let a be irrational. As $x \to a$ through rational values,
$\lim g(x) = \lim\{[x^2 - a^3]/(x-a)\}$
is not finite if $\lim[x^2 - a^3] \neq 0$ i.e. if $a^2 \neq a^3$ i.e. if $a \notin \{0, 1\}$. Hence $f'(a)$ does not exist.

There are several equivalent ways of defining such a function. For example the following function would also serve the purpose:

$$f(x) = \begin{cases} x^2(x-1) & \text{if } x \text{ is rational} \\ 0 & \text{if } x \text{ is irrational} \end{cases}$$

(4) In an $m \times n$ matrix over N the only operations allowed are multiplying a row by 2 or subtract 1 from every member of a column. Can you reach a zero matrix in finitely many steps? Justify your answer. [12]
Solution: Yes, one method is as follows: Let $A = [a_{ij}]$ be the matrix. Let m be the minimum element in the first column C_1. In fact, let m occur s times i.e. let $m = a_{i_1 1} = \cdots = a_{i_s 1}$. We may *assume* that $m = 1$. For if $m \geq 2$, subtract 1 from each element of C_1 $m-1$ times so that the minimum element in C_1 is 1.

Multiply each of the s rows $i_1, i_2, \ldots, i_s$ of A by 2. This forces the minimum element in C_1 to be 2. Subtract 1 from each element of C_1. The effect of these steps on C_1 is this: the s elements $a_{i_1 1}, a_{i_2 1}, \ldots, a_{i_s 1}$ of C_1 are still equal to 1, but the remaining elements of C_1 have all become *smaller* though they are all still ≥ 1. Hence in a finite number of steps *all* elements of C_1 will become 1. Then subtracting 1 from each element of C_1 makes C_1 a column of zeros.

Next make the second column C_2 a column of zeros as in the above. Note that the operations on C_2 have no effect on C_1 and C_1 *remains a column of zeros*. Hence in a finite number of steps A becomes the zero matrix.

10.5 Nurture Camp

The organisation of a nurture camp for prize winners of the competition is a distinctive feature of the competition. The duration of nurture camp varies from one week to ten days. The purpose of having a nurture camp is to provide insights to talented undergrads to advanced mathematics in the presence of distinguished mathematicians of the country. The camp also provides a unique opportunity to students to interact with their peers. The emphasis of the camp is on discussions and problem solving. The lectures in the camp are highly interactive and students are prompted to ask relevant questions. Some of the topics discussed in the camp are

(1) Killing-Hopf Theorem on the characterization of locally Euclidean surfaces
(2) Poncelet's Theorem in Geometry

Fig. 10.1 Nurture Camp 2016

(3) Rational Points on Quadratic Forms
(4) Curvature of Curves and Surfaces
(5) Parametrization of Conics and Quadrics

Also information about pursuing a serious career in mathematics is given to students and they are told about various options in India and abroad for graduation in mathematics. Eminent mathematicians, Professor C.S. Rajan and Professor Raja Shridharan have been regularly participating in the nurture camp as resource persons. In the next section a feedback from Professor Rajan, especially about nurture camps has been recorded (Fig. 10.1).

10.6 Feedback

Any new initiative, especially a competition, needs to have a continuous feedback mechanism. The feedback helps in improving the academic as well as organizational matters associated with the event. Here, we record a sample feedback of Madhava Competition from both students and teachers involved in the competition.

We shall also include comments of two eminent mathematicians, Professor S.G. Dani and Professor C.S. Rajan who have been closely monitoring all the activities of Madhava Competition.

10.6.1 Feedback of Students

Here we record the feedback of two talented students who participated in the competition and won a prize.
Aditya Garg from Mumbai says:

> The exam is a really good platform for talented, motivated and passionate students to showcase their talent in Mathematics. The problems were generally on the tougher side but really very interesting and well-designed and I thoroughly enjoyed solving them. I think this is one of the top-notch Mathematics competitions in the country and I really feel there should be more such competitions to motivate and find out talented students in Mathematics. And the nurture camp was even better. I got to interact with so many eminent professors, a number theorist and so many brilliant students. We had really good Mathematics sessions of problem solving and also gave presentations. We traded our knowledge of Mathematics with each other. I love Mathematics very much. And the camp was really very good. I still remember all the interactions and discussions we used to have with the teachers and the students in the camp.

Adway Gupta from Mumbai quotes:

> As I spoke during the day of the felicitation, I believe that Madhava Mathematics Competition is a genuinely good medium for both the professional and amateur mathematician. I, as a student graduating with Physics, still found Madhava to be a very engaging paper. To be honest, the difficulty of MMC is definitely higher than your usual mathematics examinations. however, I believe that it is by design and that is exactly what makes it an extremely fun exam to sit for. The questions are designed in a way that knowing hardcore mathematical formulations isn't a prerequisite for doing well, a general mathematical intuition is enough. I think for the same reason I enjoyed taking the exam probably more than I did doing well in it.

10.6.2 Feedback from Teachers

A competition of this magnitude cannot be run without the support and help from colleagues across the country. In all the twenty one regions where the competition is being conducted, a dedicated teacher shoulders the responsibility as a Regional Coordinator. The comments of two regional coordinators have been included here.

My student and colleague Geetanjali Phatak has been closely associated with the competition right from its inception. Her reaction about the competition has been quoted below:

> My relation with Madhava Mathematics Competition (MMC) has been multidimensional. I have contributed to MMC as a member of paper setting committee, coordinator of Pune Region and as a Tutor at Madhava nurture camp. As a mathematics teacher, I sincerely feel that students deserve an opportunity to face challenging problems in mathematics and MMC certainly fulfills this need. The group of students appearing for MMC is heterogeneous. However, majority of the students appearing for MMC are studying in colleges and as such do not have much of exposure. These students find the level of difficulty of problems posed at MMC as very high when compared to their university examinations. With the emergence of MMC, our undergraduate students have now started referring additional books other than the prescribed text books. We have also observed progress in terms of mutual discussions between the students as well as with the teachers. MMC helps students to generate interest in Mathematics. From the funds generated through registration fees, we can arrange various activities like guest lectures, summer workshops, etc. Nurture camp provides students an opportunity to discuss Mathematics with students from various parts of the country and experts from different institutes. I am sure that MMC will increase interest of students in Mathematics and I wish very best for the success of MMC.

Needless to mention that Madhava competition takes place in Kerala, the land where Madhava himself lived. The regional coordinator of Kerala region Dr. Aparna offered her comments:

> Though Madhava Mathematics Competition began in 2010, center in Kerala region was established in the year 2012. In 2012, 869 students wrote MMC in 10 different centers in Kerala. Through years MMC became popular not only among Mathematics students in Kerala, but students from our neighboring state Tamil Nadu is also writing exam at various sub-centers in Kerala. Last year the number of students who applied for the examination at various sub-centers in Kerala reached nearly 2000 and due to the increasing demand from the Mathematics community, we have increased the number of sub-centers to 14. For all registered students in Kerala region, a free training programme to enrich

their Mathematical abilities is organized from last year onwards. Also we are giving motivational prizes to the state level winners.

We shall now present the views of two reputed mathematicians of the country who took keen interest in Madhava Competition and helped in enhancing the quality of the competition. It is important to record their opinion about this new initiative.

10.6.3 Feedback from Mathematicians

Professor S.G. Dani, a renowned mathematician, retired from TIFR, Mumbai and currently working at IIT (Powai) is well known for his contributions in Ergodic Theory. He is also a scholarly figure in the field of History of Mathematics. Professor Dani was the Chairman of NBHM when the proposal for the Madhava Competition was sent to NBHM for the financial assistance. He took keen interest in the proposal and recognized the potential in it. In all these years, he has been a constant source of encouragement and support. The author immensely values the comments offered by him. Professor Dani's thoughts are given below:

> Competitions on a wider scale than in a limited learning group like in a school or college facilitate in generating interest in the subject and invigorate the studies as a whole. This applies especially to mathematics and has been fruitfully applied in various contexts and in various ways around the world. In India mathematical competitions at school level have flourished in the form of Olympiads, both in the main channel connected with the International Mathematical Olympiad and also outside it. A need was felt to introduce a similar activity at the undergraduate level, in colleges, in order to vitalize the interest in mathematics, especially among students with mathematical aptitude. A major step in this direction occurred when the S. P. College of Pune presented a proposal along the, spearheaded by Prof. Sholapurkar, to the National Board for Higher Mathematics for financial support. The Board recognized the immense potential in the proposal and after some tweaking of the original scheme, and gradual expansion the Madhava Mathematical Competition took shape. There has been a rapid expansion of the activity, both in numbers and the geographical spread around the country, within the short span of its existence. One would hope that it would become a defining feature of undergraduate studies in mathematics, and also expand to other countries.

Professor C.S. Rajan is a well known mathematician from TIFR, Mumbai. He works in Algebraic Number Theory. Prof. Rajan showed interest in the Competition right from the beginning. Especially, he liked the idea of having nurture camps for students. He participated in the nurture camps and gave deep insights to students. The points emphasized by Prof. Rajan have been recorded here.

(1) It helps in recognizing mathematical talent from the vast pool of undergraduate students in India.
(2) The nurture camp lets the selected students come into contact with their peers from across colleges and universities in India. Such contacts enable them to learn mathematics better; to encourage them and to maintain their interest in mathematics. Also it has the possibility of setting up long term (mathematical) friendships, which can be useful in future for collaborative work.
(3) The Indian undergraduate teaching is done mainly in colleges, many of which lack resources in terms of personnel, access to books and modern material.
(4) The nurture camp gives an opportunity for the students to come in contact with mathematics faculty drawn in from some of the best places in the country.
(5) The structure of the nurture program has been designed so that it can illustrate the use of the undergraduate mathematics they are learning in their curricula to interesting mathematical problems accessible with the material they have learnt.
(6) Thus the camps give exposure for the students to some advanced mathematics and material that are not readily accessible; to expose them to the way mathematics is done by working mathematicians.
(7) It also helps in molding the tastes and interests of the students, naturally dictated by the faculty involved in the nurture camps. This is an extremely important but subtle aspect of the nurturing process.
(8) For the faculty, it allows them to come into contact with promising undergraduate students from across the country. This is quite satisfying for a faculty to be able to reach across to a wider cross section of students than he/she normally faces in their respective places.

10.7 Epilogue and Future Plans

The Madhava competition is a relatively new initiative in the area of Mathematics Competitions. Looking at the size of the country and heterogeneity in terms of language, academic background, geographical diversity, etc. it will take another few years to reach out to all parts of the country. Opening a center in the country mainly involves the appointment of a devoted mathematics teacher who would be willing to coordinate the event in the respective region. So far, the teachers have showed tremendous interest and enthusiasm in the activity and have extended their wholehearted support to the activity.

The competition has already reached major cities in India like Delhi, Mumbai, Kolkata, Ahmedabad, Hyderbad, Pune, Cochin, etc. This year the competition has reached to North-Eastern part of the country. Next year, we shall add a few more centers such as Tamilnadu, located in the southern part.

The setting of question paper for the competition is the most crucial aspect of the competition. We have tried our best to maintain a very high standard in setting up of questions. Though, so far we are borrowing questions from sources (and try to modify them wherever possible) that are not available to students, we would like to improve the situation by designing totally new questions.

The result analysis reveals that we need to improve the problem solving abilities of the students to a great extent. We plan to bring out a consolidated report on the competition and bring it to the notice of the government and university authorities. We hope that the curriculli of the universities would be designed so that the conceptual understanding is strengthened and as a result, students would be in position to tackle tough mathematical problems.

The nurture camp has been very useful for the students. We plan to extend the duration of the camp for better results. We shall also invite more students in the camp in coming years and propose to conduct more such camps in different parts of the country.

All in all, we have received a very positive feedback from students, teachers, mathematicians and math lovers in the country. We are sure that the activity will further flourish in the days to come and in turn, benefit the students community in the country. Organising such a competition has been a very rewarding experience and we sincerely hope that the activity would help in enhancing the mathematical aptitude of students of undergraduate level. On a larger sphere, the competition would certainly contribute, in its modest way, in generating a mathematically and logically strong human resource.

References

Diwakaran, P. (2007). The first textbook of calculus: Yuktibhasa. *Journal of Indian Philosophy*, *35*, 417–443.

Joseph, G. G. (2009). *A passage to infinity*. Los Angeles, CA: Sage Publications.

Larson, L. (1984). *Problem solving through problems*. New York: Springer
Plofker, K. (2009). *Mathematics in India*. New Delhi: Hindustan Book Agency.
Zeitz, P. (1999). *The art and craft of problem solving*. Wiley.

Part VI
Thoughts About Competitions' Present and Future

Chapter 11
From the Lifetime Experience of a Seasoned Math Educator—Thoughts, Hopes, Views and Impressions

Romualdas Kašuba

Abstract This paper traces the author's evolution as a problem poser, especially for younger students, from crisp, abbreviated sentences to long, embellished artistry and, in the telling, takes the reader on a journey through a mathematical fantasy land.

Keywords Number imagination · Reformulation of problems · Small data analysis · Fantazy in mathematics

The author of any paper showcasing remarkable mathematical content ought to be exact; consequently, the materials proposed ought to be well-structured and easily accessible. This is how things should be arranged and this author will try to achieve that. Still, on the other hand, when speaking about such a vivid matter as problem solving, it is almost impossible to avoid some influence of a subjective or humanist component. That is, according to the author's personal view, not always bad. Sometimes it may even add additional charm and be quite attractive for students and readers.

If you succeed in understanding this after a while, then the door to the world of math problems will almost surely open wide for you

Let's start with the so-called classroom problem. In a classroom, one or more pupils always tell the truth. The other pupils sometimes do, and other times do not. The pupils were asked how many of them always tell the truth. The answers were: 5, 6, 2, 3, 4, 6, 3, 6, 3, 4, 6, 5, 4, 3 and 6.

R. Kašuba (✉)
Vilnius University, Vilnius, Lithuania
e-mail: romualdas.kasuba@mif.vu.lt

A. Soifer (ed.), *Competitions for Young Mathematicians*,
ICME-13 Monographs, DOI 10.1007/978-3-319-56585-9_11

So, in reality, how many pupils do always tell the truth?

(A) 2 (B) 3 (C) 4 (D) 5 (E) 6

This problem is taken from the Dutch Mathematical Olympiad AD (2011). It would be difficult to overestimate the usefulness of a problem of this kind.

What formulation of a problem might be regarded as nice or otherwise attractive?

First of all, problems must be short and, at the same time, not exactly 100-percent known. All non-standard problems that might be formulated in a single sentence are nice in that way. For instance, such is Problem 1, taken from the Cayley Paper Olympiad (UK 2013).

What is the smallest non-zero multiple of 2, 4, 7 and 8, which is a square?

Another example of that type might be the Problem B1, taken from the Dutch Mathematical Olympiad (2013).

What is the smallest positive integer, consisting of the digits 2, 4 and 8, such that each digit occurs at least twice and the number is not divisible by 4?

Anyone who reads the text of that problem more or less carefully might notice that all the numbers mentioned in the text are even, but this still does not guarantee divisibility by the second smallest even number, namely—4.

The reader should not think that problems formulated in a few words are automatically easy to solve. Below is an example.

Find an integer that is divisible by 10 and such that the sum of its digits decreases by 10% when the number itself increases by 10%.

Needless to say, the beauty or the attractiveness of the problem stems from the fact that all the numbers explicitly mentioned in the text of the problem clearly coincide.

Sometimes the text of a problem is expressed in one sentence, and in another one, which is usually much shorter, the task, or what we expect from the problem, is defined. We might say that such problems are formulated in "one and a half" sentences. Some such problems might be rather challenging or, at least, demand careful counting. An example of this might be a problem formulated while looking at problem B4, taken from the aforementioned Dutch Mathematical Olympiad:

We write down the numbers from 1 to 30,000 one after the other to form a long string of digits:

$$1\,2\,3\,4\,5\,6\,7\,8\,9\,10\,11\,12\ldots 30000.$$

How many times does 2016 occur in this sequence?

This is a shortly formulated, yet attractive problem, since the task is non-trivial, or unexpected. It would be easy to add an embellishment to such a problem. Here are two examples of that—the first of which the author has recently used as a problem proposed for the Lithuanian competition for grades 7 and 8 AD2015 (based on Problem 320, taken from one of the marvelous Ukrainian problem books (2013/14), edited by the famous composer, B.V. Rublyov (2015)).

A natural number is said to be *round in the Thai way* if it is greater than 10 and is divisible by the product of its digits. What is the maximum number of consecutive natural numbers such that each of them is *round in the Thai way*?

It is absolutely clear what kind of embellishment is used; it is enough to mention that the International Mathematical Olympiad of the corresponding year happened to take place in Thailand. The same might be equally well seen in the following problem from the same contest, proposed as the last problem, Problem 6 (based on the Problem C2, taken from the Dutch Mathematical Olympiad (2014)).

Problem 6 The natural integer N is said to be extraordinarily smart, or, in short, *a Thai pearl,* if it is possible to find another integer k having at least two digits, which are all the same (like, say, 999, or 222222), and such that the product $N \cdot k$ also consists of equal digits. For instance, the natural number 3 is clearly a Thai pearl, because $3 \cdot 222 = 666$. (A) Show us a 10-digit number which appears to you to be *a Thai pearl* and prove that this is indeed the case. (B) Prove that the number 11 is not *a Thai pearl.* (C) Prove or disprove the same question concerning the number 143 and clarify your answer.

Concerning the length of an interesting formulation

Not all problems which we regard as formulated in a challenging way are that short. Some of them are expressed in much longer texts. First, we will present such problems; again, an attempt to add some Thai flavor is undertaken. The reader may judge how successful that attempt is. After that presentation the author will explain what kind of psychological advantages this kind of formulation might bring to the solver.

You the reader surely won't believe or may otherwise express your doubts, but for the author, despite that, one thing is completely clear and will not be argued about. That summer in Thailand we saw an elephant that was able, in spite of the large the number of spectators, always successfully to mark some fields on a 6×6 table in such a way that in any row there were always exactly three marked fields and, in any column, always either one or four. We would like to repeat that we understand that for some this would be too difficult to believe; if we hadn't seen it, we also wouldn't believe it.

But we have seen it, and we really enjoyed it, frankly speaking, many times. There are two things that make this situation so challenging: firstly, it is amazing that this might be realized, and secondly, that this might be achieved, and repeatedly, by an animal so modest and peaceful looking.

We should, and willingly do, repeat what is required: some fields of the square table 6×6 are to be marked in such a way, that in any row there are always exactly three marked fields and in any column—always either one or four.

If you come to the conclusion that this is possible, then we will expect you to present us the picture of the table with the fields marked so that we could, if there is some need for it, check your table.

We would like to explain why such a formulation might not be very short. There may be extenuating circumstances that make its length right and acceptable.

When proposing a challenging problem, we create quite difficult circumstances for the solver. He may be stressed. He may be afraid that he will not be able to deal with the problem properly. He may be hesitant to attempt it. So, he must become somewhat interested. He must decide that this is a problem worth mastering and that the process in itself may be beneficial for him. In short, he must believe that the proposed problem is worth the effort.

How might a possible future solver be challenged, or otherwise engaged?

Not being in the easiest of situations, we may need some help. Frankly speaking, when we are in any non-standard situation, any help is precious and it makes sense to take it. Help may come in the form of nice words, characters, structure and many other things and circumstances that we will try to mention later.

Let us review some examples of the problems where some elements, which we often encounter in the telling or retelling of stories, are used. All of them are taken from the recent Lithuanian Olympiad for younger grades. Below are several examples showing how we have tried to achieve attractive wording. First let us see the wording of a rather easy problem (Grade 5 and 6).

Simple counting

In order to enumerate the most admirable elephants in Thailand, all the integers which are greater than 111 and smaller than 222 were used. All that were connected with that marvelous activity were deeply astonished when it became known that those elephants, and only those, from the enumeration tagged with integers with two repeating digits, could freely fly. For instance, elephants numbered with 119, 181 and 211 could fly freely. What is the number of freely flying elephants amongst all these most admirable elephants in Thailand?

(A) 30 (B) 29 (C) 28 (D) 27 (E) 26

Dividing into groups

There are 36 elephants and each of them holds a balloon on which exactly one of the numbers from 1 to 36 is painted. Unrest was anticipated in such a large group of otherwise so peacefully acting animals, so the need to divide them somehow effectively into groups was commonly recognized. For the division of the elephants to be well thought out and effective—otherwise no one could guarantee their behavior, since it is easy to imagine what wilds elephants could do when enraged—it was decided to divide them into groups based on the following two rules:

(a) The sum of any two numbers in any of these groups must be divisible by 3;

(b) There must be at least two elephants in any group.

Let us answer the following technical, but also philosophical, question: What is the smallest number of groups the elephants will form, if we divide them according to the rules mentioned above?

A simple, but obviously true, remark

The author has written or adapted such problems for approximately 20 years and enjoyed many compliments and remarks when speaking about both the advantages and shortcomings of such problem posing. There have been some remarks that should be regarded as completely correct. One of them is the following: the telling of the story might even be the best imaginable, but in the end, if the problem only has that going for it, it would be wise to rewrite the task presented in the text in a shorter fashion.

How did all that process of posing and embellishment start?

The author feels it is now his duty, as someone who often poses and embellishes problems, to explain how he happened to start using such a form of presentation. In the beginning, the author was a simple University teacher with no real connections to any math contest. That involvement started to grow after the author had successfully completed his Ph.D. studies in Greifswald (Germany) and, as the Doctor rerum naturalium Universitatis Greifswaldensis with estimation magna cum laude, returned to his home alma mater in Vilnius, the capital of Lithuania, this new evolution, which was participation in the jury of the Lithuanian MO, took up only a few days each year.

After several years, the author had become even more involved in various Mathematical Olympiads and other contests by undertaking organizing duties in the Lithuanian team-contest. That essential step was connected to the fact that in 1995 the author was appointed leader of the Lithuanian team in the Baltic Way team contest. That contest was inspired by the Lithuanian team contest and, after Lithuania regained independence, had quickly become the competition of all the countries surrounding the Baltic Sea. The internal Lithuanian team contest remained, however, and in 1999, by the initiative of the author, evolved to encompass a competition for younger grades as well. Ever since then the author has been selecting problems for that competition.

About the influence of the long term selection, preparation and embellishment of problems from the perspective of someone involved in that process

Looking back, it is clear that although the sets of the problems for the first newly initiated contests were not bad, the fact remains that the problems selected sometimes appeared to be rather hard for the solvers. As years passed, the wording of problems lengthened considerably during the process of selection. The main reason for this development was probably the simple fact that the author devoted much more time to the selection, which is no surprise, since his understanding of the needs and abilities of the audience had become considerably higher. The first milestone better: feature was the fact that in the texts of proposed problems some names of heroes appeared. Nothing more, at first it was just first names. Then the role of these heroes expanded; you might say that the heroes managed to win a more central role in the actions described in the texts of problems. It wasn't the original intention of the author, rather it was, as the author has come to understand recently, essentially influenced by the desire to increase the comprehension of the audience, the strong wish to come closer to the solvers involved and attempt to touch the audience more deeply. That is a common situation of proper teaching when you try to make an impression upon your listeners. The situation is the same in problem posing, adapting and embellishing, except that, when you are teaching, you have much better means of grasping the attention of your listeners. When proposing problems for competitions and the like, you have more or less only the text of a task proposed as the unique tool to influence your audience.

De gustibus non est disputandum or, in matters of taste, there can be no dispute. Is this definitely so?

As mentioned before, not all persons approve of presenting problems in such a way. At first, the author was surprised to hear that he should undertake concrete investigation into the validity of the new methods applied. First, he was quite astonished, because he was convinced without any scientific corroboration that his way was typical, meaning that it was of remarkable use for the audience. On the other hand, the author would like to stress strongly that he is always glad and thankful for all who express their views and attitudes, even if they strongly disagree with what the author regards as valuable and otherwise useful for so many students despite their age.

It is probably similar to the situation in which you ask a writer to prove that his way of expression or his approach to connect with his audience is proper.

Such demands are understandable and natural. To prove that his way is important, the author may base his answer not only upon the fact that he has been providing problems for contests for 20 years, but also on his lectures at Vilnius University available to students of all departments. For formulation of problems similar methods are applied, let us mention humbly, with quite remarkable success.

The lector is like an artist when he believes that his art is still alive and current and his methods are suited to the essential part of the problems he proposes to his students—despite where it may take place: at university, high school or (especially) kindergarten. Because everyone knows that the memories of small children may last forever.

Using foreign languages

Many things might be said about the ability to speak well in languages that are not the mother-tongue of the person who intends to formulate a problem using fictitious elements. There are persons who are against it because they say that it is, in general, impossible. That is, if you write in the language that is not your own, then you cannot achieve excellent results. They also say, that if the author of such problem writes using embellishments, as a real fiction writer does, he must only write in his mother-tongue. Afterwards, another person must translate it.

The author doesn't agree with such an opinion. If the author is posing the problem in a language that is not his mother-tongue, then it is better that he, only if he is really well versed, should translate his texts himself.

But afterwards, it is absolutely necessary to give the text to a native speaker for the final check of the wording.

This adheres to the motto: nice posing shouldn't be almost nice, but really nice.

Having said that, the author feels he knows Russian well enough and, having a book (Kašuba 2012) written in Russian and printed in Moscow in such a well-known publishing house as "Prosveshchenie", always welcomes anyone to check the language in any of his texts.

More remarks concerning translating various problems with embellishments into other languages

First of all, the author prepares the wording of a well-adapted problem in his mother-tongue, that is, in Lithuanian. Only then the translation to English follows. So far, the author has been able to edit five books and booklets in English. All these aforementioned books and booklets were either translated by the author ("What to do when you do not know what to do", Parts I and II, published in Riga (Latvia)), or originally compiled in English (these were devoted especially to attractive representation of problems as well as their solutions and appeared in Latvia under the title "Once upon a time I saw a puzzle" in a well-known LAIMA series.) (Kašuba 2006, 2007, 2008a, b, 2009).

Then, there usually are the translations of problems adapted for the competition for youngsters (that was initiated by the author in 1999 and linked to the Lithuanian team contest). Those problems are usually translated into English as well as Russian, German and Polish after the competition.

Needless to say that the posing and decoration of problems has its own specifics in every language; please do not forget that success in creative posing, design and any kind of decoration is heavily dependent on the skills in the particular language, which are never absolute, even with the mother-tongue.

It ought to be stressed that the author's adventures with writing (not translating, as one might suppose looking at the similar titles) did start as a kind of translation. But, after some time, a better understanding of the situation followed—thank God—that may be expressed by saying that one should avoid the strict translations of one's own books into other languages. He should simply write another, although related, book, if only he is skilled enough to accomplish it. That especially applies when the author is eager to paint the problems in a different light by using a wide range of the embellishments that may vary from language to language.

The series of adapted problems which are devoted to the Bremen City musicians

As deputy leader of the Lithuanian IMO team, the author was quite happy to be present at the 50th IMO in Bremen. As a child, he enjoyed the marvelous fairy tales about the adventures of the four smart and brave animals of Bremen lore. Upon returning home, the author tried to compose a problem with an impression of what he had seen in Bremen, as well as using all of his sentiments and childhood memories. This was one of the first times when the texts of problems proposed for the Lithuanian competition for youngsters became considerably longer.

This is also one of the reasons that the representation of the essential parts of the whole sample exist in five languages. The author feels obliged to say that although the Bremen musicians are of clearly German origin, thanks to various historic and cultural circumstances, they are more than well established in the Russian common conscience.

The immortal Bremen heroes mentioned by name

The immortal quartet of Bremen, namely the Donkey, the Dog, the Cat and the Rooster, after becoming absolute classics of hard-beat music, could no longer perform all together. On very rare occasions when three of them performed, it was treated as an occasion commanding absolute respect. Such performances and only these were called *the Bremen ecstasies.* When in a recent summer the World Session of Beautiful Young Math Minds took place in Bremen, the immortal four performed several sensational *Bremen ecstasies* in its honor. The soul of the city, Roland, who, naturally, took part in all the *Bremen ecstasies,* certified that Maestro Rooster participated more times than any other of them, 8 times, and Maestro Donkey participated less than any other, only 5 times. Then, without uttering a word, Roland made them believe that any clever mind, if it was able to concentrate at least a bit, could calculate how many *Bremen ecstasies* were provided by the members of the immortal quartet of Bremen. Would you be able to explain how many *Bremen ecstasies* were performed at the World Session of Beautiful Young Math Minds by the members of that immortal quartet of Bremen?

Above, you see an attempt to decorate a classic, simple, but nevertheless challenging problem using Bremen lore.

The eternal Roland meets Maestro Cat in a struggle of minds

On silent winter nights, when the last lonely pedestrians have disappeared from the streets, Roland, the eternal patron of Bremen city, climbs down from his monument and together with maestro Cat they arrange what they call *the silent Bremen-17 game.* For it 7 cards numbered 0, 1, 2, 3, 4, 5, 6 are necessary (with each number written on exactly one card). Roland and the Cat take one card each in turn; Roland usually starts first. The player who is able, by using only his own cards, to present a number divisible by 17 sooner than his opponent is declared the winner. On the news portal ihaha.com, there were furious quarrels about whether any of them would be able to select his cards in such a way that he would always be sure to win, in spite of what his opponent selected.

There appeared a wiseacre named Rex, who kept claiming that

(A) If either of them would really be able to win despite what his opponent chose, then that person was the person who started first. Is the wiseacre Rex right? Explain your answer.
(B) So, how about that? Was one of them indeed able to win, in spite of what his opponent selected?

Explain your answer.

This is a simple, yet challenging, problem from the famous Russian mathematical and linguistic educator I.S. Rubanov, exhibited after attempts to dress it in new wardrobe.

Careful counting of the paths in Bremen

On their infrequent leisure time, the immortal quartet of Bremen divided the usual chess board into four equal parts and started to examine one of these parts containing 16 fields (8 white and 8 black fields, colored in the usual chess order), i.e., a 4×4 square.

A zigzag path, consisting of 4 white fields, one from each row and such that any two neighboring fields shared *only a common corner* was called by them a *Bremen path*. A heated discussion immediately began about how many versions of *Bremen paths* could be made in that small 4×4 square. Roland, the patron of Bremen City, testified that they talked well into the night and couldn't come to a common conclusion about how many *Bremen paths* could thus be made.

Can you explain in an understandable way to that immortal quartet of Bremen how many *Bremen paths* could be formed in that (rather small) 4×4 square?

Here you see an attempt at posing the well-known and typical representation of a problem called "Careful Count" in a new way.

Soccer tournament

Yesterday, the Cup of Nations, a soccer tournament in which each team played exactly once against every other team, came to an end in Bremen. The matches were played according to rules that generated a health hazard: 3 points were awarded for a win, 1 point for a tie and no points for a loss. After all the matches were over, it was mentioned that all the teams together were rewarded a total of 21 points. The troubadour of the Cup, Maestro Rooster, spent the whole 3 days absolutely confident that, knowing only what has been said right here, it was still impossible either to conclude how many teams participated in that Cup of Nations or to establish how many points each team was awarded (according to its final classification). Was the troubadour of that Cup, Rooster, right in his belief? Is it possible to detect how many teams participated in that Cup of Nations and to establish how many points each team was awarded? (Compare (Mazanik et al. 2005), Problem 49.)

Tournament problems, as we all know, are one of the cornerstones of challenging mathematics.

Some things that many of us, if not all, know

Let us start from a very simple situation or an attractive notion, which is widely known and called a "Magic Square". Most have heard of them because in life, as well as in arithmetic, a coincidence is a rather curious matter which is mentioned almost everywhere, and in the case of Magic Squares, we enjoy an extreme number of coincidences of sums.

You may recall that a Magic Square is a square where, roughly speaking, the sums of the numbers in all rows as well as in all columns and even on the two main diagonals do nothing more than simply coincide.

In a classic case, the first nine natural numbers 1, 2, 3, 4, 5, 6, 7, 8 and 9 are located in the nine cells of the 3×3 square. If you deal with a freshman who is at least a bit curious, it may be interesting not only to make him believe that such a curious thing indeed exists, but also to demonstrate for him quickly how it may be detected.

To persuade such a person, you have two possibilities: either you immediately show such a table while asking to check out whether that table indeed has all the announced properties, or choose another option. Of course, this other option takes more time, but it is also more useful. Before acting in the manner "Deus ex machine", it is always worth trying to take at least one intermediate step and ask the following question, or even two that are closely linked. These questions are: assume that the numbers from 1 to 9 can be arranged in the Magic Square. How big then may the sums of elements in all rows, columns, and on these two diagonals be? Waiting until the natural and quick answer, 15, appears, one can then ask another question about what number might appear in the very center of that table? A couple of minutes later, the audience will come up with the answer 5. After that one can simply propose to take the 3 × 3 table with 5 in the very middle of the table and locate the rest of numbers in a timely manner.

What kind of attitude may even a good student often possess?

Even good students sometimes show an unseemly attitude, which in our case might be expressed thus: what is so valuable about knowing what happens in the 3 × 3 table when everything is so well established and has been discussed more than once? As a partial answer to what might happen, some concrete tasks might be suitable, representing some new aspects of such seemingly well-known a situation. Here are some attempts concerning how this might be developed. The author deeply believes that there are many ways to enrich and develop this kind of situation. Remember the Latin phrase "Verba docent, exampla trahunt"—"Words instruct, illustrations lead".

Some related examples

1. Here are some possible examples for the enrichment of the situation taken from the literature. The first problem we mention is taken from the Dutch Mathematical Olympiad 2011.

	2	
?		
		9

In this Magic Square, the three rows, the three columns, and the two diagonals all have the same sum (so this is a Magic Square, but the reader may notice that there is no requirement that the material used for that square be a number from 1 to 9!). Which number is represented by the question mark? (A) 5 (B) 6 (C) 7 (D) 10 (E) 16.

For any effective solution it is important to remember that, although in the Magic Square all cells are important, there may be one that is even more important than all the others.

Some modifications to the notion of being or not being Magic might be proposed and developed, as often happens in literature. Below is another example taken from South African sources (Laurie and Merry 2000).

2. A 4 × 4 "anti-magic" square is a square containing the numbers from 1 to 16, such that the sum of the numbers in each of the four rows and four

columns and two diagonals are ten consecutive numbers in some order. The diagram shows an incomplete "anti-magic" square. Complete the "anti-magic" square shown below.

			14
	9	3	7
	12	13	5
10	11	6	4

Instead of the usual requirements of equality of sums we can investigate a slightly different situation that might occur and we see that a great variety of questions can arise even in these simple-looking situations.

Sometimes instead of a table, the gathering place for our numbers in question might also be the dial of a traditional clock.

3. The dial of a clock has been cracked into three pieces so that the sum of the numbers on each piece is the same. Given that none of the cracks separates the digits of a number, which of the following statements is correct? (A) 12 and 3 are not on the same piece (B) 8 and 4 are on the same piece (C) 7 and 5 are not on the same piece (D) 11, 1 and 5 are on the same piece (E) 2, 11 and 9 are on the same piece.
4. Nine light bulbs are put in a square formation. Each bulb can be either *on* or *off*. We can make a move by pressing a bulb. Then, the pressed bulb and the bulbs in the same row and column change their state from on to off or vice versa. Initially, all light bulbs are on. What is the minimum number of moves needed to turn off all the light bulbs? (A) 3 (B) 4 (C) 5 (D) 9 (E) This is impossible (compare Problem 5 (Kašuba 2010)).

☺	☺	☺
☺	☺	☺
☺	☺	☺

An attempt at a nice posing of a problem, as may be done in fiction

Below we present an attempt to pose a simple problem nicely, which seems very natural in the 4×4 dimension. It was proposed in 2011 in a Lithuanian contest for youngsters in lower grades in the exact formulation as is presented below.

The White Horse, although he was rather seldom seen in the company of other animals, was in fact highly regarded as a devoted partner and trusted friend of the Hedgehog in the Fog. Also the Horse liked to appear as if descending from high above by always bringing strange problems of complicated origins. Solemnly speaking, from first glance the Hedgehog in the Fog was usually not that enthusiastic about solving them or even reading their formulations.

But in time he got used to getting involved in the process of solving, was always doing his best, but if he wasn't able to achieve at least slight progress, he would get very irritated and lose his temper with outbursts of anger that were not always predictable.

Today in the daily post the White Horse also received a puzzle that looked quite difficult. It consisted of 16 pieces of letters joined with numbers. They looked exactly as indicated below:

$$a1, a2, a3, a4, b1, b2, b3, b4, c1, c2, c3, c4, d1, d2, d3, d4.$$

In the fog, the call of the Owl was clearly distinguishable and, at the same time, an invitation for all who were of some scientific importance in that forest of clever animals, at least to attempt to fulfill the following challenging scientific task: put these letters joined with numbers into the 16 cells of 4×4 square in such a way that in each row as well as in each column all these four letters

$$a, b, c, d,$$

as well as all these four integers

$$1, 2, 3, 4$$

were represented exactly once.

The Grizzly Bear, who appeared immediately, expressed his sincere doubts concerning the possibility of solving it. The Hedgehog in the Fog, on the other hand, eagerly believed in the possibility of solving it, although without any scientific basis. He simply said that the puzzle was too nice not to be solvable. Only imagine: in every row, as well as in every column, all the digits and all letters without any repetition in any row or in any column.

Is that really possible? Or it is too good to be true?

Some changes to the situation, helpfully illustrating that not everything is so simple

One should not think that the situation in low-dimensional tables is always so easy to deal with. Here we present some examples taken from the literature and used in the mathematical contest in Lithuania. The first problem is taken from the Dutch Mathematical Olympiad and the second one was apparently used in the UK Mathematical Olympiad but rewritten in the South Africa MO.

1. We consider 5×5-tables containing a number in each of the 25 cells. The same number may occur in different cells, but no row or column may contain five equal numbers. Such a table is called *pretty* if in each row the cell in the middle contains the average of the numbers in that row, and in each column the cell in the middle contains the average of the numbers in that column. The *score* of a pretty table is the number of cells that contain a number that is smaller than the number in the cell in the very middle of the table. What is the smallest possible score of a pretty table?
2. A 5×5 square is divided into 25 unit squares. One of the numbers 1, 2, 3, 4, 5 is inserted into each of the unit squares in such a way that each row, each column and each of the two diagonals contains each of the five numbers once and only once. The sum of the numbers in the four squares immediately below the diagonal from top left to bottom right is called a score. Show that it is impossible for the score to be 20. What is the highest possible score?

If we wished, this might be made to appear as a Kangaroo problem. Then, instead of the question about the impossibility of the score to be 20, one might gradually finish by asking: What is the highest possible score: (A) 20 (B) 19 (C) 18 (D) 17 (E) 16

To be or not to be and if so how?

In his genial way he proceeded to say
(Forgetting all laws of propriety,
And that giving instruction, without introduction,
Would have caused quite a thrill in Society).
Lewis Carroll, Hunting of the Snark.

The phrase "to be or not be" in early-childhood education

The famous saying is deeply connected with the main question of what exactly a child is able to understand and especially what he does not. It demands constant verification of a child's understanding and achievements. Having such sure notions, or at least some commonly acceptable ideas, we may move to the structuring of the situation. We must also give some explanation about the way that child should be instructed in order to help manage the situation. And an optimist would also immediately add the condition "sine qua non" that the child ought to learn with pleasure.

And what the author means by "pleasure" in this case is not the typical meaning, which we try to achieve simply by saying "oh, how brave you are", but the more profound pleasure that you, for instance, must unavoidably feel after you have finished building a house (or cracking the problem, that you were not able to do yesterday, or a week or even a year before).

The problem of what exactly a child is able to achieve and how to prepare him for it, is a most subtle question, because we may so quickly come to a conclusion that even in early years a child can achieve practically everything and understand even more—when properly instructed, naturally.

There is a common view that the child can understand, even invent, everything that might be expressed using very few words, avoiding special terms and tricky ideas. For tricky ideas mankind has founded the universities.

First of all, we strongly believe that a child, just like any other normal person, especially likes to do what at least for a time appears to him as especially difficult and, most importantly, attractive. Many problems of that kind are problems dealing with numbers, arrangements of numbers and their combinations.

Instructing a child about solving something that he finds attractive is not very difficult

The next step is assisting him, or maintaining the situation. It helps a lot that a normal child is used to operate with numbers, even with large ones. Just as many people, it seems, believe even more that, consciously or not, a child is eager to demonstrate somehow for all those concerned, that the difference between him and any adult is considerably smaller than many try to believe or have experienced it to be.

So let us assume that a normal child is not afraid of any numbers and, what may be even more important, he is very eager to demonstrate that he is not that young. Let us also assume that he is able to proceed constructively, quite often in an astonishingly logical way, and that instead of making long theoretical reflections he prefers concrete actions.

We would like to discuss with the reader how to deal with some problems. When speaking about solving them, we strongly believe that we will be able to give some insight into effectively stimulating the challenging and inspiring process of reaching a solution.

The author has spent many years first learning how to solve problems and then afterwards learning specifically how to help the process. He has written a few books, which will be mentioned in the list of references.

The first problem the author will present, could, we believe, be presented for practically any age, after some instruction, of course. We humbly admit that we possess quite a lot of experience with instructing and preparing the process of solution for a child in his tenth year.

Besides the original formulation, we will also try to present the reader with a problem that he will find more attractive and then we will report on our attempts to achieve the solution. We believe that attempts to formulate the problem more attractively, so that more people will be interested, are also good for people of other age groups, not just children.

As the author has witnessed many times, if something is suitable for a normal child then the same problem will be suitable for any normal university student as well.

An example of the usual representation of a problem and how it may be embellished

We are asked to find the minimal positive integer that can be represented in two different ways as a sum of three addends in such a manner that the six addends mentioned would all be different.

Find means detect, and detect is something that Mr. Sherlock Holmes with Dr. Watson used to do every day, with great success.

This is not a difficult problem, but nevertheless let us look for some "more attractive representation", as we just did by adding two popular names that for a normal child may not add anything worth mentioning.

What object could be attractively proposed to the child as a multitude that can be split into three parts in two different ways?

What multitudes does the child see and enjoy every day? We may choose and propose some possibilities, e.g., it could be the clouds in the sky. Any natural number may be represented as a separate group of clouds. So we may have the whole set of clouds in the sky divided into separate groups of clouds. And then we may ask how to apply that language in order to present another representation of the very same natural number or the same multitude of clouds. How could we do that? The answer is simple, because the winds blow and when wind is blowing the skies change their configuration. We might point out to his attention the fact that we silently assume that no cloud disappears, but the multitude of clouds in the sky may still easily change its configuration.

As mentioned before, we strongly believe—because of our experience—that any child who is alert is able to count exactly and, moreover, is eager not only to add numbers, but to split them as well. Also, he will be very fond of us if we are able to present a task in good style.

So, in order to begin looking properly for a solution, we may ask him for an example and the child will immediately respond. A child of any age is strongest when answering a concrete question. So, we may ask him: do you know such a number, which possess the miraculous ability and can be split in two different ways into the sum of three different addends, so that all six numbers of these numbers are different? The examples are not difficult to find, in fact, they are everywhere; for instance, such is the very regular number 111 and just as regular, but more usual because of its size, is the number 21:

$$111 = 100 + 10 + 1 = 104 + 4 + 3,$$
$$21 = 1 + 9 + 10 = 2 + 4 + 15.$$

Now we just have to detect the smallest number among all such "chameleonic" integers.

Children are natural investigators

Because of their constructive nature, children are always curious to detect what integer will lose that property first.

The author has gained some experience by presenting and explaining this problem, especially at the very end of its solution. There it is rather interesting, because some kind of abstract reasoning or something of proof-like thoughts would be highly welcomed. But, according to the author's experience in a recent year in Grade 3, that abstract end was taken more as the bare statement and as the dominion of the teacher. Because the teacher said that the process must naturally stop at some number and, moreover, the teacher indicated we would find that number when we wouldn't be able to find constructively that double split into three addends, all different. Frankly speaking their reasoning ran along the lines of "it must be because it is". And nothing more needed to be said in Grade 3. Yet, in Grade 4 of the very same school on the very same day it was possible to achieve so much more; evidently, approximately a third of the pupils present were able to understand even that abstract philosophical part of the proof that 10 is that number. After experimenting with numbers, the proof that 10 is the number followed. The usual words of proof or abstract reasoning were pronounced. These words were pronounced and they were understood by a good third or possibly even a half of the audience—in Grade 4.

These words were: take the six smallest positive integers
Namely

$$1, 2, 3, 4, 5, 6.$$

Because their sum is

$$1+2+3+4+5+6=21,$$

we can see that this is more than twice as big as 10, so the number 10 clearly can't possess such a double-thrice split or representation. Following this train of thought, we tentatively state that 11 may be the smallest integer with the double split. And 11 is indeed such a number, because it can be written either as

$$2+4+5$$

or

$$1+3+7.$$

So, in conclusion, 11 is indeed the smallest such integer and the rest follows.

After experiencing the natural joy of the teacher that his students are so smart

The saying "repetitio est mater magistrum" followed. Because these pupils of Grade 4 also understood that they have achieved something. Then it was high time to repeat that main sentence, which at the same time is the short version of the proof, several times. For the first time it was done immediately after finding the solution and for the second time, after some 5 min. Then, again, the clouds were mentioned and the wind that scatters clouds remembered.

The Second Example is About the Magic of Suitable Instruction

The following ought to convince the reader that subtle advice at the right time may indeed appear to be extremely magical.

Let us take an apparently standard problem which might for a time be connected with our plans to set 9 numbers

$$15, 16, 17, 18, 19, 20, 21, 22 \text{ and } 23$$

into the 9 cells of 3 × 3 table in such a way that for each pair of numbers whose cells happen to share a common side, the sum of both integers in the pair is always different.

It ought to be mentioned that this is a problem from the Saint-Petersburg competition. Please, do not ask us the grade it was meant for because, knowing the grade, you may be afraid that the problem may not be suitable for early grades, or that it might even be presented as a pre-school problem, after suitable instruction of course.

Assume that we would like—and this is what the author, as any teacher, is always potentially trying to aspire to—to give some effective support in the form of subtle and short advice. The intended advice must be so perfectly worded, that afterwards each student in each grade would be able to achieve what the task has asked him to achieve.

Actually, after reading the text of the problem

It is not immediately apparent why this problem might be proposed, and was proposed, in the Saint-Petersburg Math Olympiad (Berlov et al. 2011). What's more, it was not presented in the lowest grade either. What could be the reason for that?

Looking back, we must mention that it is completely clear that this problem is not too difficult. We can start simply with uninspired attempts to place the numbers; we might even succeed without having any kind of a system. Now, we could proceed as a smart child would: do something, enjoy what happens and simply gather experience. After several attempts, each child would have an unavoidable feeling that he must proceed by applying some system of placement and not letting things happen chaotically.

Let us not be, however, still uninspired and write down the numbers in the most common way—by starting from the smallest one upwards like this:

15	16	17
18	19	20
21	22	23

Remembering that we must pair wisely in order to obtain different answers when summing all pairs of numbers sharing a common side, we start by adding neighboring integers, consequently getting:

$$15 + 16, 16 + 17, 15 + 18.$$

We notice that we are not able to fulfill our requirement immediately—we have just started with calculations and already have gotten the number 33 twice.

So the way of placing numbers in the most common manner proved itself to be unsuitable.

Now, in the spirit of the everlasting dance, or the mutual, careful assistance between construction and instruction, we would like to ask: is it possible and how should we proceed in order to provide effective support for our pupil, yet formally do very little? What kind of impulse might we give? Are there some wonderful means for that?

The effect of achieving much by doing very little is quite memorable; the child at each age is very fond of rapid success and, moreover, he never forgets it. Rapid success and how to deal with it might be a really valuable chapter in the theory of effective and, especially, joyful teaching.

In our situation all that effectiveness

Is due to the idea of a slight rotation of the whole table around its center. Nothing more is needed at this time. After that very small rotation the idea comes to mind: help the diagonals to be at least a bit more like the horizontals. Then the same standard placement in this new setting will prove itself to be the solution. But before we begin, let us help the formal diagonals look like the horizontals—the best means for that is the 45° rotation.

That brings us almost immediately to the desired solution presented in the table below:

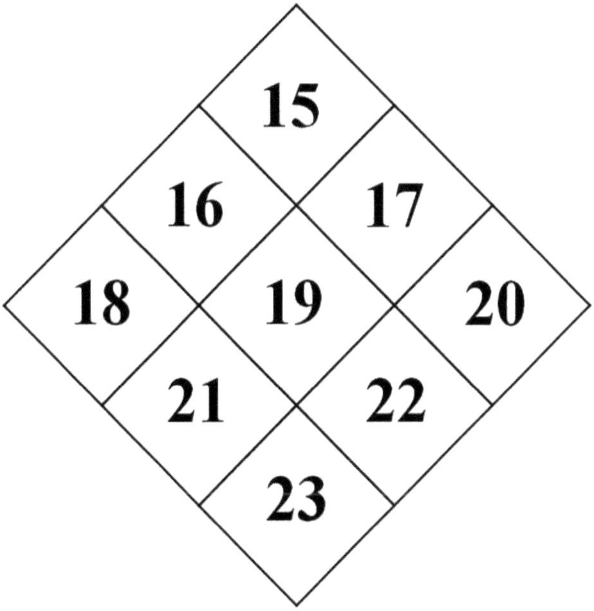

After rotating backwards, we'll have the answer presented in the usual form, or

15	17	20
16	19	22
18	21	23

Let us note that now all 12 of the neighboring pairs do indeed provide different sums:

$$15+17=32, 17+20=37, 15+16=31, 17+19=36,$$
$$20+22=42, 16+19=35, 19+22=41, 16+18=34,$$
$$19+21=40, 22+23=45, 18+21=39, 21+23=44.$$

So, once again, we must state that in the eternal dance of instruction and construction or, in other words, in the everlasting mutual assistance between the "normal reality" and "possibilities which we possess to shape that reality", vast, inexhaustible potential remains. According to the understanding of the author, it is one of the most powerful resources of any skilled teacher. It enforces the hold of such teacher at school and serves as one of the main resources for his pedagogical and natural human enthusiasm.

You might sometimes be lucky enough to express infinity by saying practically nothing, as we have just been rewarded with in our case with the rotation. And that was enough for complete success in dealing with the problem proposed.

Early childhood with all its infinite possibilities and boundaries is a constant challenge for instruction and construction

In the extremes, this statement may lead to the crucial question: how difficult can the problems be so that a child will still be able to deal with them effectively? Here we would like say that to deal effectively might not necessarily mean to solve them. It is enough to be able to achieve some progress and, of course, to feel the satisfaction of being really involved. Experienced math educators know that the feeling of being involved is the reason that makes math education such a precious and sought-after subject.

How inexhaustible are the real possibilities for a child? What are the effective boundaries of an alert child with some seemingly difficult task? What might make even the most exacting problem accessible and even attractive?

There are no definite answers to these questions. There also are no simple answers to the question of what makes a good problem into a piece of the irresistible art of human challenge and possibilities.

Let us try to illustrate some of our thoughts and impressions while, at the same time, discussing an astonishing Kangaroo problem

First, we'd like to present the text of the problem (Kangaroo problems for the Year 2012).

In our opinion, this is a problem that may be shown to any child who can read the text in order to illustrate the following truth: a math problem can be ready for reading, yet not necessarily ready for solving. There are some problems which are as nice as fairy tales are in literature. According to the author's opinion, this proves the following: a mathematical problem may be more than just a mathematical problem. Such a concept might seem unusual in math education but, on the other hand, expressions of that sort are well enough known in art, literature and other fields (including sport).

Every cat in Wonderland is either wise or mad. If a wise cat happens to be in one room with 3 *mad ones it turns mad. If a mad cat happens to be in one room with* 3 *wise ones it is exposed by them as being mad. Three cats entered an empty room. Soon after the* 4*th cat entered, the* 1*st one went out. After the* 5*th cat entered, the* 2*nd one went out, etc. After the* 2012*th cat entered, it happened for the first time that one of the cats was exposed as mad. Which of the cats could both have been mad after entering the room?*

(A) *The 1st one and the 2011th one* (B) *The 2nd one and the 2010th one* (C) *The 3rd one and the 2009th one* (D) *The 4th one and the last one* (E) *The 2nd one and the 2011th one.*

That is a typical answer in Kangaroo style, 5 alternatives are proposed with exactly one of them being correct.

From first glance, the proposed task does not appear to be easy

Yet, it does introduce some structure into the world, borrowing from a fairy tale. That structure is exotic, but the way the problem plays out is arranged following some predictable rules. From that point of view, the presentation of the problem, that may seem lengthy for the Kangaroo competition, is not so because of the magic in the presentation. As a whole, it looks more like a fairy tale than a mathematical problem. The heroes named wise and mad cats immediately presuppose tension. The possibilities of transition and exposition described guarantee additional excitement for the fantasy of a child. So, let us try to mention all the remarkable circumstances that could inspire us to recommend these problems for radically young minds. As stated before, simply by reading the text of a problem we might be predisposed to think that the problem possesses a mystical flavor. At the same time, the action lasts for quite a while, for a remarkably long time one cat enters and another leaves. So, as we are told, cats are not only filling the scene, but also performing by following defined rules. These cats, a lot of them, form a long line and act under strictly fixed rules thus making the magical situation partly predictable and, consequently, more attractive. But we must mention that, just like in fairy tales, the circumstances that allow the hero to win, or for the situation to take a suitable turn, must be analyzed rather carefully.

That special turn of our situation is the moment when, for the first time, a child must understand that he is dealing with a situation where, if the first three cats are mad, the fourth must be so as well.

In many such cases, when dealing with a problem with a bit of magic, attractive and yet not that easy to solve, we may raise the following question: what is the first remarkable observation that a child can make regarding the particular situation? The first observation of a child quite often later proves to be the main idea of solution, or at least a big part of it.

That is, the first thought of a child when dealing with good problems is usually excellent or otherwise worth mentioning.

And, like in every fairy tale, the child is hoping, sometimes even unconsciously, for a happy end

Which in our case also includes the essential understanding of the whole situation. From that moment the desired solution usually follows. Sometimes this is where we, who believe that we and the child can both clearly grasp the whole process of solving, may start to think that, after suitable instructions, this problem could be accessible for all those interested or involved.

So, what could be the first remarkable observation of the child in our dance of understanding and explaining the problem? What could be the first words that the child might utter after some consideration? Sometimes the first words are those that describe or investigate some of the "peculiar cases".

What might the worst case be in our situation? The one which eliminates the situation presented in the problem.

Such a situation might of course be the situation with three consecutive mad cats happening somewhere in that long line of 2012 cats. The situation with 3 "consecutive" mad cats would completely and definitely change everything. These 3 cats would make all other cats mad even if they weren't before. We believe understanding that might be the first serious thought of the child and the first remarkable stone upon which the solution will be built. The situation with three consecutive mad cats is clearly the "drastic" case after which it would be impossible to reach the outcome described in the text.

We dare to imagine, that a clear concept of that situation is not too complicated for any sensible child to reach.

So let's assume that the child clearly understands that three consecutive mad cats anywhere in that long line of cats eliminates the outcome described, and this is what the child must never forget.

The child is also expected to be aware of the following circumstance, which is similar to every good fairy tale: that the most important things are often mentioned in the end

In our case, the most optimistic reality occurs, if the first mad cat was exposed only after the 2012th cat entered. What are the circumstances that would allow delaying this for so long?

Returning to the initial situation, the child sooner or later decides to examine the situation carefully with the first three cats (immediately before the fourth one appears). Three (or even four) cats won't be too much for any child, so long as they do not all live in the same flat.

So, let the child examine the situation with the first 3 cats. If all of them had been mad, oh, this would bring the bright child to the worst case, which we have already examined above.

We will repeat the important observations again: if the 4th cat is mad, then everything will remain as bad as it was; if the 4th cat isn't mad when entering, he will very quickly become mad. Afterwards it is of no importance which cat leaves. So, it is quite obvious that the first cat that leaves has to be mad. Then, if the situation with three mad cats occurred, it would remain the same until the end of the line.

Because of that, it is clear that not all 3 of consecutive cats can be mad.

Now, the child could regard the "opposite situation"

What if all three initial consecutive cats were all wise ones? If all of them were wise ones, then the fourth cat, when entering or before entering, couldn't be mad, because otherwise, immediately after entering, he would be exposed as mad. But we are informed in the text of problem that this happened much later. So, according to the fable of the whole story, if three initial cats are all wise and if no case of madness is exposed until the 2012th cat enters, then this means that all "intermediate" cats that enter have to be wise. In that situation, when the supposedly mad 2012th cat entered, everything would happen as described.

And what then? The situation is indeed possible but none of the five alternatives proposed, none of them, suit the situation. But that is not necessarily an error, because in the text of problem we read: "which of the cats could both have been mad after entering the room?" "Could" doesn't mean "must be".

It allows both us and the child to cross off both extreme situations, when all three initial cats are either “mad” or “wise”. In the latter case the situation is possible, but none of 5 choices mentioned in the answers suits it completely.

Now we are expected to regard the “intermediate situation” with the initial cats being a mix of mad and wise. That means that among these 3 first cats there is at least one mad and at least one wise cat. We have to admit that after the 4th cat entered (and the 1st left) the situation couldn’t be allowed to go to extremes, for reasons we already examined.

Consequently, let us regard these last two possible cases. In the first case, 2 cats of the 3 are mad and in the second 2 of the 3 are wise.

In the first case the child understands that the 4th cannot be mad when entering, because, as shown and mentioned, this would create again the situation with 3 mad cats among consecutive 4, and this would make all remaining cats mad after entering thus determining the whole happening “ad infinitum” and not as expected.

Again, if 2 cats among the first 3 are wise, then the 4th cat, when entering, can’t be wise, because then 3 of the 4 cats would be wise and the exposition of madness would immediately follow—and this contradicts the fable of our problem.

Having investigated both cases, the child may conclude that when the 4th cat enters he must create a “fifty-fifty situation” with two cats being wise and another two cats being mad. Keeping this in mind, the child could start to list all possible variations. These are the following (where W naturally denotes a wise cat and M a mad one):

$$MMWW, MWMW, MWWM, WMMW, WMWM, WWMM.$$

Now in all of these situations, it is high time for the 1st cat to bid farewell to the society of cats. When the 1st cat is leaving then the possibility for the 5th one to introduce itself appears. Now the main concern is whether we realize that if the cat leaving the room is wise, then mad cats will form a majority and that majority is what the 5th cat must neutralize. To put it simply, if the 1st cat leaving the room is wise, then the one entering will have to be wise too. On the other hand, if the 1st cat leaving the room is mad, then the cat entering has to be mad too.

All of these deductions prove that the situation is circling itself, or behaving in a cyclic fashion, or as arithmeticians would say, can be thought of as being modulo 4, so we have 6 possible sequences:

1. *MMWW MMWW MMWW* . . .
2. *MWMW MWMW MWMW* . . .
3. *MWWM MWWM MWWM* . . .
4. *WMMW WMMW WMMW* . . .
5. *WMWM WMWM WMWM* . . .
6. *WWMM WWMM WWMM* . . .

Finally, the solution

Once again, according to the fable of the problem, after the 2012th cat enters, *M* must be in the room. What's crucial is to realize that this mad cat is not necessarily the one who entered. Exposing a mad cat takes 3 wise ones. 3 mad ones would have eliminated wise ones completely and destroyed the fable. But we already spent a lot of time proving there must be a cycle of any four consecutive cats and why. But the 2012th cat has to break a pattern, because it exposes madness for the first time. So, a situation must be achieved where before 2012th cat enters, there must be 2 wise cats and 1 mad cat in the room. Because of that, we can eliminate the 1st, 2nd and the 4th sequence.

Only these remain:

MWWM MWWM MWWM . . . *MWWW*
WMWM WMWM WMWM . . . *WMWW*
WWMM WWMM WWMM . . . *WWMW*

So now we can easily check that only in second of remaining cases the only answer *B* is possible.

Some simple words practically before the end

The author believes that a young child is able to understand the first three sentences which may express a solution of a not very complicated mathematical (or arithmetical) problem. For this to be so, the problem, of course, could not require specific methods or formulas, just logic and common sense.

So, if the problem is not overspecialized, then practically all that can be expressed in the first three clever sentences is potentially understandable for any alert child, yet the dance of inspiration and time is necessary for the solution to be properly arranged.

The child must have enough time. Then he may achieve much more than he imagines and even more than we, as educators, might expect.

The final part, or some story about what you might gain from the problem and what you might not

This is a fragment which, under the title of "Lawn tennis or the bright musketeer D'Artagnan", is taken from the author's book "What to do when you don't know what to do?"

Once these three musketeers were training on the tennis court—don't forget that they all were warriors of King Louis XIII, the counterpart of Cardinal Richelieu.

Meanwhile they all were involved in a tennis game. The game was arranged in the following manner: two were playing one set and the third was their judge. After the game was over the loser became the judge and the former judge would then play with the winner, and so the whole affair ran on.

After this long time for play was over, it was stated that Athos had played 15 sets or 15 times, Porthos 10 and Aramis 17. D'Artagnan, who had just arrived, seeing that the entire statistics consisted of 3 numbers, declared his ability to determine the losers of many sets.

- But how on earth can one determine the loser of so many sets having these 3 numbers only?
- Many facts may be gleaned from these 3 numbers.
- For instance, which facts exactly?
- For example, I could tell you who lost the second set you played.
- Might you be able to tell also who won the second set?
- I beg your pardon; I'm not speaking about who **won** the second set but I'm speaking now about who **lost** it.
- Does it make any difference?
- Yes, it does. I repeat that I'm not claiming to be able to determine everything in France having only these 3 numbers, but I insistently repeat with all my might that I can say for sure who lost the second set.
- Can you say also who lost the third set?
- No, I'm not talking about that.
- What else could you say having these numbers 10, 15 and 17?
- Many things. For instance, I can tell who **lost** the 16th set.
- And maybe you can tell us also who **won** this 16th set?
- Again I am not speaking about who **won** it but about who **lost** it.
- So what is this? You who are such an excellent fighter and perhaps the first amongst the king's warriors suddenly became a specialist on losing?
- I do what I can.
- Please explain to us how you are doing it?
- Let us sit down on the bank and, please, listen for a while.

Dear reader, we understand that not all of you nor I would be very fond of such artfulness in the popularization of so simple a problem. Not everyone likes such adaptations. But still it's always worth trying.

On the other hand, if I'm willing to deal with a problem, I'm expected to know what the problem I chanced to meet is about and also to be able to keep in mind all essential circumstances of the deal.

The adaptation can also be slightly irritating and not convincing when I'm running around with a serious face repeating that one jump of a kangaroo can be 100 km long or that a horse can leap over a fence that is 15 m high. Or it's a poor fantasy that is still at least a bit attractive?

Still the author would never give up his opinion and will go on repeating that any at least somewhat successful adaptation or more vivid presentation of the possible task makes the sense of the whole happening more attractive to the possible future solver. It simply and naturally makes the possible future solver cleverer and keeps up his spirit, and that, together with all other useful things, is the most valuable thing that education may bring and propose.

To make the solver cleverer than he'd been before, and so to keep up his spirit—this is the most honorable duty of every human art, especially mathematics

Sometimes we are too shy to say it **expressis verbis**, or express it in words.

Let us go back to explanation of D'Artagnan. In a curious way he was speaking about the simplest things. Let's listen to him and figure out what he meant.

First he suggested adding up these 3 numbers, 10, 15 and 17. Summing up we've got

$$10 + 15 + 17 = 42.$$

What can we extract from this prosaic, banal sum 42? It gives us the common number of all persons involved in all provided sets.

Now we ought to emphasize strongly that one and the same person from any 3 of them will be counted several or many times.

Two players make a set. Two players mean 1 set. So 42 players mean

$$42 : 2 = 21$$

sets exactly. So from that prosaic sum we extract the fact that there were 21 sets to play.

Take a glimpse at the 3 given numbers again.

Athos had played 15 sets or times, Porthos 10 and Aramis 17. There were 21 sets as has been said and repeated. Who was the most successful player? Again the most natural thing is to assume that the person who played most is the strongest. He was not eliminated so often. There remains practically no doubt that the person who played most is the strongest one among them. So according to the numbers Aramis seems to be the strongest, Athos almost as strong as Aramis, and finally Porthos is the weakest. So perhaps he, as the weakest, is that one who lost the second set. Why must it necessarily be him? What's the reason? Everyone, even the strongest player, can lose.

Only God Can't lose

Indeed, will it really turn out that it was Porthos?

Look again at the number of games Porthos played. Porthos played 10 times or sets. There were 21 set. It is clear that each musketeer was playing at least every second time. That means that in two neighbouring sets each of them must have played at least once. But only 10 participations in 21 sets leaves Porthos only one possibility to realize such a game configuration. This possibility is necessarily to have played the second set, then the fourth, the sixth, the eight, the tenth, the twelfth, the fourteenth, the sixteenth, the eighteenth and the twentieth set.

And unfortunately the only possibility to obtain such a game configuration or to play 10 times in 21 sets, means also that he has lost all these sets.

Otherwise he would have participated in some 2 neighbouring sets. But he did not. He lost them all.

We remember now that D'Artagnan spoke only about losers and not about winners. He spoke also about who'd lost the 16th set. He avoided any talk about possible winners.

Nothing more can be figured out with the exception that in the other $21 - 10 = 11$ sets, Athos was playing versus Aramis and that Aramis should have won more games.

Question to the reader:

Can you deduce from these three numbers 10, 15 and 17 how many times Athos won and how many times Aramis did?

For deep studies see Vygotsky (1978).

References

Berlov S. L., et al. (Eds.). (2011). *Zadachi Sankt-Peterburgskoj olimpiady shkolnikov po matematike* (*Problems of Sankt-Petersburg Mathematical Olympiad*), Nevski Dialekt Ed. ISBN 978-5-7940-0147-1.

Kangaroo problems for the Year 2012. http://www.kengura.lt.
Kašuba, R. (2006). *What to do when you don't know what to do?* (127 pp.). Riga: Macibu gramata. ISBN 9984-18-234-7.
Kašuba, R. (2007). *What to do when you don't know what to do?* Part II (166 pp.). Riga: Macibu gramata. ISBN 9984-18-284-3.
Kašuba, R. (2008a). *Once upon a time I saw a puzzle*, Part I (59 pp.). Riga: University of Latvia. ISBN 978-9984-45-045-2.
Kašuba, R. (2008b). *Once upon a time I saw a puzzle*, Part II (67 pp.). Riga: University of Latvia.
Kašuba, R. (2009). *Once upon a time I saw a puzzle*, Part III (88 pp.). Riga: University of Latvia. ISBN 978-9984-45-133-6.
Kašuba, R. (2010). Some remarks on possible components of advanced components of mathematical thinking. *Mathematics Competitions, 23*(1), 12–26.
Kašuba, R. (2012). *Kak reshat' zadachu, kogda nie znajesh kak*, (*How to solve the problem when you do not know how*) (2nd ed., 2014). Moscow: Prosveshchenie Publishing House. ISBN 978-5-09-023583-9.
Laurie, D., & Merry, B. (Eds.). (2000). *South Africa and the 41st international mathematical Olympiad.* University of Cape Town. ISBN 0-7992-2068-x.
Mazanik, S. A., et al. (Eds.). (2005). *Zadachi minskoj gorodskoj olimpiady mladshykh shkolnikov* (*Problems of the Minsk city math Olympiad for younger grades*), Minsk, "Konkurs" ass.
Rubloyv, B. V. (Ed.). (2015). *Matematichni olimpiadni zmagannia shkoliariv Ukrainy*, 2013/14 (Mathematical Olympiad contests of students in Ukraine). Kharkov, "Gimnaziya". ISBN 978-966-474-255-6.
United Kingdom Mathematics Trust. (2013). *Intermediate Mathematical Olympiad and Kangaroo* (IMOK).
Vygotsky, L. (1978). *Mind in society: The development of higher psychological processes.* Cambridge, MA: Harvard University Press.
We eat problems for breakfast. Preferably unsolved ones… 50th Dutch Mathematical Olympiad 2011, Stichting Nederlandse Wiskunde Olympiade, Amsterdam, 2012.
We eat problems for breakfast. Preferably unsolved ones… 52nd Dutch Mathematical Olympiad 2013, Stichting Nederlandse Wiskunde Olympiade, Amsterdam, 2014.
We eat problems for breakfast. Preferably unsolved ones… 53rd Dutch Mathematical Olympiad 2014, Stichting Nederlandse Wiskunde Olympiade, Amsterdam, 2015.

Chapter 12
Future Directions for Research in Mathematics Competitions

Peter James Taylor

Abstract In this chapter I outline two major areas for future research on competitions and mathematics enrichment generally. The first is a development of better classification of problems, and an evaluation of their effectiveness, particularly those at the level which can lead a student with classroom knowledge to a deeper mathematics experience. The second area is the gathering and analysis of more data on competition and Olympiad alumni, in order to help gauge the effectiveness of competitions on choice of and success in future careers.

Keywords Competition · Olympiad · Problem · Classification · Alumni

12.1 Introduction

Competitions have a unique role in the education system. Because the problems are usually composed externally to any particular school, and can test a student's ability to use the mathematics they have learned in the classroom in new contexts, they can help a student to be increasingly useful in later studies and in ultimate careers.

The modern existence of competitions for schools commenced in 1894 in Hungary (Romania may have started earlier), Olympiad style competitions started in the Soviet Union in the 1930s (Leningrad and Moscow), and large inclusive-style competitions first appeared in the US in the 1950s, and spread from then to Canada, Australia, Europe and elsewhere.

P.J. Taylor (✉)
University of Canberra, Bruce, Australia
e-mail: pjt013@gmail.com

A. Soifer (ed.), *Competitions for Young Mathematicians*,
ICME-13 Monographs, DOI 10.1007/978-3-319-56585-9_12

By 1984 there were so many competitions in various forms nationally, regionally and internationally that a professional society, the World Federation of National Mathematics Competitions (WFNMC), was established at a meeting held at ICME-5 (International Congress on Mathematics Education, Number 5) in Adelaide. A detailed history of WFNMC can be found in Kenderov (2009). The role of competitions in mathematics education generally was part of ICMI Study 16, which was reported on in Barbeau and Taylor (2009).

12.2 Research

There has been some research on competitions, and some has been published in the WFNMC's Journal *Mathematics Competitions*. What I will discuss in this chapter are two broad areas which would be useful on a larger and systematic way, to help practitioners have a clearer understanding of what they are doing and help in their planning. To this end I will quote known results from projects on smaller scales.

12.2.1 Competition Syllabus

There are many programs around the world for gifted students in mathematics, but those with greatest commonality would be competition-related activities, so I feel that a study of student performance in competitions can give us the best insight into what sort of mathematics is best for extending students' experience beyond the classroom. At the highest end the International Mathematical Olympiad (IMO) has a defined set of 4 topics (Algebra, Number Theory, Geometry and Combinatorics, with Calculus as the main omission) but does not attempt to define a more detailed syllabus below this.

If we assume that this syllabus is an ultimate goal (and it is about the only international one available) then it is a long way from classroom experience and I am interested in identifying the first steps on this path, which can be used in the classroom, the starting point for competitions.

In my paper Taylor (2015) I attempted to identify and define a suitable set of mathematics topics, all used in school competitions. I have also examined this idea on my web site Taylor (2014). I will now consider my breakdown of competition topics, with typical examples and solutions, as well as discussion on some pedagogical aspects.

1. Diophantine Equations

Diophantine equations are linear equations in which we look for integer solutions. Finding such solutions under various restrictions can be a very accessible task for school students. Even if students have not seen Diophantine equations per se in the classroom, students have shown they can use intuition to adapt to these situations.

An example, from the Australian Mathematics Competition (AMC) 2010 is

Problem 1

Eric and Marina each wrote two or three poems every day. Over a period of time, Eric wrote 43 poems while Marina wrote 61. How many days were in this period of time?

Note: This problem, as is the case in some others below, were set as multi-choice. But large inclusive competitions use multi-choice not because they like the format, but it is the only practical way of assessing hundreds of thousands of scripts in a short time. In this and the cases below I will not pose as multi-choice.

Solution 1

Suppose the number of days is z, with Eric writing three poems on x days and Marina on y days. Then we have the equations $3x + (z - x)2 = 43$ and $3y + (z - y)2 = 61$. These simplify to $x + 2z = 43$ and $y + 2z = 61$. This leads to $y = x + 18$. From the first equation it is apparent that x is odd. For $x \geq 3$, $y \geq 21$ but $z \leq 20$, which is not possible. This leaves the only solution $(x, y, z) = (1, 19, 21)$.

Note: This problem was not too difficult. A total of 41% of Australian Year 8 entrants in the Competition were successful. Whereas the above solution is probably the most direct, an important part of mathematics is that you get the same solution no matter which valid method is used and it is my view we need to demonstrate mathematics as a rich subject with integrity by providing alternative solutions when known. The above problem has two nice alternatives, given below.

Solution 1, Alternative 2

It could not be less than 21 days, as in 20 days no one could write more than 60 poems.

It could not be more than 21 days, as in 22 days each one of them would have written at least 44 poems.

So the time period must be 21 days. It is possible if Eric writes three poems on just 1 occasion and Marina on 19 occasions.

Solution 1, Alternative 3

Since Marina wrote 18 more poems than Eric, there were at least 18 days when Eric wrote 2 poems and Marina wrote 3.

Now $43 = 18 \times 2 + 2 + 2 + 3$ and $61 = 18 \times 3 + 2 + 2 + 3$.

So there were 21 days.

2. Pigeonhole Principle

Also known as Dirichlet's Principle, it is very easy to explain and provides a powerful tool for simplifying an argument.

Often a problem lending itself to a Pigeonhole proof will have some form of constraint, with phrases somewhere like "at most", "at least", "more than" or "less than". If so, one looks to identify pigeons and pigeonholes.

An excellent example is the following Tournament of Towns problem, in which the problem is attributed to Hungarian High School student M. Vora, and solution to Canadian mathematician Andy Liu.

Problem 2

In a football tournament of one round (each team plays each other once, 2 points for win, 1 point for draw and 0 points for loss), 28 teams compete. During the tournament more than 75% of the matches finished in a draw. Prove that there were two teams who finished with the same number of points.

Note: The proof below starts in the normal way by calculating the number of games and hence the minimum number of games which can end as draws. For each team the difference between the difference between the number of wins and defeats is seen to be a device as they must be different. This enables the Pigeonhole Principle to be employed as the number of positive or negative differences must then be at least half.

Solution 2

There are ${}^{28}C_2 = 378$ games in all, at least 284 of which end in draws. Thus there are at most 94 victories, and at most 94 defeats. For each team, compute the difference between its number of victories and its number of defeats. In order for all teams to have distinct scores, these 28 differences must be distinct. At most one of them can be 0.

Of the remaining 27, the Pigeonhole Principle dictates that either at least 14 are positive or at least 14 are negative. By symmetry, we may assume the former is the case. The number of victories of these teams must be at least $1 + 2 + \cdots + 14 = 105$. Since there at most 94 victories in all, this situation is impossible. Hence at least two teams finish with the same score.

3. Discrete Optimization

This is a method of proof with two parts, one showing an upper or lower bound, and the display of an example satisfying this bound. Students in most countries will not normally learn this method in the classroom, but it is a nice structural idea which can be easily explained to an able student.

This method is often needed in solving Olympiad problems, but the Australian experience shows it can well be used in inclusive classroom competitions. The following problem was set in the 2007 AMC and 17% of the entrants successfully solved it.

Problem 3

A $1 \times 1 \times 1$ cube is cut out of a $10 \times 10 \times 10$ cube. Then a $2 \times 2 \times 2$ cube is cut from the remainder followed by a $3 \times 3 \times 3$ cube and so on. What is the largest cube which can be cut out?

Note: Such a solution will need to prove optimality, and also find an example to show existence of the optimal. Quite often optimality is the more difficult part of the proof, but here, optimality is fairly obvious and to tidy up it is a little work to show how it can be done.

Solution 3

Part 1: Optimality

We cannot cut out a $6 \times 6 \times 6$ cube and a $5 \times 5 \times 5$ cube at the same time, as there would have to be overlap on at least one layer, as $5 + 6 > 10$, the number of layers in the large cube.

Part 2: Existence

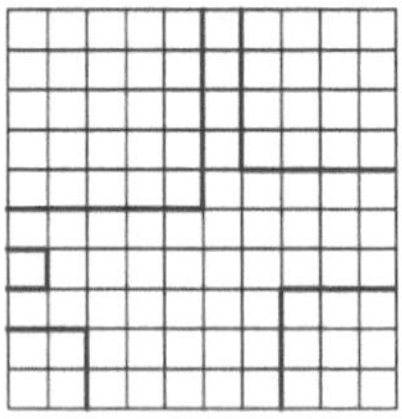

The end on view shows how cubes $1 \times 1 \times 1$, $2 \times 2 \times 2$, $3 \times 3 \times 3$, $4 \times 4 \times 4$ and $5 \times 5 \times 5$ can be cut from the larger cube, so the largest possible is $5 \times 5 \times 5$.

Note: This shows a $5 \times 5 \times 5$ is *possible*, but it does depend on the smaller ones being taken out in certain ways, such as the one shown.

4. Proof by Cases

This is an exhaustive method of proof involving breaking down a situation to mutually exclusive cases, where separate proofs apply within each case. The mutually exclusive cases must include each possible outcome. The following is an example set in the AMC in 2010, which 6% of the entrants were successful.

Problem 4
An *ascending* number is one in which each successive digit is greater than the one before. A *descending* number is one in which each digit is less than the one before.

Find the 3-digit descending number which is the square of an ascending number.

Note: In this solution we resort to finding each case of a possible 3-digit descending number which can be a square and identifying the one from these cases which is the square of an ascending number.

Solution 4
A 3-digit descending number will be at least 210 and at most 987, so it will be the square of an ascending number n which is at least 15 and at most 31. Also n^2 cannot end in 9, so n cannot end in 3 or 7. So, the possible values for n are:

n	15	16	18	19	24	25	26	28	29
n^2	225	256	324	361	576	625	676	784	841

We note that $29^2 = 841$ is the only three-digit descending number which is the square of an ascending number.

5. Proof by Contradiction
This is a well-established method of proof using logical reasoning used in several popular proofs, such as that the square root of 2 is irrational, and that the number of primes is not finite. School students are familiar with this method in some countries.

I illustrate the method with a problem set in the very first Tournament of Towns, in 1980, by Agnis Andjans, of Riga, and solution by Jordan Tabov, of Sofia. The solution does require some graph theory, which will be discussed later.

Problem 5
In an $N \times N$ array of numbers, all rows are different (two rows are said to be different even if they differ only in one entry). Prove that there is a column which can be deleted in such a way that the resulting rows will still be different.

Note: Whereas there are other methods of solution, contradiction presents as an obvious option, by assuming the result is not possible.

Solution 5

Suppose the contrary, i.e. that if we delete any column of the given array then among the resulting rows there are two identical rows.

Consider a graph G, whose vertices are the rows of the given array, and whose edges are determined by the columns of the array, each column determining exactly one edge according to the following rule: after deleting any column c_i ($i = 1, 2, \ldots, n$), there is at least one pair of equal resulting rows; choose one of these pairs, and connect the rows in this pair by an edge; the edge so obtained corresponds to c_i.

Note that an edge cannot correspond to different columns c_i and c_j, because otherwise the rows connected by this edge must be equal, which is not true.

So we obtain a graph G with n vertices and n edges. But every such graph contains a cycle, i.e. a sequence of vertices r_1, r_2, …, r_m, r_1 (pairwise distinct), each two consecutive of which are connected by an edge.

Let c_i be the column corresponding to the edge $r_i r_{i+1}$, $i = 1, 2, \ldots, m - 1$, and let c_m be the column corresponding to the edge $r_m r_1$. Then r_1 and r_2 differ only in the number lying in c_1, and hence their numbers in c_m are equal. Similarly the numbers of r_2 and r_3 in c_m are equal, …, the numbers of r_{m-1} and r_m in c_m are equal. Together with the fact that $r_m r_1$ corresponds to c_m this leads to the conclusion that r_1 and r_m are equal, which contradicts the given property of the array.

Consequently our assumption is false, i.e. there is such a column that after deleting it from the array all the resulting rows are still different.

Note: This is a method of proof and can't be tested properly in an inclusive competition with multiple choices, but the method is accessible to students showing talent in the classroom.

6. Counting by Exhaustion

Counting problems are commonly used in competitions, especially if the number of cases is small enough to not make the process too laborious as to lose the structure of the situation. They are not normally discussed in the classroom, but can be used in inclusive multiple choice problems, as they do not normally require much formal instruction. Rather, they can test a student's intuitive ability to organize and find a systematic pathway. Certainly the Australian Mathematics Competition has a history of setting problems of this type, with reasonable success rates. The following is an example.

Problem 6

In the school band, five children each own their own trumpet. In how many ways can exactly three of the children take home the wrong trumpet, while the other two take home the right trumpet?

Solution 6
Suppose the students taking home the wrong trumpet are called *A*, *B* and *C*. These can take the wrong trumpets in *two* ways, e.g. *A* takes trumpet *B*, *B* takes trumpet *C* and *C* takes trumpet *A*, or *A* takes trumpet *C*, *B* takes trumpet *A* and *C* takes trumpet *B*.

We need also to know how many ways *A*, *B* and *C* can be chosen from the five. This is the same as the number of ways in which the two with the right trumpets can be chosen, this being ten (e.g. if the students are called *A*, *B*, *C*, *D* and *E* these are):

$$\begin{aligned}&A \text{ and } B, A \text{ and } C, A \text{ and } D, A \text{ and } E,\\ &B \text{ and } C, B \text{ and } D, B \text{ and } E,\\ &C \text{ and } D, C \text{ and } E, \text{ and}\\ &D \text{ and } E,\end{aligned}$$

or a student familiar with combinatorics would know the number of ways of choosing 2 objects from 5 is

$$\binom{5}{2} = \frac{5!}{2!(5-2)!} = \frac{5!}{2!(3)!} = \frac{120}{(2)(6)} = 10.$$

Thus the answer is $10 \times 2 = 20$.

Notes: There are various types of systematic counting. One in which we try to count the number of ways in which every member of a set is placed in the wrong order is called a derangement problem. In the above problem only some of the members are in the wrong order, so it is called a partial derangement problem. There are many other types of problem in which systematic counting will work when the numbers are small enough to be manageable. I explore this further in the next discussion.

7. Counting Systematically
The above problem was of manageable size. But could we solve the problem if there were 100 students and trumpets and 79 exactly took home the wrong trumpet? Clearly we get quickly to a complicated situation. It would be convenient if we could explore for patterns and find a formula. Well, a full derangement formula can be found via the inclusion-exclusion (see Niven 1965), and associated with this there is a formula for solving partial derangement problems as above.

The full derangement formula is, if $D(n)$ is the number of ways a set of n numbers can all be placed in the wrong position,

$$D(n) = n!\left(1 - \frac{1}{1!} + \frac{1}{2!} - \frac{1}{3!} + \cdots + (-1)^n \frac{1}{n!}\right).$$

For example, the number of ways of placing 4 numbers in a way that none are in the right position is

$$D(4) = 4!\left(1 - \frac{1}{1!} + \frac{1}{2!} - \frac{1}{3!} + \frac{1}{4!}\right) = 24 - 24 + 12 - 4 + 1 = 9.$$

Alternative Solution 6

The formula for a derangement problem, giving $N(n, r)$, the number of ways of arranging n numbers with r placed in the wrong order, is

$$N(n, r) = \binom{n}{r} \times D(r).$$

In the case of the trumpet problem, n = 5 and r = 3, and so the answer is

$$\binom{5}{3} \times D(3) = \frac{5!}{2!3!} \times 3!\left(1 - \frac{1}{1!} + \frac{1}{2!} - \frac{1}{3!}\right) = \frac{120}{12} \times (9 - 9 + 3 - 1) = 10 \times 2 = 20,$$

as before.

Comment

In fact the inclusion-exclusion principle is a rich way of generalizing counting problems to large numbers. There are other cases of counting generalizations we have used in the Australian Mathematics Competition is the necklace formula of George Pólya, where the number of ways of making a necklace with n beads of k colors is $k - (k^n - k)/n$.

8. Inverse Thinking

Sometimes in dynamical problems, where the state of a system changes step by step to a new state, it can become simpler to identify the final state and identify the change mechanism in reverse.

The following problem, taken from the Mathematics Challenge for Young Australians, illustrates this.

Problem 8

A Fibonacci sequence is one in which each term is the sum of the two preceding terms. The first two terms can be any positive integers. An example of a Fibonacci sequence is 15, 11, 26, 37, 63, 100, 163, …. We will assume that terms must be positive.

1. Find a Fibonacci sequence which has 2000 as its fifth term.
2. Find a Fibonacci sequence which has 2000 as its eighth term.
3. Find the greatest value of n such that 2000 is the nth term of a Fibonacci sequence.

Notes: This problem clearly invites inverse thinking, that is starting with the last term in the sequence and finding terms which can lead to it. The term before can be anything, but once that has been chosen the preceding ones are fixed. So all depends on the choice of the term before 2000.

Solution 8

Part 1. One can easily find second last terms which lead backwards to a negative term rather quickly. But it is not too difficult to find one, such as choosing 1250 as the fourth term, leading to 250, 500, 750, 1250, 2000.
Part 2. It is now a much tighter choice for the fourth term. Systematic trialing will show that only the choice of a number between 1231 and 1249 will yield a sequence with 8 positive terms.
Part 3. The only way of getting 10 terms is the following (showing sequences in reverse).

$$2000,\ 1236,\ 764,\ 472,\ 292,\ 180,\ 112,\ 68,\ 44,\ 24$$
$$2000,\ 1237,\ 763,\ 474,\ 289,\ 185,\ 104,\ 81,\ 23,\ 58$$
$$2000,\ 1238,\ 762,\ 476,\ 286,\ 190,\ 96,\ 94,\ 2,\ 92$$

Note that the last two cannot be extended as the next term would be negative. But the first one can be extended 20, 4, 16 before going negative. So the answer is $n = 13$.

Note: This problem had an added bonus in that it can lead to the golden ratio. This can be found by looking at the solution of the recurrence relation $a_{n+2} = a_{n+1} + a_n$.

9. Invariance

Sometimes in a dynamical system where the state changes step by step, a useful insight into the problem can be found by looking for a condition which does not change as the system progresses. An example might be noting that parity of a number, or remainder when divided by a given number, might be unchanged.

The following well-known problem typifies the idea in this method.

Problem 9

On the island of Camelot live 13 grey, 15 brown and 17 crimson chameleons. If two chameleons of different colors meet, they both simultaneously change color to the third color (e.g. if a grey and brown chameleon meet they both become crimson). Is it possible they will all eventually be the same color?

Comment
When I tried to solve this problem I experimented working forwards from the beginning data but made little progress. Then I tried inverse thinking, starting with a successful end point, and this helped me finally to identify the pattern leading to the invariant.

Solution 9
The starting configuration has populations 0, 1, and 2 modulo 3. This situation remains invariant no matter which two chameleons touch noses. So the desired configuration, which has all three equal to zero modulo 3 is not possible.

10. Coloring
Sometimes, such as in a tiling problem, an insight can be obtained by coloring tiles according to a set of rules (some might see it as a similar thing to finding a construction line in geometry). A very simple example of such a problem is the following well-known one.

Problem 10
An 8 by 8 checkerboard has it top-left and bottom-right squares removed. Can 31 dominoes (1 by 2) can be placed over the remaining 62 squares?

Comment
One can try this problem for some time trying to get dominoes to fit and not succeed nor necessarily get a reason for why not. But coloring provides a very simple explanation.

Solution 10
Each domino necessarily has one of each color in the normal checkerboard coloring scheme. However the two squares removed are of the same color, leaving an imbalance.

Further Comment
This example provides a coloring with two colors on a chessboard, but in fact the idea can be extended to more colors in richer situations.

11. Geometry
Geometry is an important part of the syllabus, not because of an expectation that various theorems will really be used by the average person in later life, but because it provides accessible methods for developing systematic thought processes. It is often unpopular with teachers who are not

well-trained, and hence it has declined in many syllabi, but once getting past a certain point, good students can be trained in it.

Geometry is a very common component of competitions, and in most it is expected a student has a reasonable knowledge of line and circle geometry. It provides problems for instance which are successfully solved by students at International Mathematical Olympiads. The following problem, composed by Bob Bryce, of the Australian National University, was set in the Australian Mathematics Competition, and shows how the line and circle geometry can be used to solve useful and practical problems.

Problem 11

The latitude of Canberra is 35°19′S. At its highest point in the sky when viewed from Canberra the lowest star in the Southern Cross is 62°20′ above the southern horizon. It can be assumed that rays of light from this star to any point on the earth are parallel.

What is northernmost latitude at which the complete Southern Cross can be seen?

Comment

The solution follows fairly readily after drawing a diagram and recognizing a cyclic quadrilateral.

Solution 11

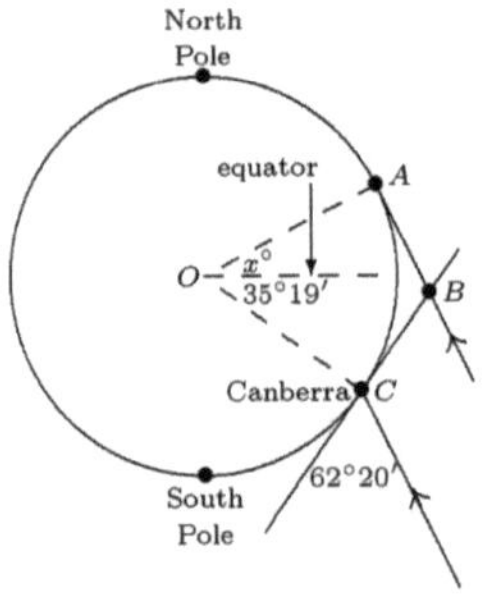

The northernmost point is where a ray of light from the star is tangential to the earth's surface, i.e. point A in the diagram. The object is to determine the value of x in the diagram.

Consider the quadrilateral $OABC$ in the diagram.

Because the angles at A and C are 90°, the remaining two angles must add to 180°.

Therefore $x° + 35°19' + \angle ABC = 180°$.

Because the rays reaching points A and C are parallel it is clear that $\angle ABC = 180° - 62°20'$. Therefore $x° + 35°19' + 180° - 62°20' = 180°$, giving $x° = 62°20' - 35°19' = 27°01'$.

Comment

It is worth noting that this means the Southern Cross can be seen from just over 27° north and this was used by French navigators pioneering air routes over the Sahara desert to South America.

12. Graphical Methods

When an algebraic problem looks messy to solve as such, the problem solver can often gain great insight by attempting to sketch components in a problem. Sometimes a function with several terms can have its graph built up in stages by graphing its various components and combining them. This is well illustrated by the following problem.

Problem 12

Find all the solutions of $x^2 - [x] - 2 = 0$.

Comment

Initially algebraic methods look complex, while even drawing the graph of the function on the left hand side is difficult. However the solution follows more quickly if we note that the equation is equivalent to $x^2 - 2 = [x]$ and draw the graphs of the functions on each side, looking for where they cross.

Solution 12

We have the following diagram.

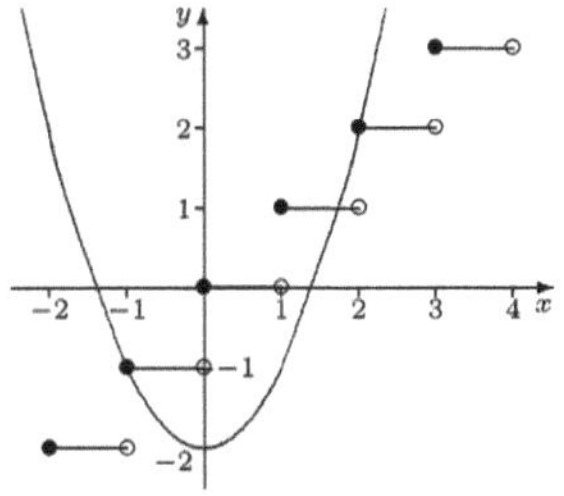

This leads to a clear discovery that there are three solutions, two of which are integer, $(-1, -1)$ and $(2, 2)$ and the one where $y = 1$ and it turns out using symbolic calculation that this root is at $(\sqrt{3}, 1)$.

13. Probability

Probability is a very important, and neat branch of mathematics, but our experience is that great care needs to be taken in formulating the problem statement, as such problems can easily be worded in a way to lead to double meaning, especially if there is a Bayesian context. Many good problems can be formulated without broaching the Bayesian and solutions often require skills overlapping with other methods, such as finding mutually exclusive cases and counting methods.

A good example of an AMC question with Bayesian flavor, which was set in multi-choice mode, is the following.

Problem 13

A deck of 16 cards contains the four aces, four kings, four queens and four jacks. The 16 cards are thoroughly shuffled and my opponent (who always tells the truth) draws two cards simultaneously and at random from the deck. I ask him if he has drawn at least one ace. He looks at the cards simultaneously and replies ‘yes’. What is the chance that he holds two aces in his hand?

Comments

This was a difficult question, beyond the student with average classroom experience, and set as a challenge to the talented student. The number of cards, and outcomes, is too large to count the outcomes exhaustively, so a structural solution is needed.

Solution 13

Let C_1 be card 1 (drawn first) and C_2 be card 2 (drawn second).

Then the total number N_1 of pairs which include exactly one ace equals the number of pairs for which C_1 is one of the four aces and C_2 is one of the 12 non-aces or C_1 is one of the 12 non-aces and C_2 is one of the four aces. This equals $4 \times 12 + 12 \times 4 = 96$.

The total number N_2 of pairs for which both cards are aces equals $4 \times 3 = 12$.

Therefore the probability of 2 aces given that at least one is an ace is the number N_2 of pairs with two aces divided by the number $N_1 + N_2$ of pairs with at least one ace, i.e.

$$\frac{12}{96+12} = \frac{1}{9}$$

or 1 in 9.

14. Logic

Inclusive competitions like the Australian Mathematics Competition often pose problems which require finding strings of logical implications from given statements to finding a broader one. They can be rather intuitive. An example of one which was correctly answered by 55% of Year 8 students is

Problem 14

If, in the Republic of Utopia, the statement ‘all citizens have two legs’ is false, which of the following statements must be true?

I. All citizens have more than two legs.
II. Not all citizens have two legs.
III. No citizen has two legs.
IV. There are some citizens with more than two legs.

Solution 14

It is self-evident that Statements I, III and IV need not be true.

15. Graph Theory

This is a popular subject, but does need some instruction. However students can be quickly led into some rather powerful concepts, such as those Euler discovered when solving the Königsberg bridges problem. The following problem, based on Muscovite folklore, is a nice example from the Tournament of Towns, set in 1984.

Problem 15

A village is constructed in the form of a square, consisting of 9 blocks, each of side length l, in a 3×3 formation. Each block is bounded by a bitumen road.

If we commence at a corner of the village, what is the smallest distance we must travel along bitumen roads, if we are to pass along each section of bitumen road at least once and finish at the same corner?

Comment

Whereas the solution relies on Euler's discovery, the following proof, by Andy Liu, of Edmonton, is also a classic example of a discrete optimization proof, with an optimization component and an existence component.

Solution 15

(Existence) The diagram shows a closed tour of length 28 and we claim this to be a minimum.

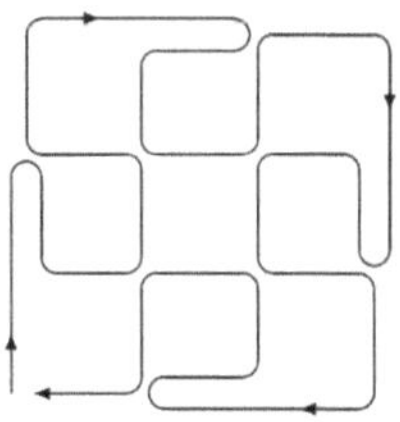

(Minimality) Each of the four corners is incident with two roads and requires at least one visit. Each of the remaining twelve intersections is incident with three or four roads and requires at least two visits.

Hence the minimum is at least $4 + 12 \times 2 = 28$.

16. Mechanics

This is important, but not popular. Also, of the various methods referred to here, we found it to be the one topic in which boys do significantly better than girls. Gender issues are discussed in more detail below.

17. Cutting Edge

Contrary perhaps to some perceptions, problem solving is an evolving branch of mathematics. Sometimes long-forgotten methods re-emerge. The most remarkable of these is the re-emergence, in Western mathematics of the use of Barycentric Coordinates, developed by Möbius, which can simplify solution of some geometry problems such as proving collinearity.

Also, new problem solving methods and methods of teaching problem solving, evolve, and the process of solving or composing problems can lead to new discoveries in mathematics. I note that in Olympiad training the emergence of Muirhead's Inequality has streamlined the required knowledge, and members for example of the Problems Committee of the Mathematics Challenge for Young Australians have written a number of refereed papers as a result of discoveries during problem composition.

New methods can especially evolve via national school problem-solving journals such as *Kvant* in Russia and *KöMal* in Hungary, sometimes taking advantage of research discoveries in mathematics.

To the author the nicest such development in recent years was that of the method of moving parallels, which enables the solution of certain problems when tessellating within polygons. It is illustrated here by the following problem from the International Mathematics Tournament of Towns. It was used in 1983, composed by VV Proizvolov of Moscow, and the solution is by Andy Liu, of Edmonton.

Problem 17

A regular $4k$-gon is cut into parallelograms.

(a) Prove that among these there are at least k rectangles.
(b) Find the total area of the rectangles in (a) if the lengths of the sides of the $4k$-gon equal a.

Comment

This is a highly unconventional type of geometry problem. Students with classroom geometry tools might try some of them but traditional Euclidean methods are not really available. The solution below illustrates how moving parallels inside the polygons can be used.

Solution 17

(a) Let the regular $4k$-gon be dissected into parallelograms.
Let x_1 and x_2 be a pair of opposite sides. The set of all parallelograms with one side parallel to x_1, starts from x_1 and eventually reaches x_2, possibly subdividing into several streams. The diagram illustrates the case of a regular octagon.

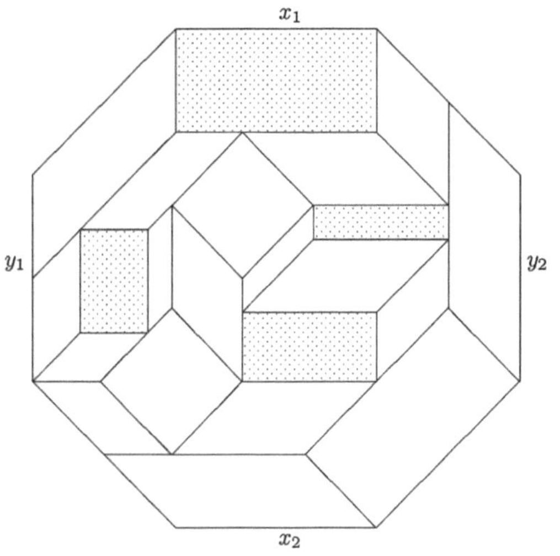

Since the regular polygon has $4k$ sides, there is a pair of opposite sides y_1 and y_2 perpendicular to x_1 and x_2. The set of parallelograms with one side parallel to y_1 starts from y_1 and eventually reaches y_2, again possibly subdividing into several streams.
Now these two sets of parallelograms must cross each other.
This is only possible at parallelograms with one pair of opposite sides parallel to x_1 and the other to y_1.
Since x_1 and y_1 are perpendicular, this parallelogram is actually a rectangle (due to subdividing into several streams, four such rectangles based on x_1, x_2, y_1, y_2 in the diagram exist and are shaded).
In the regular $4k$-gon, there are k sets of mutually perpendicular pairs of opposite sides. Hence there must be at least k rectangles in the dissection.
Note that in the diagram we can also identify a rectangle (in fact three exist) based on the two other pairs of opposite sides.

(b) Since the sides of the $4k$-gon are all of length a, the width of each set of parallelograms in (a), in the direction of the side of the $4k$-gon defining the set, is equal to a. It follows that the sum of the areas of all rectangles in the set is a^2.

It follows that the total area of the rectangles is ka^2.

12.2.2 Research Directions

The above journey through a number of problem solving methods and topics is intended to ask research questions and answer them when possible. Before doing this I make one general observation, and that is that these methods increase the student's ability to think structurally and have nothing to do with dealing with calculation intensity. I also believe that proficiency with these methods which can be implemented beyond the classroom better equip the student for their experiences in later life. I feel there is a great role for researchers, most likely through the WFNMC as the relevant international professional society in developing this list and studying the effects on students, both by analyzing results in competitions, but also interviewing or surveying students not only in the wake of their competition experiences, but on reflection later in life.

The more such material can be developed, the more it will help future people design better competitions and better problems. But it should not imply there is only one way of doing things. On the contrary, it should be in an innovative setting, with the understanding that mathematics is not a dead and immovable object, but in a discipline in which new things are discovered. This purpose is reflected in "17. Cutting edge" where examples from Tournament of Towns and corresponding discussions in student Journals such as *Kvant* have seen problem solving methods evolve.

There are a number of other similar areas where WFNMC should build up more formal knowledge about the effect of different types of problems on different categories of students. One of the most recent papers analyzing these sort of issues is Leder and Taylor (2010). This contains some gender and some topic analysis.

12.2.3 Gender Issues

It is a well-documented fact that boys tend to outnumber girls in IMO teams and competition medal lists. This tends to happen despite girls being equal in number in the initial entry group. It is true that our previous studies showed slight (not statistically significant, normally) average scores in favor of boys, we did an analysis of standard deviations, and discovered boys' standard deviations to be consistently higher than that for girls across the board.

In this paper we also looked at gender scores by topic, including that at primary schools. This study is of scores in the Australian Mathematics Competition for the years 2004–2008. We categorized questions by arithmetic, algebra and geometry, each subdivided into basic (single-step) and advanced (multi-step). We also looked at problem solving, where a situation was described in words, both in familiar and unexpected situations. With geometry we also looked at problems with and without diagrams provided and those in 2 dimensions and others in 3 dimensions. We also looked at special other categories such as mechanics, enumeration and ratio. The data involves several hundreds of thousands of students.

We found that at the secondary levels boys were ahead in all categories but in all except mechanics not significantly ahead. With mechanics we found the results for boys were significantly better than for girls.

The results were similar in most topics at primary schools, but girls narrowly outscored boys in primary schools in most geometric topics, probably against conventional wisdom.

12.2.4 Risk-Taking Patterns

A further area in which competitions have helped provide useful information is in measuring risk taking. This can, and has been done in situations where a penalty is applied for incorrect response. The Australian Mathematics Competition applied such penalties until 2002, under the assumption that it would discourage guessing and encourage responding only when the student believed they had solved a problem. Atkins et al. (1991) used the Ziller statistic and other measures devised by the authors to analyze risk taking in the 1988 and 1989 AMC papers. They did deduce that risk taking declined as students grew older, but left an open result on gender, contrary to popular belief that boys take more risks than girls. If anything girls were more likely to take more risks when younger, and boys more likely when older.

12.3 Alumni

I have found it very useful to have statistics of what careers competition winners take up in later life, but also to have their own reflections on how their competition experience influenced their decisions. Olympiad programs are intense and expensive, and often involve travel and living away from home expenses. Many families cannot afford this, especially when a talented student is likely to be attending two long camps away from home for domestic training, and then international airfares. In Australia we particularly have felt this, because of distances within our own country, and our distance from most other countries.

So it was important for us to obtain government funding, to enable participation by all selected students and not just those from wealthy families. Whereas we had a lot of information for the last 3-year funding cycles while I was Director of the Australian Mathematics Trust, in order to justify funding the Australian Government provided extra funds for significant surveys, the first just on alumni, the second a major external review which was conducted by international accounting firm PricewaterhouseCoopers. No one should fear such reviews. Alumni enjoy their experience and there is no doubt all such experience will improve their skill and knowledge base, and their confidence.

12.3.1 National Competition Winners Alumni—Australia

I will come to them but first I look at our own follow-ups on alumni from our national competition. Many papers have been written on the medalists from the Australian Mathematics Competition (AMC), which started in 1978 and attracts several hundred thousands of entries each year, mostly from Australia, but also from about 40 countries, mainly in the Asia Pacific region and is now held under the auspices of the Australian Mathematics Trust. In Australia, and in some other countries, this event, which is easy to organize in schools, and involves multi-choice questions, is used as a first identifier of students who might enter invitational Olympiad training.

We have conducted research analyzing many features of student activity in the AMC, including gender issues, risk taking behavior and student interests and ambitions. The key one which reflects on the students as alumni would be Leder (2011), in which we conducted a large scale survey of former medalists. In the period between 1978 and 2006 there were 420 different students who received medals, the top award. Of the 420 letters sent out 52 were returned as undeliverable, leaving 368 possible respondents. In all we received 90 useable responses, which at over 25% of the possible population we felt was good. A small subset of these who were mathematicians and indicated willingness to be interviewed further were followed up with more in-depth questioning.

About 40% of the respondents said they were in professions which used their mathematics. Responses to questions such as how they felt about the experience was overwhelmingly favorable, with comments such as

- *A source of pride—we were immensely competitive in a good-natured way at school and there were 3 or 4 students in my year who won AMC medals in various years. We still get together every year to do the...AMC competition paper over dinner (our 15th year this year).*
- *Selection into the Mathematical Olympiad training programme, with many flow on benefits, including: learning much more mathematics and at a higher level, meet like-minded people many of whom are now good friends, encouragement to continue with mathematics.*
- *I think (it) got me an invitation to participate in the Tournament of the Towns—which in turn meant regular exposure to (a) more challenging mathematics, and (b) other extremely talented students. I gained a great deal from this program. At school, I got somewhat embarrassed by the fuss and teacher pride, but on the other hand my teachers were happy to let me do other things in class once I finished class work.*

These types of response definitely indicate that students felt their experience in the competition added value to their educational profiles. I do note that the students discussed above are the elite. My own personal interest though goes to the wider class of average standard student who we were unable to survey. I accept that below-average students will not enjoy a competition experience, but I would hope that average students can, and that some discover talent that the classroom was unable to detect.

The Leder (2011) paper was also one of a number in which we built a profile of the gifted student. When winning a medal students complete a survey which canvases their other interests and their aspirations. We found that there was a very high correlation between AMC medalists and students competent with a musical interest, and also with sport (some appearing to hold sporting records).

12.3.2 Olympiad Alumni—Australia

I now move on to our former Olympians. The best record I have of Australian IMO alumni from 1981 to 2012 is at Taylor (2013). I just consider those to 2007, as the later ones have not had an opportunity to establish a career. Some are difficult to classify as the have changed careers in mid-stream or they have worked over more than one discipline at the same time.

Altogether there are 118 different students who have participated at IMO for Australia. Of these we are not aware of what happened to 5 of them, mainly from the earliest three years. But 99 of them undoubtedly moved into seriously good careers, involving mainly mathematics, but sometimes informatics, physics, a career in finance or technology specially using their mathematics knowledge, or in a couple of cases full professors of philosophy at prestige universities. A further 12 moved into medicine or law, although in some cases they combined the professions with some form of quantitative work (mathematical biology, still training IMO students etc.). Just 2 are known to have moved out of mathematics for most of their careers, but both had mathematical components in their earlier careers.

Some might say this is not surprising, and that such students would expect to go on to mathematics anyway. But in the follow up studies it becomes apparent they were not headed to mathematics until the out of school opportunities arose, and always when asked they are emphatic that their career choice was positively influenced by their Olympiad experience. Example comments from our 2006 (government funded) survey:

- *Significantly. It was one of the influences that eventually resulted in my pursuing a Pure Mathematics major in my science degree.*
- *I was probably always going to do Maths & Computer science, but the amount I learnt at the training camps meant I was able to accelerate my subject choices.*
- *It showed me how there is much more intellectual depth to Maths than what is taught at school.*

12.3.3 Olympiad Alumni—Germany

Another country which has a similar alumni data base is Germany. Germany has published a book Engel et al. (2009) which lists all its IMO alumni, in a similar way to Australia, from the inception of IMO in 1959–2008. If later activity is known it says whether or not the alumnus studied further mathematics at University, whether or not this led to a doctorate, it will state later employment if known also. I did not count the ones after 2002 because then it is not reasonable to expect a doctorate before 2008 where the data ceases. It seems that the German (East and West, and unified) programs have been a bigger pathway to a doctorate in mathematics or related subject in physical sciences or engineering or philosophy than Australia.

I counted 158 East German alumni, but a large number (30) where unknown afterwards, these being in particular from earlier years. Of the 128 known students no less than 93 went on to receive doctorates in mathematics or closely related disciplines, and a further 28 studied for a mathematics or related degree and then had a degree based on that, such as in the teaching or finance professions. Of the remaining seven, 5 had worthy professions such as medicine or other non-mathematical areas.

It is similar for those in West German or later unified German teams. Of these there are 95 alumni (West Germany started at IMO much later), of whom 15 later careers are unknown. Of the known 80 alumni, 55 achieved mathematics-related doctorates, 22 obtained mathematics related degrees and went on to appropriate later careers such as teaching or finance, and all of the remaining 3 went into respectable professions such as medicine.

One observes from both the Australian and German experiences that IMO training has at least held these students in high level mathematics and if the attitudinal Australian surveys are indeed reflective, the experience of IMO training and involvement influenced these career choices. I am sure these data can be extrapolated to show that IMO programs are good for a country's economy. In fact in the PricewaterhouseCoopers report for the study commissioned by the Australian Government referred to above, which is confidential, it is assumed that building up a higher level knowledge base in mathematics is good for a country's economy.

12.4 Concluding Remarks

I have reflected on the achievements and current status and operations of WFNMC and outlined two major areas where large scale systematic research can benefit the quality of competitions, enhance their image and justify support from government and other sponsors. Building up a large body of knowledge on the effects of various types of problems on various classes of students, and development of a similar body of knowledge on the effects of competition experience on later careers of students can lead to a better understanding on why we run competitions and how we might do it better.

References

Atkins, W. J., Leder, G. C., O'Halloran, P. J., Pollard, G. H., & Taylor, P. J. (1991). Measuring risk taking. *Education Studies in Mathematics, 22*, 297–308.

Barbeau, E. J., & Taylor, P. J. (2009). *Challenging mathematics in and beyond the classroom: The 16th ICMI study*. New York: Springer.

Engel, W., Gronau, H.-D., Langmann, H.-H., & Sewerin, H. (2009). *The German teams at the international mathematical Olympiads 1959–2008*. Bad Honnef: KH Bock.

Kenderov, P. (2009). Competitions and mathematical education. *ICM Bulletin and WFNMC*. http://www.wfnmc.org/history.pdf.

Leder, G. (2011). Mathematics taught me Einstein's old cocktail of inspiration and perspiration: Mathematically talented teenagers as adults. *Canadian Journal of Mathematics, Science and Technology Education, 11*(1), 29–38.

Leder, G. C., & Taylor, P. J. (2010). Are Raelene, Marjorie and Betty still in the race? *The Australian Mathematics Teacher. 66*(2), 17–24.

Niven, I. (1965). *Mathematics of choice—How to count without counting*. Mathematical Association of America.

Taylor, P. (2013). *IMO students*. https://sites.google.com/site/pjt154/home/a2-imo-students.

Taylor, P. (2014). *Mathematics content*. https://sites.google.com/site/pjt154/mathematics-content.

Taylor, P. (2015). Classifying methods of problem solving—And my favourites. *Mathematics Competitions, 28*(1), 7–27.

Part VII
A Bridge Between Competitions and 'Real' Mathematics

Chapter 13
Are Mathematics Competitions Changing the Mathematics that Is Being Done and the Way Mathematics Is Done?

María Falk de Losada

Abstract The relation between mathematics competitions, as a branch of mathematics education, and mathematics is examined with the aim of establishing a particularly strong influence of competitions over mathematics itself. It is argued that mathematics competitions have generated a school that spread from Hungary all over the world and changed the face of mathematics, especially in the last half of the twentieth century, and that competitions such as the IMO have helped to form mathematicians who have rethought mathematics itself.

Keywords Mathematical competitions · Problem solving · Contemporary mathematics · Hungarian school · IMO-International Mathematical Olympiad · Epistemology · Methodology

13.1 Introduction

Parallel to our 35 years of experience with mathematical olympiads, at the national, Iberoamerican and international levels we have witnessed changes to mathematics itself. In this paper we wish to explore the question of whether or not mathematics education can have an impact on mathematics. The particular branch of mathematics education that we will examine is that related to mathematics competitions, competitions of solving original and challenging mathematical problems.

M.F. de Losada (✉)
Colombian Mathematics Olympiads, Universidad Antonio Nariño, Carrera 38 #58 A - 77, Bogotá, D.C, Colombia
e-mail: mariadelosada@gmail.com

A. Soifer (ed.), *Competitions for Young Mathematicians*,
ICME-13 Monographs, DOI 10.1007/978-3-319-56585-9_13

Mathematical problem solving competitions are not new. For example, in the twelfth century on the occasion of the visit of the Holy Roman Emperor to Sicily, the local Norman king launched a competition to solve a particular problem, a cubic equation, as part of the celebration. Leonardo de Pisa, also known as Fibonacci, took part in the competition. The equation (in modern notation) was

$$x^3 + 2x^2 + 10x = 20$$

Now cubic equations have been solved at least since the classic era of Greek mathematics, and the original methods involved finding the intersection of two conic sections. Similar methods were employed by Arab mathematicians in the tenth and eleventh centuries. And although Lconardo learned his mathematics from Arab scholars, his approach to this particular problem was novel. First, he factored 10 to obtain

$$10\,(x + 1/10\,x^3 + 1/5\,x^2) = 20 \quad \text{or} \quad x + 1/10\,x^3 + 1/5\,x^2 = 2.$$

Since mathematics only worked with positive numbers at the time, this last equation implies that $x < 2$. But if we substitute $x = 1$ in the original equation, we obtain $1 + 2 + 10 = 13 < 20$, so that $x > 1$. Leonardo observes that the equation must have a root between 1 and 2, a thoroughly modern observation in which notions of continuity and the intermediate value theorem are implicit. He then proves that the root cannot be rational, and that it does not belong to any of Euclid's categories for numbers constructible by ruler and compass. Leonardo then goes on to approximate the value of x using sexagesimal fractions as $x \approx 1°22'7''42'''33''''4'''''40''''''$.

The implication is that in algebra it is possible to encounter numbers that are not encountered geometrically (de Manrique and de Losada 1997). Or to put it more broadly, the Greek's identification of positive real numbers with magnitudes of segments is correct, but the concomitant supposition that these segments can be constructed with ruler and compass is incorrect. Leonardo published his results in his *Flos*.

The impact of Leonardo's work on the solution of this competition problem was felt for centuries. It includes the impulse given to work in algebra and solution of the cubic and quartic equations by the great Italian algebraists of the sixteenth century, the shift toward an algebraic foundation for mathematics as opposed to a geometric one, the notion of algebra as a source for numbers that are not geometrically constructible, the consideration of the idea of a polynomial function and its continuity, the representation of functions by power series ("infinite polynomials"), among many other features of renaissance and early modern mathematics. This shift occurred over centuries and, clearly, was not at once noticeable.

Coming closer to our topic in today's world, what was the relationship of mathematics to mathematics education at the time mathematical problem solving competitions reappeared in the second half of the nineteenth century in the setting of school mathematics? We believe that the most relevant feature was the increasing specialization in mathematical research, the necessity of studying a specialized branch of mathematics deeply in order to do research and solve original mathematical problems. In other words, mathematics itself had gone beyond the scope of what both *aficionados* from the general public and students could understand and work on. We believe this is what led both mathematicians and mathematics teachers to look for interesting, genuine and challenging problems that young minds could understand and solve, and become involved, sometimes passionately involved, in mathematics.

What we wish to argue in this paper is that, in a similar way to the case of Leonardo de Pisa, mathematical problem solving competitions are changing the mathematics that is being done and the way mathematics is being done, and we anticipate that their impact will extend into the future.

13.2 A Word About the Hungarian School

Mathematical problem solving competitions, as a branch of mathematics education, has a feature that distinguishes the work being done from every other initiative in the field. And this has its roots in Hungary in the Eötvös and Kürschák competitions and the journal of problems in mathematics and physics, *Középiskolai Matematikai Lapok* or KöMaL. With common roots in these pioneering competitions, a school was formed that produced outstanding figures in mathematics, in methodology and in epistemology.

Beginning with the work and leadership of Lipót Fejér (Leopold Weiss) who grew up solving problems from KöMaL and who placed second in the Eötvös competition of 1897, a school was formed that came to include, in varying degrees, Paul Erdős, George Pólya and Imre Lakatos, the great mathematician and collaborator with mathematicians around the globe, the influential thinker on problem solving and method, and the philosopher–epistemologist who dared to question formalist mathematics proposing an alternative interpretation of the character, origins, structure and justification of mathematical knowledge and its historic evolution. These three stand out among the many great Hungarian mathematicians whose mathematical formation began in or was intimately related to the competitions, especially because they migrated to England and the United States and worked and published in English, thus opening their ideas and results and bringing them to bear on the worldwide community of mathematicians and mathematics educators. In what follows we outline their contributions.

Lipót Fejér (1880–1959), precursor

Once competitions devoted to solving challenging mathematical problems were well established in Hungary, a new school of mathematics began to take form in that country rooted in the competitions. One of the first winners of the Eötvös Competition was Leopold Weiss (Lipót Fejér), and he was destined to become one of the mathematicians who were highly instrumental in forming the new generations of the Hungarian school.

Fejér's attitude towards mathematics changed dramatically in secondary school when he began solving problems from KöMaL and in 1897, the year in which he graduated from secondary school in Pécs, Fejér won second prize in the Eötvös Competition. That same year, Fejér began his studies in the Polytechnic University of Budapest where he studied mathematics and physics until 1902. Among his professors in Budapest were József Kürschak and Lórand Eötvös, whose names are well known to all who have worked in mathematics olympiads.

Pólya said the following of Fejér (Albers and Anderson 1985).

> Why was it that Hungary produced so many mathematicians in our time? Many people have asked this question which, I think, cannot be fully answered. However, there were two factors whose influence on Hungarian mathematics is clear and undeniable, one of these was Leopold Féjer, his work, his personality. The other factor was the combination of a competitive examination in mathematics with a problem solving journal.

In what follows we offer some considerations to answer the question: What is the relationship between Erdős, Pólya and Lakatos and why is it important to the ideas we wish to express?

Paul Erdős (1913–1996), the great mathematician, problem poser and solver

Erdős got his first formation in mathematics from his parents who were mathematics teachers. He won a national competition in problem solving (Jószef Pelikán has informed us that it was not the Eötvös), which allowed him to study mathematics at university. He wrote his thesis under the direction of Fejér. He was awarded a postgraduate scholarship in Manchester and then in Princeton.

Asked if he believed that his mathematical development had been influenced by *Középiskolai Matematikai Lapok (KöMaL)*, Erdős answered (Freud 1993):

> Yes, of course, you really learn to solve problems in KöMaL. And many good mathematicians realize early on that they have mathematical ability.

Asked to what he attributed the great advance in Hungarian mathematics, Erdős said (Albers and Anderson 1985):

> There must have been many reasons. There was a journal for secondary school, and the competitions, that began before Féjer. And once they began, they self perpetuated up to a certain point … But such things probably have more than a single explanation …

Erdős' contributions were great in quantity and importance, and cover a great range of topics. Erdős was primarily a problem solver, not a theory builder. He was attracted principally by problems in combinatorics, graph theory and number theory. For Erdős a proof must provide insight into why the result is valid and not be only a sequence of steps that lead to a formal proof without providing understanding.

Several of the results associated with Erdős had been proved previously by other mathematicians. One of these is the prime number theorem: *the number of primes not exceeding x is asymptotic to x/ln x.*

The theorem had been conjectured in the eighteenth century, Chebyshev got close to a proof, and it was proved in 1896 independently by Hadamard and de la Vallée Poussin using complex analysis. In 1949 Erdős and Atle Selberg found an elementary proof, one which did indeed provide insight into why the theorem was true. The result is typical of the kind of mathematics that Erdős worked on. He proposed and solved problems that were elegant and simple to understand, but very difficult to prove.

In 1952 Erdős received the Cole Award from the American Mathematical Society for his several results in number theory, but in particular for his article *On a new method in elementary number theory which leads to an elementary proof of the prime number theorem* which contained this very proof.

George Pólya (1887–1985), mathematician, educator and methodologist

At university in Budapest, Pólya learned physics with Lórand Eötvös and mathematics with Fejér. Pólya stated (Albers and Anderson 1985).

> Féjer was a great influence on me, as he was on all of the mathematicians of my generation, and, in fact, once or twice I collaborated with Féjer on small things.

Pólya was awarded his doctoral title in mathematics in the academic year 1911–1912 having studied, essentially without supervision, a problem in geometric probability. In the problem-solving book in analysis that he wrote jointly with Gabor Szegö, Pólya had the idea of grouping the problems according to the method of solution used rather than the usual grouping by topics. He explained why he approached mathematics in a way that differed from the usual treatment in the following terms (Albers and Anderson 1985):

> I came to mathematics very late … as I got close to mathematics and began to learn something about it, I thought: Well, this is true, I see, the proof seems conclusive, but how is it that people can find such results? My difficulty in understanding mathematics was how was it discovered?

Although the book of solutions to problems of analysis that he wrote with Szegö was a masterpiece that would make both authors famous, Pólya continued to look for answers to this question, publishing his three well-known works: *How to solve it*, *Mathematics and plausible reasoning* (1954), and *Mathematical discovery* in two volumes (1962, 1965).

Pólya maintained that in order to work on problem solving it is necessary to study heuristics, stating (Pólya 1945) that

> the purpose of heuristics is the study of the rules of discovery and invention… As an adjective heuristic means 'that which allows one to discover' … its purpose is that of discovering the solution to a problem that one is studying…. What is good education? It is systematically giving the student the opportunity to discover for himself.

Speaking in general about teaching, Pólya said (http://www-history.mcs.st-andrews.ac.uk/Biographies/Polya.html):

> Teaching is not a science, it's an art. If it were a science there would be a best way of teaching and everyone would have to teach that way. Since teaching is not a science, there is room to accommodate different personalities…. Let me say what I think teaching is. Perhaps the first point, which is widely accepted, is that teaching should be active, or better that learning should be active…. The main point of teaching mathematics is developing problem solving strategies.

Imre Lakatos (1922–1974), philosopher and epistemologist

In 1953 Lakatos was supporting himself translating mathematical books into Hungarian. One of the books he translated at that time was Pólya's *How to Solve It.*

In the Hungarian Revolution of 1956 and the Soviet repression that followed, Lakatos realized that he was about to be imprisoned and escaped to Vienna, going from there to England where he began a doctoral program in philosophy. The ideas of Popper and Pólya were greatly influential in his work and his thesis *Essays on the Logic of Mathematical Discovery* was completed in 1961. It was following a suggestion of Pólya that the thesis took its theme from the history of Euler's formula $V - E + F = 2$. Lakatos never published his thesis as a book because he intended to improve it. In 1976, after his death (1974) the book was published by J. Worrall and E.G. Zahar (eds.), *I. Lakatos. Proofs and Refutations: The Logic of Mathematical Discovery.*

Worrall (Lakatos 1976) described this work in the following terms.

> The thesis of 'Proofs and Refutations' is that the development of mathematics does not consist (as conventional philosophy of mathematics tells us it does) in the steady accumulation of eternal truths. Mathematics develops, according to Lakatos, in a much more dramatic and exciting way - by a process of conjecture, followed by attempts to 'prove' the conjecture (i.e. to reduce it to other conjectures) followed by criticism via attempts to produce counter-examples both to the conjectured theorem and to the various steps in the proof.

Of high importance to the present analysis with regard to *Proofs and Refutations* is its attack on formalism in the style of Hilbert, although it is worthwhile noting that Hilbert himself always recognized the importance of singular, unique problems in attracting young minds to mathematics, and his famous list of problems was made known precisely with that objective in mind.

An article that Lakatos wrote and that was originally published in *The Mathematical Intelligencer* (Lakatos 1978) entitled "Cauchy and the Continuum: The Significance of Non-Standard Analysis for the History and Philosophy of Mathematics" shows, in Hersh's interpretation (Hersh 1978), the objective that Lakatos pursued in his approach to the history of mathematics.

> The point is not merely to rethink the reasoning of Cauchy, not merely to use the mathematical insight available from Robinson's non-standard analysis to re-evaluate our attitude towards the whole

history of the calculus and the notion of the infinitesimal. The point is to lay bare the inner workings of mathematical growth and change as a historical process, as a process with its own laws and its own 'logic', one which is best understood in its rational reconstruction, of which the actual history is perhaps only a parody.

With these three figures, formed in one way or another in the Hungarian school of mathematical problem solving competitions, the key ingredients have been readied that constitute the theoretical framework of the mathematics education related to competitions and that would allow a change in the way that mathematics is being done or, at least, a change in the mathematics that is being done. First, a prolific mathematician, foremost a problem solver rather than a theory builder, who worked his entire life with mathematicians throughout the world. That is to say, a view of the nature of mathematics. Second, an epistemologist who theorized about the nature of mathematical knowledge and broke with the formalist tradition that had dominated mathematics for much of the twentieth century. That is to say, a view of the nature of mathematical knowledge. Third, a methodologist who led change on the level of education. That is to say, a view of how such mathematics can (and should) be learned.

These are the underpinnings of the work done in creating the mathematical challenges of competitions.

13.3 The Role of the International Mathematical Olympiad

Rumania has also been highly influential in mathematical problem solving competitions, its competition activity beginning on the level of primary school even before the Eötvös and Kurschak competitions in Hungary. And in 1959 Rumania founded the International Mathematical Olympiad which has spread mathematics competitions all over the world. In particular, it is not only the IMO itself, but the selection process of the participants representing each country, which is usually done vía the realization of a national Olympiad, that gives thousands and even millions of students the experience of solving Olympiad problems, problems that are original, challenging, many of them beautiful, and that open the student's horizons to include experiences in which his/her own thinking is considered a highly valued contribution.

Thus the IMO and the different national contests and Olympiads have been the means for attracting young scholars to the field of mathematics, although not all participants choose to follow mathematics. Many of these young scholars have been attracted to mathematics, as was Erdős, by the challenge of solving very beautiful and very difficult problems. While pursuing their advanced studies in mathematics, most experience the rites of initiation into a mathematical establishment in which theory building, heavily influenced by thinkers in the tradition of Bourbakí, is predominant. Is this predominance fading, or at least being offset?

The question has to be answered cautiously. There are other important factors at work, key among them the use of computers in doing mathematics and in producing mathematical proofs, leading to new emphases and entirely new fields, as well as new philosophical perspectives.

13.3.1 Distinctions Made by Timothy Gowers and Freeman Dyson

Several protagonists have emerged who have unequivocally attempted to legitimate the work of those who emphasize the solving of beautiful and difficult problems, and especially the beautiful solving of beautiful problems, above (or course, not to the exclusion of) the work of building theories. Among these is Timothy Gowers, IMO gold medalist and winner of the Fields Medal, who over the first decade of the present century has championed in several scenarios the necessity for taking a new look at the doing of mathematics. Gowers in a chapter that he wrote for the book *Mathematics: Frontiers and Perspectives* published by the AMS in 2000, spoke clearly about the important changes that have taken place in the way mathematics is being done.

In his writing Gowers (2000), following C.P. Snow, talks about the two cultures he perceives in the mathematical community and maintains that, on one side, a group of mathematicians continue to work in the formalist tradition, whereas there are others whose primary interest is in problem solving.

> Loosely speaking, I mean the distinction between mathematicians who regard their central aim as being to solve problems, and those who are more concerned with building and understanding theories.... it involves a certain oversimplification, but not so much as to make it useless. If you are unsure to which class you belong, then consider the following two statements.

(i) The point of solving problems is to understand mathematics better.
(ii) The point of understanding mathematics is to become better able to solve problems.

Most mathematicians would say that there is truth in both (i) and (ii).

So when I say that mathematicians can be classified into theory-builders and problem-solvers, I am talking about their priorities.

In the material he prepared for the AMS Einstein Lecture in 2008, Freeman Dyson (2009) makes a similar, though clearly different, distinction as follows.

> Some mathematicians are birds, others are frogs. Birds fly high in the air and survey broad vistas of mathematics out to the far horizon. They delight in concepts that unify our thinking and bring together diverse problems from different parts of the landscape. Frogs live in the mud below and see only the flowers that grow nearby. They delight in the details of particular objects, and they solve problems one at a time.... Mathematics needs both birds and frogs. Mathematics is rich and beautiful because birds give it broad visions and frogs give it intricate details. Mathematics is both great art and important science, because it combines generality of concepts with depth of structures.... The world of mathematics is both broad and deep, and we need birds and frogs working together to explore it.

Dyson goes on to classify historical figures such as Francis Bacon and René Descartes, and notes that

> For me, as a Baconian, the main thing missing in the Bourbaki program is the element of surprise. The Bourbaki program tried to make mathematics logical (In the Bourbaki scheme of things, mathematics is the abstract structure included in the Bourbaki textbooks. What is not in the textbooks is not mathematics. Concrete examples, since they do not appear in the textbooks, are not mathematics. The Bourbaki program was the extreme expression of the Cartesian style. It narrowed the scope of mathematics by excluding the beautiful flowers that Baconian travellers might collect by the wayside.).

13.3.2 Timothy Gowers

Gowers (2000) considers himself to be a mathematician whose priority (in the tradition of Paul Erdős) is problem solving, and he comments on some of the areas in which one or the other of the two cultures he had identified predominates. In particular, he mentions

> … graph theory, where the basic object, a graph, can be immediately comprehended. One will not get anywhere in graph theory by sitting in an armchair and trying to understand graphs better. Neither is it particularly necessary to read much of the literature before tackling a problem: it is of course helpful to be aware of some of the most important techniques, but the interesting problems tend to be open precisely because the established techniques cannot easily be applied.

He continues, saying

> My main purpose here is to defend some of the less fashionable subjects against criticisms commonly made of them. I shall devote most of my attention to combinatorics, since this is the area I know best. However, what I say applies to other areas as well. I often use the word "combinatorics" not quite in its conventional sense, but as a general term to refer to problems that it is reasonable to attack more or less from first principles. … Such problems need not be discrete in character or have much to do with counting. … Nevertheless, there is a considerable overlap between this sort of mathematics and combinatorics as it is conventionally understood. Why should problem-solving subjects be less highly regarded than theoretical ones?

Appraising the situation, Gowers cites Sir Michael Atiyah, who takes the opposing position

> that so much mathematics is produced that it is not possible for all of it to be remembered. The processes of abstraction and generalization are therefore very important as a means of making sense of the huge mass of raw data (that is, proofs of individual theorems) and enabling at least some of it to be passed on. The results that will last are the ones that can be organized coherently and explained economically to future generations of mathematicians. Of course, some results will be remembered because they solve very famous problems, but even these, if they do not fit into an organizing framework, are unlikely to be studied in detail by more than a handful of mathematicians.

Gowers' objective is to answer this criticism. He maintains that, according to Atiyah,

> … it is useful to think not so much about the intrinsic interest of a mathematical result as about how effectively that result can be communicated to other mathematicians, both present and future. Combinatorics appears to many to consist of a large number of isolated problems and results, and therefore to be at a disadvantage in this respect. Each result individually may well require enormous ingenuity …, and future generations of combinatorialists will not have the time or inclination to read and admire more than a tiny fraction of their output.

Gowers affirms that the difference is to be found in that

> The important ideas of combinatorics do not usually appear in the form of precisely stated theorems, but more often as general principles of wide applicability. An example will help to make this point more clearly. One form of Ramsey's theorem is the following statement. *Theorem. For every positive integer k there is a positive integer N, such that if the edges of the complete graph on N vertices are all coloured either red or blue, then there must be k vertices such that all edges joining them have the same colour.*

Gowers proceeds to present a proof and speak about a result due to Erdős, that places a lower bound on *N*. Now the line of reasoning that Erdős follows uses probability arguments in this problem of combinatorics. The crux of the argument is the idea of metric concentration, an idea that turns out to be fertile and that is later capitalized on in various other situations.

Gowers comments that

> My main point about such principles is .. that they play the organizing role in combinatorics that deep theorems of great generality play in more theoretical subjects.

And he sums up his points saying

> I have been trying to counter the suggestion that the subject of combinatorics has very little structure and consists of nothing but a large number of problems. While the structure is less obvious than it is in many other subjects, it is there in the form of somewhat vague general statements that allow proofs to be condensed in the mind, and therefore more easily memorized and more easily transmitted to others.

Note here how this echoes the organization that Pólya gave to his book coauthored with Szegö on the solution of problems in analysis.

Gowers has continued to sustain this idea, notably as editor of an outstanding compendium, *The Princeton Companion to Mathematics* (Princeton University Press, 2006). Gowers characterizes himself as a vigorously interventionalist editor. In the Introduction of the book, he states

> The most obvious way of classifying mathematics is by its subject matter.... Another approach is to try to classify the kinds of questions that mathematicians like to think about. This gives a usefully different view of the subject.

In a section titled "The general goals of mathematical research", Gowers concentrates on "what mathematicians do with concepts and the kinds of questions they ask about them". The point he wishes to make once again is that certain thinking strategies cross artificial frontiers erected between mathematical topics and theories, organizing mathematical thought in a new and valuable way.

13.3.3 IMO 50, Bremen 2009 and Beyond

In 2009 Gowers was one of several invited speakers at a special event organized in Bremen to celebrate the 50th anniversary of the IMO. Other speakers included Terry Tao and Lászlo Lovász, also IMO gold medalists as well as winners of the Fields Medal. We contacted both of them concerning the question that is the title of the present paper, and received very different responses. We also contacted two combinatorialists ex-members of Colombian IMO teams, and also received two very different responses.

Terry Tao and Federico Ardila. Terry Tao and Federico Ardila seem to coincide in their thinking, and that thinking seems to address the direct influence of participation in the IMO on future mathematicians. We give their views. Tao, referring to his blog, says:

> My impression is that the impact of mathematics competitions on mathematical research is primarily indirect, in that it helps encourage bright students to develop their mathematical interests; in a few cases (notably in combinatorics) some Olympiad training is directly helpful in research, but more for solving individual subproblems in a project (e.g. proving a tricky lemma) rather than guiding the overall research from a high-level perspective.

From the context of the blog that Tao maintains, it appears correct to claim that the position he has taken is that a young mathematician must learn all the basics of the great corpus of mathematical knowledge that has been built up and study very deeply those parts that are pertinent to a certain specialized area in order to write his or her doctoral thesis. His appreciation concerns a direct link between problem solving training for participation in the IMO and the solving of the problem of a doctoral dissertation, while Gowers is analyzing the situation from a much ampler perspective.

Ardila, also speaking about olympiad training, states (translation ad hoc)

> that it is excellent for developing technique and confidence. Without reference to how profound the mathematics is that one is doing, I believe that problem solving teaches a great arsenal of techniques that can later help to solve small parts of big problems. … On the other hand, the Olympiad participant comes to have great ability to solve problems without necessarily understanding them. Observing and studying the nature of mathematics, letting the mathematics reveal itself, carefully and without haste nor trying to be ingenious, teaches a lot. … Some ex Olympiad participants find it difficult to have such patience.

Lászlo Lovász. Lovász answers that he does not have time to write to the question although he considers it a very interesting one. But in several of his statements, he has made it clear what the entire mathematical olympiad experience, including his personal preparation for competitions, meant to him as a mathematician and how it influenced him. Lovász at age 14 encountered an article written by Paul Erdős in KöMaL and was so taken with it that he read it at least 20 times. The following year he met Erdős personally and has this to say about that meeting:

> I had the great fortune to meet Paul Erdős as a high school student in 1963. In those days the cold war was quieting down a little, and he began to visit Hungary more and more often. … It is an understatement to say that I have learned a lot from him, not only mathematics in the technical sense, and not even only elements of the fine art of problem solving, but also his way of pursuing knowledge.…

Luis Serrano. Luis Guillermo Serrano, another IMO medalist from Colombia, states that "in the olympiads we take really simple concepts, such as high school algebra and Euclidean geometry, and do really complicated things with them", thus echoing Gowers' description of combinatorics and problem solving in combinatorics. Serrano recognizes the division of mathematicians into theory builders and problem solvers; for him the former are those who ask the deep questions, trying to generalize or reduce hypotheses, and the latter those who help to answer them, qualifying this distinction by stating that every successful mathematician is a bit of both, but in most cases is geared towards one or the other, and classifies himself first as a problem solver. He believes that the mathematical olympiad experience attracts many problem solvers to mathematics who might otherwise have attempted other professional fields where problem solving is the principal strength required. This will, of course, be reflected in the mathematics that they do and the way they do mathematics. He also states that theory builders are rare so that the Olympiad experience naturally attracts more problem solvers and coincides with Ardila in stating that problem solvers can begin to prove theorems without fully understanding the objects involved, and to actually learn by working on things.

Finally, he believes

> there is a bias in looking at contemporary mathematicians and seeing a change in orientation with regard to the past because only the theory builders tend to be remembered,...that it has become harder and harder to be a generalist as are most problem solvers, and... that any engine (like the olympiads) that brings people in mass towards doing mathematics, will naturally bring more problem solvers than theory builders to mathematics. Maybe the theory builders would be the only ones who may still become mathematicians even if they don't do olympiads, ... so it is a sampling bias. And this shows a positive influence of the olympiads towards mathematics, in the sense that they discover many mathematicians who would not have become mathematicians otherwise.

A sound summary of the ideas expressed by Tao, Ardila and Serrano is that they speak to the influence of mathematics olympiads in terms of the young people they attract to mathematics. Serrano echoes Gowers in linking combinatorics to an emphasis on problem solving over theory building, but states clearly that the theory builders will be the ones most remembered. Gowers searches for a new way of organizing mathematical knowledge so that the work of problem solvers will be remembered equally; the others do not address this issue.

13.4 Following the Trail Between Competitions Problems and Mathematics

In this section we follow the trail that leads from mathematics competitions to mathematics, and also look briefly at the road leading from mathematics to competition problems. The protagonists are Serrano and Ardila. The question asked is what are the most memorable or favorite Olympiad problems for you, and do they have any relation with the work you are doing in mathematics? Although several well-known ex-Olimpians were contacted, only the responses from Serrano y Ardila actually addressed these questions and their answers indeed show close relations with their mathematical work, although in significantly different ways. However, the contexts in which this work is being done are notable and pertain directly to the questions we are attempting to answer in this article.

Luis Serrano does not remember the exact wording of the problem that he considers most memorable, but rather the general strategies of his solution to it (translation ad hoc).

> While training with Titu Andreescu in Bogotá, he proposed a problem that had to do with maximizing a sum of the form $a_1x_1 + \ldots + a_nx_n$ with some constraints. It looked hard. The next day I was sick and did not go to training, but I spent the entire day trying to solve the problem. None of the known methods worked, but after trying for a long time, I noticed that with a small change in the a_i there was a shift from the family $(a_1, \ldots, a_n)$ to a new $(a_1', \ldots, a_n')$ that increased the weighted sum of the a_ix_i, and in some way, repeating this process, a maximal solution was attained and there the process stabilized. This method was completely new to me and, in fact, it seemed to me to be a bit strange, so I tried for a long while to find a more classic solution without success. The next day I showed my solution on the blackboard and Titu applauded. Others told me that they had liked my solution a lot, and ever since then I acquired confidence with inequalities, and whenever I saw one, I would immediately set myself to solve it.
>
> This has a lot to do with what I am doing now! Twenty years later, I came to learn that the crux of machine learning optimization, and other related topics is finding the correct parameters to maximize a function that contains them (for example, the a_i in the sum of the a_ix_i). And what is one of the methods most used to maximize (or minimize) machine learning? The answer is gradient descent which is precisely that, beginning with any parameters, iterate and iterate in such a way

that the function continues to increase, until a point is found where it stabilizes and the solution is (or is very close to being) a maximum.

Federico Ardila answered by showing a problem he met with in sixth or seventh grade and which he worked on in different ways for over 10 years. The problem, originally posed on a math Olympiad paper in China in 1962, is stated in the following way.

1. *Given* $2n + 1$ *points in general position in the plane, prove that there is at least* one *"halving circle", that passes through 3 of the points and has* $n - 1$ *of the points in its interior and* $n - 1$ *of the points in its exterior.*

   Another version of the same problem appeared in *Crux Mathematicorum* in 1995 (and on the IMO 1999 short list).

2. *Prove that any set of 5 points in general position in the plane has exactly 4 halving circles.*

   Ardila writes that his solution to the problem in *Crux* revealed another related problem that he proposed in the APMO (Asian-Pacific Mathematical Olympiad) in 1998 and that states:

3. *For any set of* $2n + 1$ *points in general position in the plane, the number of halving circles has the same parity as* n.

He says (translation ad hoc).

However, I still didn't see the relation between problems 2 and 3. I kept on thinking about it in Colombia in the summer of 1998 during training for IMO 1998 and finally I understood. The answer appeared as "The number of halving circles" (Amer. Math. Monthly 111 (2004) 586–592).

For any set of $2n + 1$ *points in general position in the plane, the number of halving circles is exactly* n^2.

Continuing, he says.

It seems to me that this is a very surprising result. In problems in discrete geometry of this sort it is habitual to find bounds but not exact results. For example, the number of "halving lines" of a set of $2n$ points is not constant, and the problem of finding the best upper and lower bounds for this number has received a lot of attention.

Although I know of several proofs of this theorem, and I have discussed it with many experts, I do not know of any truly satisfactory and transparent explanation. Why is the number constant? Why is it so simple?

Here we have a fine example of the impact of competition problems on research.

Proceeding along the same trail in the opposite direction, Ardila then shows how a problem proposed for IMO 2006 stemmed from a result obtained in research he did with Sara Billey.

The research of Ardila and Billey was looking to clarify a construction given in an article in algebraic geometry written by Billey and Ravi Vakil concerning "Schubert calculus of the flag manifold". They needed to understand the following problem.

A flag is a sequence of subspaces $\{0\} = F_1 \subseteq F_2 \subseteq \cdots F_{d-1} \subseteq R^d$ *with dim* $F_i = i$.

Consider 3 generic flags E, F, G in R^d *and all of the lines* $E_i \cap F_j \cap G_k$ *that are determined by them when they are intersected. There are* $C(d, 2)$ *of these lines.*

Question: Which d-subsets of these lines generate R^d*? Let's call them E-F-G bases.*

Ardila recounts that

Curiously, the first solution found to this question depended on the solution of a problem of tessellations of a triangle that I then proposed for IMO 2006 and that states: *A holey triangle is an upward equilateral triangle of side length n with n upward unit triangular holes cut out. A diamond is a* $60° - 120°$ *unit rhombus. Prove that a holey triangle T can be tiled with diamonds if and only if the following condition holds: Every upward equilateral triangle of side length k in T contains at most k holes, for*$1 \leq k \leq n$.

The connection is the following. There is a bijection between *E-F-G* bases and holey triangles that can be tessellated.

The result can be generalized to *n* flags and tessellations of holey $n - 1$ simplices and led to a conjecture, the "Spread Out Simplices Conjecture" that, despite having been studied by experts, has remained open for 10 years.

In the context of the questions posed by this article, these accounts are remarkable precisely because they allow us to see clearly that there exists a cycle of competition problem posing and research that is impacting (and most probably changing the nature of) the mathematics that is being done and the way mathematics is being done, our supposition being that problems of this sort can captivate a young student and lead her or him to dedicate significant time to studying the problem and its possible generalizations, leading to new research results (and when it is possible to explain why it is true, perhaps leading to insights into new theoretical ideas).

13.5 Balance

Is this revolutionary? It is not. Gowers is convinced but conciliatory; for him and the others there is room for both mathematical cultures. Atiyah on the other hand is confrontational; for him there is no way to organize the thousands upon thousands of results attributed to the problem solvers; an organized body of knowledge, a system, is needed to relate these results and the arguments they contain. Systematic, for Atiyah, is tantamount to organized in a theory. Systematic for Pólya and Szegö, as well as for Gowers, can also be applied to ways of thinking and arguing.

New solution methods, new ways of thinking. Erdős' use of a probability argument in a combinatorics problem relates directly, for example, to a student's solving an algebra problem with a geometric argument, or to Leonardo de Pisa's introduction of entirely new ideas that expand or extend geometric arguments to algebra, exactly what mathematics competitions set the stage for students to do. While myriad schools of mathematics education follow the established body of mathematical knowledge, mathematics competitions open the doors continually to new connections and new arguments gleaned from new solutions given to original problems as well as new proofs from first principles of old results frequently founded on laboriously elaborate theories.

Thus Erdos' brilliant proof of the prime number theorem. What will the future hold for, say, Fermat's last theorem? A proof has been given based intricately on advanced theories. Will it too eventually yield to arguments stemming from first principles?

Each beautiful olympiad problem solved by a budding young mathematician builds the expertise to do mathematics in the way preferred by Gowers, to stress thinking strategies and arguments as being as meaningful anchors for organizing mathematics as systems and theories can be.

And perhaps this is exactly the way in which human mathematicians and human made mathematics will differentiate themselves from computer generated mathematics as the mathematical community marches confidently into the future.

The stress put by Ardila and Serrano on being able to prove something in olympiads about mathematical entities that one does not fully understand conveys a profound aspect of contemporary mathematics (and contemporary mathematics education). This is the mutual enrichment provided by addressing simultaneously the two major components of mathematical reasoning, that have been identified as being mathematical understanding and mathematical thinking. Frequently math educators suppose that it is indispensable to understand a concept fully in order to be able to confront successfully a problem that concerns that concept, that one must progress from understanding to thinking. The solution of challenging problems reveals however, that mathematical thinking about a concept—thinking autonomously, solving a problem that involves the concept—can enrich the understanding of that concept. This is especially clear when the solution is found outside of the realm of normal applicability of the concept, and reinforces Erdős' insistence that a proof should show clearly why a result is true.

This is an important observation for mathematics education, and it is essentially the claim made by Gowers when he comments that certain ignorance can be an asset when addressing a problem of combinatorics that remains open despite many attempts to solve it.

13.6 Some Conclusions

The influence of Leonardo de Pisa's ideas, developed in the context of a problem-solving competition, reverberated through at least six centuries and eventually changed the basic mathematical perspective from geometric to algebraic.

The long-termed influence of thinkers like Erdős and Gowers, formed in the tradition of addressing difficult problems whose comprehension does not require extensive previous knowledge, a tradition renewed and enriched by mathematics competitions, has yet to be gauged. Currently, the impact has been both deep and wide, nudging aside the great theory builders such as Bourbaki in order to open up some space for combinatorics as defined by Gowers. The extent of that influence in time may well depend on achieving new organizational principles of mathematical knowledge based on reasoning strategies rather than topical categories.

References

Albers, D. J. & Anderson, G. L. (Eds.). (1985) *Mathematical people: Profiles and interviews.* Boston.

Ardila, F. (2004). The number of halving circles. *The American Mathematical Monthly, 11,* 586–592.

Ardila, F. (2015, 2016). Private communications.

Ardila, F., & Billey, Sara. (2007). Flag arrangements and triangulations of products of simplices. *Advances in Mathematics, 214,* 495–524.

de Manrique, M. A., & de Losada, M. F. (1997). *Recorriendo el álgebra: de la solución de ecuaciones al álgebra abstracta.* Bogotá. Universidad Nacional de Colombia & Colciencias.

Dyson, F. (2009). Birds and frogs. *Notices of the AMS* (Vol. 56 No. 2). American Mathematical Society

Freud, R. (1993). Paul Erdős 80—A personal account. *Periodica Mathematica Hungarica, 26*(2), 87–93.

Gowers, W. T. (2000). Chapter: The two cultures of mathematics. In V. I. Arnol'd et al. (Eds.), *Mathematics: Frontiers and perspectives.* American Mathematical Society.

Gowers, W. T. (Ed.). (2006). *The Princeton companion to mathematics.* Princeton, NJ: Princeton University Press.

Harel, G. (2010). DNR-based instruction in mathematics as a conceptual framework. In B. Sriraman & L. English (Eds.), *Theories in mathematics education: Seeking new frontiers.* Berlin: Springer.

Hersh, R. (1978). Introducing Imre Lakatos. *Mathematical Intelligencer, 1* (3), 148–151.

Lakatos I. (1976). *Proofs and refutations: The logic of mathematical discovery* In J. Worrall & E. G. Zahar (Eds.) Cambridge University Press.

Lakatos, Imre. (1978). Cauchy and the continuum: The significance of non-standard analysis for the history and philosophy of mathematics. *The Mathematical Intelligencer, 3,* 151–161.

Lovász, L. (2015). *Private communication.*

Pólya, G. (1945). *How to solve it.* Princeton University Press.

Serrano, L. G. (2015, 2016). *Private communications.*

Silver, E. (1994). Mathematical thinking and reasoning for all students: Moving from rhetoric to reality. *Proceedings of the 7th International Congress on Mathematical Education 1992*, pp. 311–326.

Tall, D. (2013). *How humans learn to think mathematically: Exploring the three worlds of mathematics.* Cambridge University Press.

Tao, T. (2015). *Private communication.*

Wilczek, F. (2016). A puzzling solution for math education. *Wall Street Journal*, p C3, January 9–10, 2016.

http://www-history.mcs.st-andrews.ac.uk/Biographies/Polya.html.

https://terrytao.wordpress.com/.

Chapter 14
Classic Conjectures Allow Young Mathematicians to Commence Research

Alexander Soifer

Abstract The four classic conjectures of mathematics presented here are still open. They appear together with trains of thought that lead to them and relevant partial results. This material shows how strongly problems of mathematical Olympiads in general, and the Colorado Mathematical Olympiad in particular, are interwoven with the forefront of mathematics, and influence each other. These conjectures are "classic" because they are easy to understand and hard to prove or disprove. They are accessible to young high school and college mathematicians and allow students to commence research and creative work in mathematics.

Keywords Open problem · Conjecture · Mathematics · Research · Combinatorics · Chromatic number · The plane · Paul Erdős

14.1 Part Naught: Overture

What is creating conjectures about?

This essay was first read as the Closing Plenary Talk at the 7th Congress of the *World Federation of National Mathematics Competitions*, very successfully organized in Barranquilla, Colombia by Maria Falk de Losada in July 2014. Its early version appeared in the journal of the World Federation of National Mathematics Competitions *Mathematics Competitions* 27(1), 2014, 16–39.

A. Soifer (✉)
University of Colorado, 1420 Austin Bluffs Parkway, Osborne Center for Science, Office A423, Colorado Springs, CO 80918, USA
e-mail: asoifer@uccs.edu
URL: http://www.uccs.edu/~asoifer/

A. Soifer (ed.), *Competitions for Young Mathematicians*,
ICME-13 Monographs, DOI 10.1007/978-3-319-56585-9_14

In my opinion, it is the art of predicting the future.

Niels Bohr took such predicting jokingly:

Predicting is very difficult, especially the future.

Albert Einstein—nonchalantly:

I never think of the future – it comes soon enough.

In conjecturing, we use the power of intuition to envision the result without being able to prove it.

After the 1989 Paul Erdős' lecture here at the University of Colorado, Professor of Mathematics Gene Abrams was unimpressed. "What is a big deal about posing problems? Proving results is much more important," he declared. "Without someone posing problems, and moreover predicting results by conjectures, as Erdős has done, there will be nothing to prove," I replied. When we commence research, we do not know what is true, and let our intuition lead the way. We have to rely on insight or good luck in choosing a conjecture to prove. And if we choose a conjecture that is not true, it would take a very long time to prove it, an infinite time!

I oppose discrimination of young high school and college mathematicians based on their youth and inexperience. Thus, we ought to share with them unsolved problems and conjectures that are waiting for their conquerors. There is plenty of contemporary mathematics, dealing with an elaborate maze of definitions, and sometimes consisting of merely juggling with them. This kind of juggling does not interest me much. I prefer classic problems. By 'classic' I do not necessarily mean problems that are centuries old, but rather problems that are easy to understand by anyone, including a middle school student or a layman, but tantalizingly difficult if at all possible to solve. There are additional conditions on admission of a problem into 'classic' category: an aesthetic appeal of the problem is essential, as are expectations that the result may defy our intuition.

In this essay I will demonstrate interaction between Olympiad and research problems, and present 'live' fragments of mathematics, centered on predicting the future by formulating classic conjectures. I am offering you a journey on a mathematical train of thought through problems, conjectures, and results. I hope you will enjoy the ride!

14.2 Conjecture I. Squares in a Square (Erdős 1932)[1]

We are in Budapest, Hungary, year 1932. The 19-year old Paul Erdős poses the following problem. Inscribe in a unit square r squares, which have no interior points in common. Denote by $f(r)$ the maximum of the sum of side lengths of the r squares. (We allow side lengths to be zero.) The problem is to evaluate the function $f(r)$:

Open Problem 1 For every positive integer r find the value of $f(r)$.

In fact, this formulation came about later, in 1992, when Paul Erdős and I commenced joint efforts to settle this problem. Originally Paul formulated the following narrow but surprisingly difficult conjecture. When he shared the conjecture with me, he offered a $50 price for its first proof or disproof.

Fifty Dollar Squares in a Square Conjecture 2 (Paul Erdős 1932). For any positive integer k,

$$f(k^2+1)=k.$$

The conjecture is still open today, in the year 2017, waiting, as Paul Erdős used to say, "for stronger arms, or, perhaps, brains" to be settled. However, Paul and I reached a progress in a broader problem of describing the function $f(r)$. First of all, we observed the following lower and upper bounds for $f(r)$. Symbol $\lfloor x \rfloor$ as usual denotes the largest integer that is not greater than x.

Result 3 (Erdős and Soifer 1995). The following inequality is true for any positive integer r:

$$\lfloor\sqrt{r}\rfloor \le f(r) \le \sqrt{r}.$$

Proof

1. *The Upper Bound.* The celebrated Cauchy Inequality states that

$$\left(\sum_{i=1}^{r} a_i b_i\right)^2 \le \left(\sum_{i=1}^{r} a_i^2\right)\left(\sum_{i=1}^{r} b_i^2\right).$$

[1]This chapter is based on sections E14 and E15 of the author's book (Soifer 2011).

Setting $b_i = 1$ for every $i = 1, 2, \ldots, r$ we get

$$\left(\sum_{i=1}^{r} a_i\right)^2 \leq \left(\sum_{i=1}^{r} a_i^2\right) r. \qquad (*)$$

Let r squares of side lengths a_i, $i = 1, 2, \ldots, r$ with no interior points in common be placed in a unit square. Then the combined area of the r squares does not exceed the area of the unit square: $\sum_{i=1}^{r} a_i^2 \leq 1$, and we get from the inequality (*) above the required upper bound:

$$f(r) = \sum_{i=1}^{r} a_i \leq \sqrt{r}.$$

2. *The Lower Bound*. Surely, the function $f(r)$ is non-decreasing, therefore, $r \geq \lfloor\sqrt{r}\rfloor^2$ implies $f(r) \geq f\left(\lfloor\sqrt{r}\rfloor^2\right)$.

Now let us partition the unit square into $\lfloor\sqrt{r}\rfloor^2$ congruent squares, each of the side length $\frac{1}{\lfloor\sqrt{r}\rfloor}$, and calculate the sum of side lengths of these $\lfloor\sqrt{r}\rfloor^2$ squares: we get $\frac{1}{\lfloor\sqrt{r}\rfloor} \times \lfloor\sqrt{r}\rfloor^2 = \lfloor\sqrt{r}\rfloor$. Observe that this partition and the calculation demonstrate the inequality $f\left(\lfloor\sqrt{r}\rfloor^2\right) \geq \lfloor\sqrt{r}\rfloor$. By combining the two inequalities of this and the preceding paragraphs, we get the required lower bound:

$$f(r) \geq f\left(\lfloor\sqrt{r}\rfloor^2\right) \geq \lfloor\sqrt{r}\rfloor.$$

■

Result 3 has the following consequence:

Corollary 4 If $r = k^2$ for a positive integer k, then we get the equality $f(r) = k$.

Corollary 4 allows us to see the Erdős Fifty Dollar Squares in a Square Conjecture in a slightly different light:

Fifty Dollar Squares in a Square Conjecture, Second Version 5 (P. Erdős). At perfect square numbers $r = k^2$ (k is an integer), the function $f(r)$ does not increase:

$$f\left(k^2 + 1\right) = f\left(k^2\right).$$

Paul Erdős and I were able to prove that the function $f(r)$ is strictly increasing everywhere else. But to prove that we needed to find a much sharper lower bound for $f(r)$.

Theorem 6 (Erdős and Soifer 1995). *Any positive integer r can be presented in a form $r = k^2 + m$, where $0 \le m \le 2k$. Accordingly, the following inequalities hold:*

(A) *If $m = 2t + 1$, where $0 \le t < k$, then $f(r) \ge k + \frac{t}{k}$;*

(B) *If $m = 2t$, where $0 \le t \le k$, then $f(r) \ge k + \frac{t}{k+1}$.*

Proof Given a positive integer r, we can present it in a form $r = k^2 + m$, where $0 \le m \le 2k$. Indeed, it suffices to choose $k = \lfloor \sqrt{r} \rfloor$. If r is a perfect square, $r = k^2$, then $m = t = 0$, and Corollary 4 provides the exact value $f(r) = k$, which is a part of the required inequality (B). We can assume now that r is *not* a perfect square, i.e., $m \neq 0$. The parity of m dictates two cases.

(A). $m = 2t + 1$ and $0 \le t < k$. Let us first partition the unit square into k^2 congruent squares; we get a $k \times k$ square grid, call it G; and then replace a $t \times t$ subgrid of the grid G with a $(t + 1) \times (t + 1)$ square grid of the same total size as the removed subgrid (Fig. 14.1).

Fig. 14.1

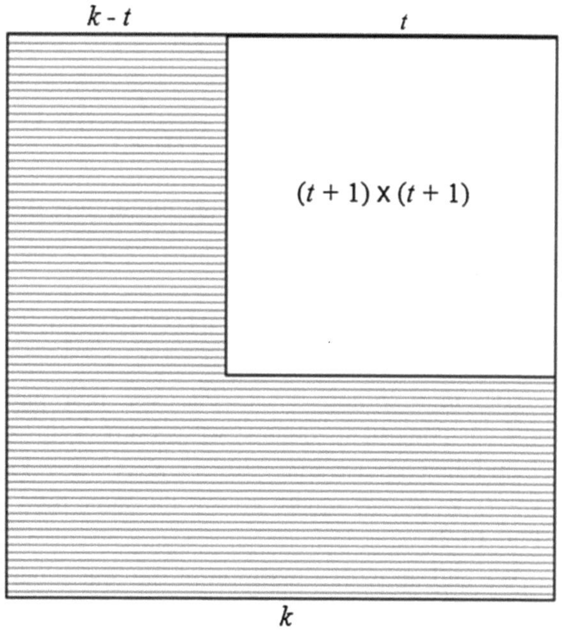

Fig. 14.2

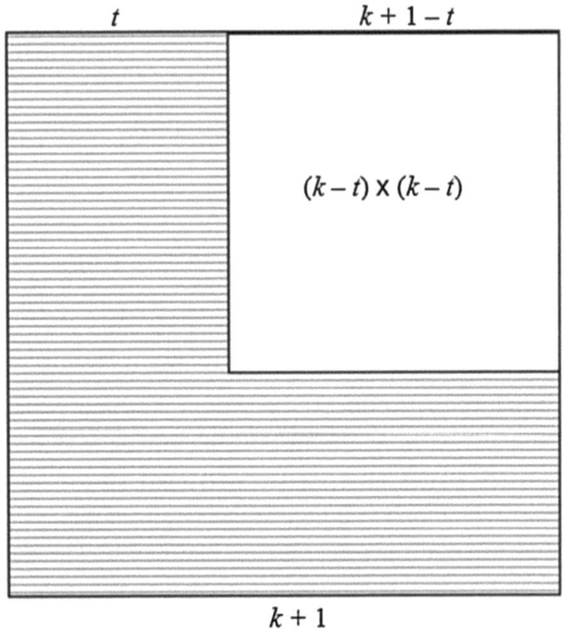

We end up with a partition of the unit square into $k^2 - t^2 + (t+1)^2 = k^2 + 2t + 1$ little squares, some of which [the original ones] have side length $\frac{1}{k}$, and others [the squares of the inserted $(t + 1) \times (t + 1)$ square grid] of the side length $\frac{t}{k(t+1)}$. Let us calculate the sum of side lengths of all these $k^2 + 2t + 1$ little squares, we get:

$$\frac{1}{k}k^2 - \frac{1}{k}t^2 + \frac{t}{k(t+1)}(t+1)^2 = k + \frac{t}{k}.$$

This partition and the calculation deliver the following lower bound for $f(r)$:

$$f(r) \geq k + \frac{t}{k}.$$

(B). $m = 2t$ and $0 < t \leq k$. We first partition the unit square into $(k + 1)^2$ congruent squares; we get a $(k + 1) \times (k + 1)$ square grid, call it G; and then replace a $(k - t + 1) \times (k - t + 1)$ subgrid of the grid G with a $(k - t) \times (k - t)$ square grid of the same total size as the removed subgrid (Fig. 14.2).

We end up with a partition of the unit square into $(k+1)^2 - (k-t+1)^2 + (k-t)^2 = k^2 + 2t$ little squares, some of which [the original ones] have side length $\frac{1}{k+1}$, and others [the squares of the inserted $(k-t) \times (k-t)$ square grid] of the side length $\frac{k-t+1}{(k+1)(k-t)}$. Let us calculate the sum of side lengths of all these $k^2 + 2t$ little squares, we get:

$$\frac{1}{k+1}(k+1)^2 - \frac{1}{(k+1)(k-t+1)}(k-t+1)^2 + \frac{k-t+1}{(k+1)(k-t)}(k-t)^2 = k + \frac{t}{k+1}.$$

This partition and calculation deliver the following lower bound for $f(r)$:

$$f(r) \geq k + \frac{t}{k+1}.$$

Done! ∎

Result 7 (Erdős and Soifer 1995). The function $f(r)$ is strictly increasing everywhere except possibly at perfect square points, i.e., if $r \neq k^2$ for an integer k, then $f(r+1) > f(r)$.

Proof Once again parity of m and Theorem 6 dictate two cases.

(A). $m = 2t + 1$ and $0 \leq t < k$. In this case $t+1 \leq k$, and by substituting $t + 1$ for t in the lower bound found in part 2 of Result 3, we get:

$$f\left(k^2 + 2t + 2\right) \geq k + \frac{t+1}{k+1}.$$

This inequality and result 1 deliver the necessary chain of inequalities:

$$f(r) = f\left(k^2 + 2t + 1\right) \leq \sqrt{k^2 + 2t + 1} < k + \frac{t+1}{k+1} \leq f\left(k^2 + 2t + 2\right) = f(r+1).$$

(B). $m = 2t$ and $0 < t \leq k$. By using result 1 and the lower bound found in part 1 of result 3 above, we get the necessary chain of inequalities:

$$f(r) = f\left(k^2 + 2t\right) \leq \sqrt{k^2 + 2t} < k + \frac{t}{k} \leq f\left(k^2 + 2t + 1\right) = f(r+1).$$

Result 7 is proven.

Note: In the proof above I omitted a demonstration of two inequalities: $\sqrt{k^2 + 2t + 1} < k + \frac{t+1}{k+1}$ and $\sqrt{k^2 + 2t} < k + \frac{t}{k}$. I hope their verification would be a welcome exercise in secondary algebra for the reader. ∎

Paul Erdős and I believed that the lower bounds in Theorem 6 were quite good, and conjectured that they just may be the best possible:

The Erdős–Soifer Conjecture 8 (Erdős and Soifer 1995). Any positive integer r can be presented in a form $r = k^2 + m$, where $0 \le m \le 2k$. Accordingly, we conjecture the following equalities:

(A) If $m = 2t + 1$, where $0 \le t < k$, then $f(r) = k + \frac{t}{k}$;

(B) If $m = 2t$, where $0 \le t \le k$, then $f(r) = k + \frac{t}{k+1}$.

We also observed that our examples in Theorem 6 completely tiled the unit square, and thus posed the following open problem:

Open Problem 9 (Erdős and Soifer 1995). Is it true that for any positive integer r, the value of $f(r)$ can be attained by a set of r squares that form a complete tiling of the unit square by themselves or with an addition of at most *one* extra square?

As I thought about Paul Erdős' problem, it appeared natural for me to pose a 'dual' problem, and thus give birth to the *New Squares in a Square Problem.*

Let □ stand for a square shape and $r > 1$ a positive integer. Denote by $S(\square, r)$ the smallest area of a square Q such that any r squares whose areas add up to at most 1, can be packed in Q (i.e., embedded in Q with no interior points in common).

In 1997 I offered this conjecture for small values of r at the 14th Colorado Mathematical Olympiad.

Squares in A Square Problem 10 (Soifer 2011).

(A) Prove that any two squares whose areas add up to 1 can be inscribed with no interior points in common in a square of area 2.
(B) Prove that any four squares whose areas add up to 1 can be inscribed with no interior points in common in a square of area 2.
(C) Prove that any five squares whose areas add up to 1 can be inscribed with no interior points in common in a square of area 2.

This Olympiad assertion shows that for any r in the range $2 \le r \le 5$, $S(\square, r) = 2$. I then formulated the following conjecture:

New Squares in a Square Conjecture 11 (Soifer 1996). For any positive integer $r > 1$, $S(\square, r) = 2$, or to simplify notations, $S(\square) = 2$.

Two years have passed since I created this conjecture. In May 1997, I was in Lincoln, Nebraska grading papers of the USA Mathematical Olympiad, together with other members of the USA Mathematics Olympiad Subcommittee. During a break, I put the New Squares in a Square Conjecture on the blackboard. Later the same day Richard Stong told me "I proved your conjecture." Indeed, he did! Richard devised a simple 'greedy' algorithm and a nice, clever proof that his algorithm works, and thus New Squares in a Square Conjecture became a theorem, which I happily published in *Geombinatorics*:

Theorem 12 (Stong 1997). *Any finite set of squares of the combined area 1 can be packed in a square of area 2.*

Later I discovered that Conjecture 11 and Theorem 12 were not new, and although Stong's proof was better, he was preceded by 30 years by J. W. Moon and Leo Moser of Edmonton, Alberta, Canada (Moon and Moser 1967). Ecclesiastes (1:9–14 NIV) comes to mind:

> What has been will be again, what has been done will be done again; there is nothing new under the sun.

On a positive side, I brought a new excitement and new players to the problem. Moreover, I was already riding toward the next station on my train of thought, the one, it seems, no one has traveled before. I conjectured (Soifer 1998a) that the identical result was true for circular discs (I will use here the word "disc" to mean a circular disc). This 1998 conjecture is still open today:

Discs in a Disc Conjecture 13 (Soifer 1998a). Any finite set of circular discs of combined area 1 can be packed in a circular disc of area 2: $\boldsymbol{S}(\mathrm{O}) = 2$.

What about triangles? In working with similar to each other triangles we encounter issues that did not exist for circular discs—limitations on the way 'clones' are embedded. We can place no limitations on embedding at all and end up with our original function $\boldsymbol{S}(\Delta)$ defined as the smallest area of a triangle T such that any r triangles whose areas add up to at most 1, can be packed in T (i.e., embedded in Q with no interior points in common). Or we can limit embedding to translations, and thus define a function $\boldsymbol{S}_T(\Delta)$ as the smallest area of a triangle T such that any r triangles whose areas add up to at most 1, can be embedded by translations in T with no interior points in common. Of course, $S(\Delta) \leq S_T(\Delta)$ as in the latter case we impose limitations on acceptable embeddings. It was not at all obvious whether these two values are equal. In 1995 T. J. Richardson calculated the easier of the two values:

Packing Triangles Theorem 14 (Richardson 1995). Any finite set of similar to each other triangles of combined area 1 can be packed in a similar to them triangle of area 2, i.e., in my notations $S(\Delta) \leq 2$.

In 1999 I pointed out the difference in triangular embedding case and posed these 'triangular problems' in *Geombinatorics* (Soifer 1999). In 2003 the Polish geometer Janusz Januszewski improved Richardson's result on the pages of *Geombinatorics*:

Theorem 15 (Januszewski 2003). $S(\Delta) = 2$ *if and only if the triangle* Δ *is equilateral.*

On January 27, 2009, Januszewski informed me that he calculated the harder value $S_T(\Delta)$ that I asked for in 1999 (Januszewski 2009):

Packing Triangles by Translations Theorem 16 (Januszewski 2009). For any triangle Δ, $S_T(\Delta) = 2$, i.e., any finite set of similar to each other triangles of combined area 1 can be packed in a similar to them triangle of area 2 *by translations alone.*

Let us roll the time back 19 years. By 1998 I felt it was time to generalize these observations to include *all* geometric figures in our 'games.' I got busy.

Definition (Soifer 1998b). Given figures f and F; it is convenient to call a figure f an *F-clone* if f is homothetic to F.

Definition (Soifer 1998b). Given a figure F. Let $S(F)$ be the minimum real number such that any finite set of F-clones of the combined area 1 can be packed in an F-clone of area $S(F)$.

Theorems 11 and 13–14 can be written in these notations as follows:

$$\text{For a square } \square,\ S(\square) = 2;$$
$$\text{For any triangle } \Delta,\ S(\Delta) = 2.$$

However, it is easy to see that numbers $S(F)$ are not even bounded if we impose no limitation on figures F in the study:

Result 17 (Soifer 1998b). For any number r, there is a figure F such that $S(F) > r$.

Proof Indeed, for any r, we can construct a cross C thin enough so that only one of the two C-clones of area $\frac{1}{2}$ can be inscribed in a C-clone of area r (Fig. 14.3). ∎

Thus, it makes sense to limit the scope of our games to *convex* figures. The main problem then can be formulated as follows:

Main Open Problem 18 (Soifer 1998b). For any convex figure F, find $S(F)$.

This is a difficult problem that in full generality may withstand centuries. However, partial solutions are possible and welcome.

Our journey is not over: in fact, it has only begun. We got a sense of packing some particular shapes, and are now ready to commence a search for its essence, a result encompassing all convex figures. In 1998 I came up with a bold conjecture:

Fig. 14.3

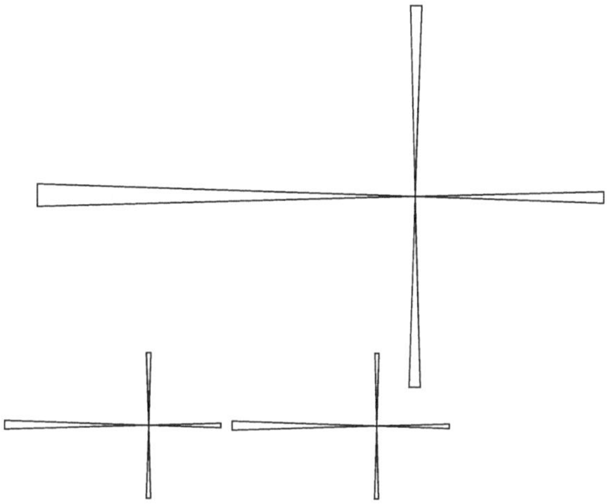

Clones in Convex Figures Conjecture 19 (Soifer 1998b). For *any* convex figure F, any set of F-clones $F_1, F_2, \ldots, F_n$ whose areas add up to at most 1, can be packed in a clone F_0 of area 2, i.e.,

$$S(F) \leq 2.$$

However, when I wrote up this conjecture for *Geombinatorics* (Soifer 1998b), I inadvertently put it as $S(F) = 2$. As the Russian proverb has it, *There is no bad without some good in it*! Three years later, in 2001, the Slovak geometer Pavel Novotný constructed a counterexample to the published equality $S(F) = 2$:

Novotný's Example 20 (Novotný 2002). For rectangle R_0 of size $\sqrt[8]{\frac{3}{2}} \times \sqrt[8]{\frac{2}{3}}$ we get $S(R_0) = \sqrt{\frac{8}{3}} < 2$.

Novotný understood my typo, as he wrote "Soifer's conjecture could be changed to $S(F) \leq 2$."

A year later, in 2002, Janusz Januszewski (2003), beautifully completed the above result of Novotný:

Januszewski's Theorem 21 (Januszewski 2003). For any rectangle R, $S(R) \leq 2$. Moreover, $S(R) = 2$ if and only if the rectangle R is a square.

Januszewski gave the main conjecture its final attribution:

The Soifer–Novotný Conjecture 22 (Januszewski 2003). For any convex figure F, $S(F) \leq 2$.

Finally, Januszewski posed a natural problem that is a particular case of problem 18:

Januszewski's Problem 23 (Januszewski 2003). Classify convex figures F for which $S(F) = 2$.

He among others noticed (Januszewski 2003) that, perhaps, the Soifer–Novotný Conjecture 22 can be generalized to n-dimensional Euclidean spaces:

***n*-Dimensional Conjecture 24** Let F be a convex body in an n-dimensional Euclidean space. Then any set of F-clones of the combined volume 1 can be packed in an F-clone of volume 2^{n-1}.

I would also like to know the minimum value of $S(F)$:

Open Problem 25 Find min$S(F)$ over all convex figures F in the n-dimensional Euclidean space E^n and classify figures F for which this minimum is attained. In particular, solve this problem for the plane E^2.

Most of these series of results appeared on pages of *Geombinatorics*, a quarterly dedicated to problem posing essays in combinatorial and discrete geometry (hence its title). Peter Winkler of Dartmouth University dedicated a section of his book (Winkler 2004, pp. 146 and 157) to the Discs in a Disc Conjecture:

> This lovely conjecture is due to Alexander Soifer of the University of Colorado, Colorado Springs. It and its relatives have been the subject of a dozen of articles in the journal *Geombinatorics*; it is known, for example, that squares of total area 1 can be packed into a square of total area 2. The generalization to higher dimension was suggested by your author, among others; the case of two balls, each of volume 1/2, shows that 2^{d-1} is best possible.

I hope you have enjoyed your ride on this train of mathematical thought. Out of the window of our train you noticed the terrain that has been continuously changing, with one problem giving birth to another. However, the original 1932 conjecture is still open 85 years later, in the year 2017. It is time to double the prize:

Conjecture I. Hundred Dollar Squares in a Square Conjecture (P. Erdős). At perfect square numbers the function $f(r)$ does not increase: $f(k^2+1) = f(k^2)$.

14.3 Conjecture II. The Happy End Problem (Erdős–Szekeres 1933)

During the winter of 1932–33, two young friends, mathematics student Paul Erdős, aged 19, and chemistry student George (György) Szekeres, 21, solved the problem posed by their young lady friend Esther Klein, 22, but did not send it to a journal for a year and a half (Erdős and Szekeres 1935).

Erdős–Szekeres's Theorem 26 (Erdős and Szekeres 1935). For any positive integer $n \geq 3$ there is an integer $ES(n)$ such that any set of at least $ES(n)$ points in the plane in general position[2] contains n points that form a convex polygon.

The authors knew only two values, obtained by the members of their group of Jewish–Hungarian friends:

Esther Klein: $ES(4) = 5$.

E. Makai and Paul Turán independently: $ES(5) = 9$.

It is fascinating how sure Erdős and Szekeres were of their conjecture. In one of his last, posthumously published problem papers (Erdős 1997), Paul Erdős attached the prize and modestly attributed the conjecture to Szekeres: "I would certainly pay $500 for a proof of Szekeres's conjecture."

Conjecture II: The Erdős–Szekeres Happy End $500 Conjecture $ES(n) = 2^{n-2} + 1$.

In 1933 Erdős and Szekeres proved lower and upper bounds for $ES(n)$; the conjectured above value is their lower bound. The upper bound has only recently been improved first by Ronald L. Graham and Fan Chung and then by others, but is still far from the conjectured value. Paul Erdős named it *The Happy End Problem.* He explained the name often in his talks. On June 4, 1992 in Kalamazoo I took notes of his talk:

> I call it The Happy End Problem. Esther captured George, and they lived happily ever after in Australia. The poor things are even older than me.

This paper also convinced George Szekeres to become a mathematician. For Paul Erdős the paper had a happy end too: it became one of his early mathematical gems, Paul's first of the numerous contributions to and

[2]I.e., no three points lie on a line.

leadership of the Ramsey Theory and, as Szekeres put it, of "a new world of combinatorial set theory and combinatorial geometry."

The personages of The Happy End Problem appear to me like heroes of Shakespeare's plays. Paul, very much like *Tempest's* Prospero, gave up all his property, including books, to be free. George and Esther were so close, that they ended their lives together, like Romeo and Juliette. In the late summer 2005 e-mail, Tony Guttmann conveyed to the world the sad news from Adelaide, Australia:

> George and Esther Szekeres both died on Sunday morning [August 28, 2005]. George, 94, had been quite ill for the last 2–3 days, barely conscious, and died first. Esther, 95, died an hour later. George was one of the heroes of Australian mathematics, and, in her own way, Esther was one of the heroines.

On May 28, 2000, during a dinner in the restaurant of the Rydges North Sydney Hotel in Australia, George Szekeres told me "my student and I proved Esther's Conjecture for 17 with the use of computer," i.e., $ES(6) = 17$. "Which computer did you use?" asked I. "I don't care how pencil is made," answered George.

14.4 Conjecture III. Chromatic Number Conjecture (Soifer 2008)

In 1986, the Third Colorado Mathematical Olympiad included the following problem from our typical category: its solution is easy to see, especially when somebody shows it to you.

Problem 27 Santa Claus and his elves paint the plane in two colors, red and green. Prove that the plane contains two points of the same color exactly one mile apart.

Solution. Toss on the plane an equilateral triangle with side lengths equal to one mile. Since its three vertices (pigeons) are painted in two colors (pigeonholes), there are two vertices painted in the same color (at least two pigeons in a hole). These two vertices *are* one mile apart. ∎

You can prove (do) the same result about a 3-colored plane:

No matter how the plane is 3-colored, it contains two points of the same color exactly one mile apart.

One may think that we know everything about the Euclidean plane. What else can there be after Pythagoras and Euclid? First of all, the Ancient Greeks did not think about these kinds of problems, where nothing is known about the coloring. Secondly, these simple problems are the starting points of a deep and still unresolved train of thought. For instance, try to push the problem to the next natural step, a 4-colored plane:

Is it true that no matter how the plane is 4-colored, it contains two points of the same color exactly one mile apart?

Imagine, nobody knows!

Chromatic Number of the Plane Problem 28 *What is the smallest number of colors with which we can color the plane in such a way that no color contains two points distance 1 apart?*

This number is called *the chromatic number of the plane* and is often denoted by χ. We can easily show (do) that 7 colors suffice, and thus $\chi \le 7$. And so we know that $\chi = 4$ or 5 or 6 or 7. Which is it?

In August 1987 I attended an inspiring talk by the member of the U.S. National Academy of Sciences Paul Halmos at Chapman College in California. It was entitled "Some problems you can solve, some you cannot." This problem was an example of a problem "you cannot solve." So far Halmos is correct.

While writing *The Mathematical Coloring Book* (Soifer 2009), in ca. 2007, I formulated the following conjecture:

Chromatic Number of the Plane Conjecture 29 (A. Soifer 2007). $\chi = 7$.

If you are familiar with 3- and more generally n-dimensional Euclidean space E^n, you will readily see that this problem straight-forwardly generalizes to n dimensions, and we can ask a more general question of the chromatic number $\chi(E^n)$ of E^n. In (Soifer 2009) I conjectured a simple formula for the chromatic number of the n-dimensional Euclidean space E^n:

Conjecture III: Chromatic Number of n-Space $\chi(E^n) = 2^{n+1} - 1$.

As Paul Erdős used to say, "If true, this conjecture may take centuries to prove, but we shall see!"

14.5 Conjecture IV. Triangular Covering (Conway–Soifer 2004)

In 2004 we held the 21st Colorado Mathematical Olympiad, for which I created the following problem (Soifer 2011):

Problem 30: To Have a Cake

(A) We need to protect from the rain a cake that is in the shape of an equilateral triangle of side 2.1. All we have are identical tiles in the shape of an equilateral triangle of side 1. Find the smallest number of tiles needed.

(B) Suppose the cake is in the shape of an equilateral triangle of side 3.1. Will 11 tiles be enough to protect it from the rain?

Solution (A). Mark 6 points in the equilateral triangle of side 2.1: its vertices and midpoints of the sides (Fig. 14.4). A tile can cover at most one such point, therefore we need at least 6 tiles.

On the other hand, 6 tiles can do the job. There are different ways to achieve it. Here is one. We can first cover the three corners (Fig. 14.5a), and then use 3 more tiles to cover the remaining hexagon (Fig. 14.5b).

(B). We can use 4 tiles to cover the top triangle of side 2, and then use the remaining 7 tiles for a bottom trapezoid (Fig. 14.6). ∎

Have you noticed that I did not ask the Olympians to prove that 11 covering tiles are necessary? At the Olympiad, I could only ask what I can prove myself!

Fig. 14.4

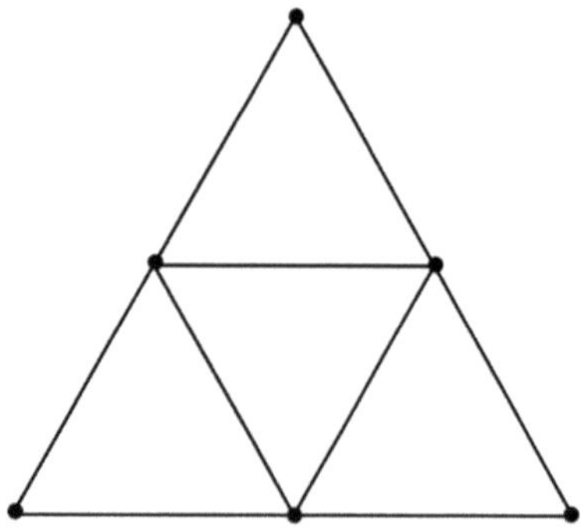

Fig. 14.5

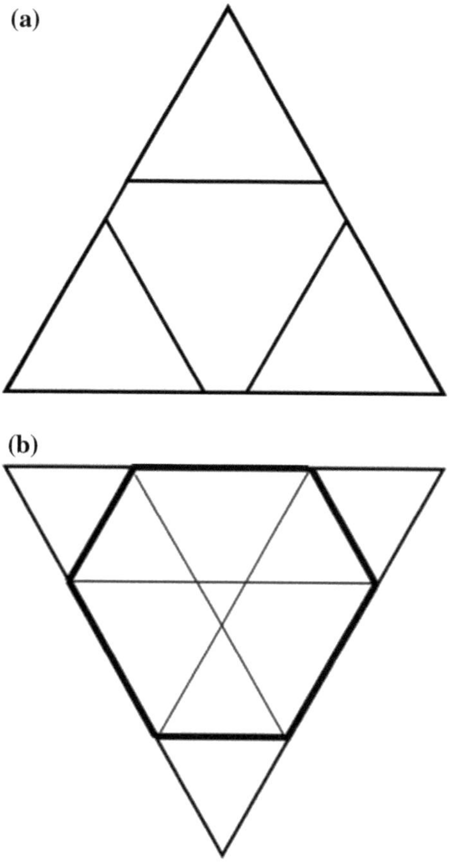

Fig. 14.6

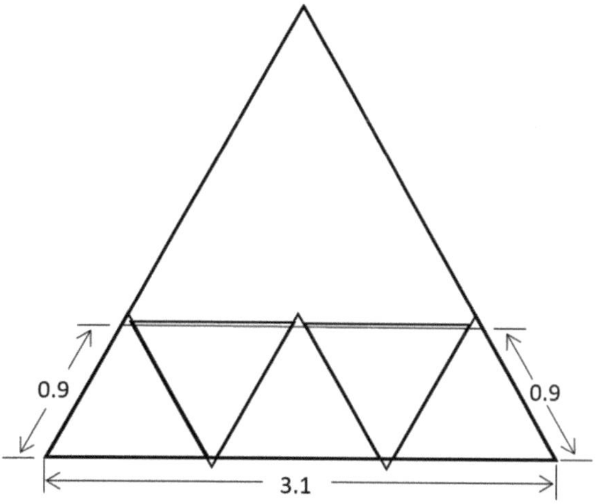

Upon my return to Princeton, where I worked at the time, I shared a more general form of this problem with John H. Conway. Imagine, we both found proofs of the sufficient condition, which were markedly different. And so John and I decided to set a world record: to publish an article containing just one word in its text. Let me reproduce here our submission to *The American Mathematical Monthly*.

Can $N^2 + 1$ Unit Equilateral Triangles Cover an Equilateral Triangle of Side > N, Say $n + \varepsilon$?

John H. Conway and Alexander Soifer
Mathematics, Princeton University, Fine Hall, Princeton NJ 08544, USA
e-mail: conway@math.princeton.edu; e-mail: asoifer@princeton.edu

$n^2 + 2$ can (Fig. 14.7):

Fig. 14.7

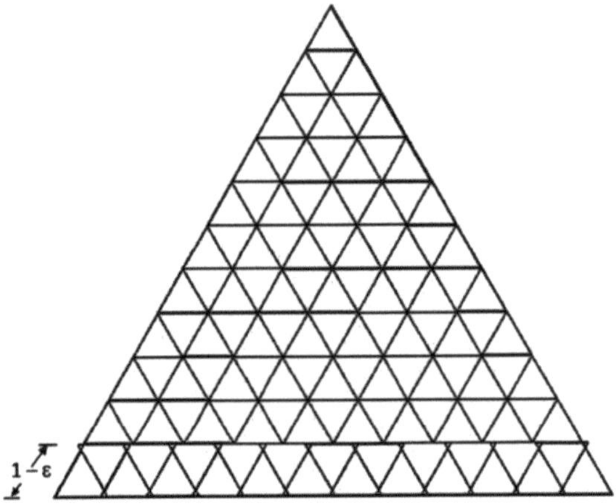

Fig. 14.8

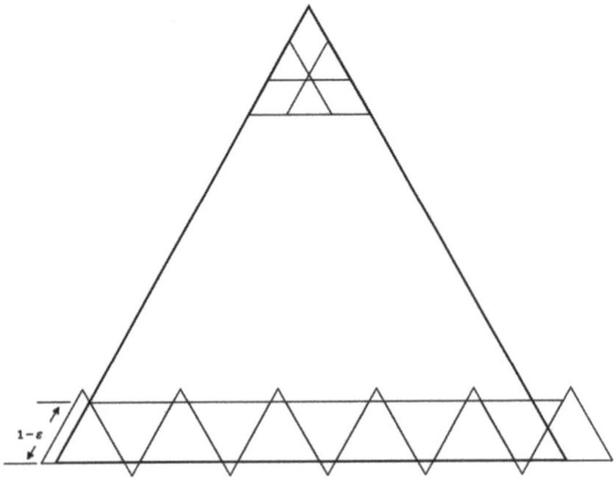

The American Mathematical Monthly was puzzled. On April 30, 2004, Editorial Assistant of the Monthly Margaret A. Combs sent me an e-mail:

> The Monthly publishes exposition of mathematics at many levels, and it contains articles both long and short. Your article, however, is a bit too short to be a good Monthly article ... A line or two of explanation would really help.

Having learned from me about *The Monthly* reply, John Conway exclaimed, "Do not give up too easily!" And so, I replied the same day as follows:

> I respectfully disagree that a short paper in general – and this paper in particular – merely due to its size must be "a bit too short to be a good Monthly article." Is there a connection between quantity and quality? ... We have posed a fine (in our opinion) open problem and reported two distinct 'behold-style' proofs of our advance on this problem. What else is there to explain?

The American Mathematical Monthly published our article (Conway and Soifer 2005), but spoiled our single-word world record by unilaterally including our title in the body of the article!

John Conway believed that since his and my coverings were so vastly distinct, the problem was too hard to continue fighting with it. However, shortly after, the Columbia University undergraduate student Dmytro Karabash joined me in working on this problem. We generalized the problem to covering an arbitrary triangle T. Tiling triangles will be similar to T and their corresponding sides will be $n+\varepsilon$ times smaller—let us call them $1/(n+\varepsilon)$-clones of T.

Result 31 (Karabash–Soifer 2005). Any non-equilateral triangle T can be covered by n^2+1 $1/(n+\varepsilon)$-clones of T.

Proof An appropriate affine transformation maps equilateral triangle on Fig. 14.8 onto T. This transformation gives a covering of T with n^2+2 tiling clones, but now we can cover the transformed top triangle (see Fig. 14.8) with 2 clones instead of 3 as shown in Fig. 14.9, thus reducing the total number of covering clones to n^2+1. ∎

Dmytro and I also generalized the problem by introducing *trigons* (Karabash and Soifer 2005). But we were unable to prove the Conway–Soifer Conjecture that the equilateral triangle requires n^2+2 covering triangles. Imagine, the equilateral triangle proved to be the hardest of all! You have a chance to prove it yourselves—sharpen your pencils!

Fig. 14.9

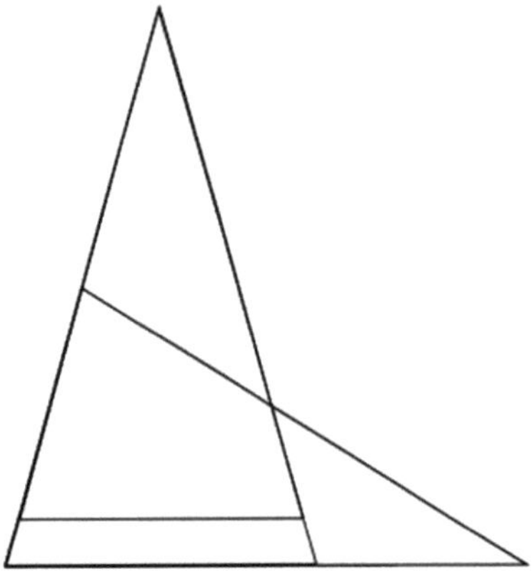

Triangular Covering Conjecture IV (Conway–Soifer 2004). An equilateral triangle of side n cannot be covered by $n^2 + 1$ equilateral triangles of side $1 - \varepsilon$.

The 'smallest' open case is the following conjecture:

The Hexagon Conjecture 32 (Karabash–Soifer 2005). Seven equilateral triangles of side $1 - \varepsilon$ cannot cover an equilateral hexagon of side 1.

You will find 10 more bridges from the problems of the Colorado Mathematical Olympiad to open problems of mathematics in my forthcoming book (Soifer 2017).

Acknowledgements I thank Col. Dr. Robert Ewell for converting my hand-drawn sketches into computer-aided illustrations.

References

Conway, J. H., & Soifer, A. (2005). Covering a triangle with triangles. *American Mathematical Monthly, 112*(1), 78.

Erdős, P. (1997). Some of my favorite problems and results. In R. L. Graham and J. Nesetril (Eds.), *The mathematics of Paul Erdős* (Vol. I, pp. 47–67). Berlin: Springer.

Erdős, P., & Soifer, A. (1995). Squares in a square. *Geombinatorics*, *IV*(4), 110–114.

Erdős, P., & Szekeres, G. (1935). A combinatorial problem in geometry. *Compositio Mathematica, 2,* 463–470.

Januszewski, J. (2003). Packing clones in triangles. *Geombinatorics*, *XII*(2), 73–78.

Januszewski, J. (2009). Optimal translative packing of homothetic triangles. *Studia Scientiarum Mathematicarum Hungarica, 46*(2), 185–203.
Karabash, D., & Soifer, A. (2005). On covering of trigons. *Geombinatorics, XV*(1), 13–17.
Moon, J. W., & Moser, L. (1967). Some packing and covering problems. *Colloquium Mathematicum, XVII*, 217–244.
Novotný, P. (2002). A note on packing clones. *Geombinatorics, XI*(1), 29–30.
Richardson, T. J. (1995). Optimal packing of similar triangles. *Journal of Combinatorial Theory, Series A, 69*, 288–300.
Soifer, A. (1996). Squares in a square II. *Geombinatorics, V*(3), 121.
Soifer, A. (1998a). Discs in a disc. *Geombinatorics, VII*(4), 139–140.
Soifer, A. (1998b). Packing clones in convex figures. *Geombinatorics, VIII* (1), 166–168.
Soifer, A. (1999). Packing triangles in triangles. *Geombinatorics, VIII*(4), 110–115.
Soifer, A. (2009). *The mathematical coloring book: Mathematics of coloring and the colorful life of its creators*. New York: Springer.
Soifer, A. (2011). *The Colorado mathematical Olympiad and further explorations: From the mountains of Colorado to the peaks of mathematics*. New York: Springer.
Soifer, A. (2017). *The Colorado mathematical Olympiad, the third decade and further explorations: From the mountains of Colorado to the peaks of mathematics*. New York; submitted, to appear in: Springer.
Stong, R. (1997). Squares inside of a square. *Geombinatorics, VII*(1), 29–34.
Winkler, P. (2004). *Mathematical Puzzles: A Connoisseur's collection*. Natick: A.K. Peters.

About the Editor and Contributors

Kiril Bankov, **Bulgaria**, a Ph.D. in mathematics, prepares future mathematics teachers as a Professor of Mathematics Education at the University of Sofia and the Bulgarian Academy of Sciences. He has been involved in mathematics competitions in Bulgaria for more than 20 years as an author of contest problems and as a member of juries. He has written many articles, made presentations, and is a co-author of books on mathematics competitions, problem solving, work with mathematically gifted students, etc. Two of his papers in "Mathematical Gazette" won "Article of the Year" citation (1995 and 1999). Bankov has an extensive experience in international large-scale studies in mathematics education. For more than a decade he has been a member of the International Expert Committee (SMIRC) for TIMSS study. He also worked as a mathematics coordinator for the International Study Center of the Teacher Education and Development Study—Mathematics (TEDS-M) at Michigan State University (USA). In 2014 he was the Chairperson of the European Baccalaureate Examining Board. Bankov was the Secretary of World Federation of National Mathematics Competitions (WFNMC) 2008–2012, and serves presently as the Senior Vice President.

Francisco Bellot–Rosado, **Spain**, served as Chair of the Mathematics Department of High Schools "Marqués de la Ensenada", Logroño (1966–70) and "Emilio Ferrari", Valladolid (1970–2006), and Associate Professor on the Faculty of Sciences of the University of Valladolid (several times during 1971–1999). He was Team Leader or Deputy Leader of the Spanish Team in the International Mathematical Olympiad (IMO) and Iberoamerican Math Olympiad (OIM), 1988–1997, Europe Representative in WFNMC 1996–present; and Member of the Joint Committee of Governors of the RSME in various periods 1997–present. He served as **the** Editor of the Journal *SIPROMA* (1996–98), published by the Organization of Iberoamerican States (OEI). He founded the Mediterranean Mathematics Competition *(Peter O'Halloran Memorial)* in 1998; and starting in 1984 organized in Spain the Mathematical Kangaroo Contest. In 2000, Bellot-Rosado was presented the

A. Soifer (ed.), *Competitions for Young Mathematicians*,
ICME-13 Monographs, DOI 10.1007/978-3-319-56585-9

Paul Erdős Award by the WFNMC. Starting in 2002, he served as the Editor of the digital journal *Revista Escolar de la OIM*. H served as the Leader of the Spanish team at the 5th European Girls Mathematics Olympiad (Busteni, Romania, 2016).

Luis F. Cáceres-Duque, **Puerto Rico**, a Ph.D. in Logic from The University of Iowa, is a Professor at the University of Puerto Rico, Mayaguez Campus. He is well recognized in the educational system of Puerto Rico, and has worked for the last fifteen years in several educational projects including professional development of mathematics and science teachers. Cáceres is well-known in the field of international mathematical competitions. He is the Secretary of the Association Kangaurou Sans Frontieres. He is also the Founder and Director of The Mathematical Olympiad Program in Puerto Rico. In 2016, Cáceres was presented the Paul Erdős Award by the World Federation for National Mathematical Competitions. He has several papers and books in Logic, Ring theory, math Olympiads and secondary school mathematics. In 2014, he published a joint paper "So Much Happening, So Much to Do," *Notices of the AMS*. He recently published the book "Math Olympiads in Puerto Rico: 2015–2016." Cáceres collaborates with the organization of math Olympiads in different countries and he is invited frequently to different countries to offer courses and conferences. His interests include geometry, math Olympiads problems, math talented students and professional development of math teachers.

Sergey Dorichenko, **Russia**, a 1995 M.A. in Mathematics from Moscow State University (MGU), has taught in the Mathematical Circles program of MGU. As a result, he published the book "A Moscow Math Circle: Week-by-Week Problem Sets," Math Circles Library, MSRI, AMS, 2012.

Multiple times he was Deputy Chair of the Organizing Committee and Head Judge of the Moscow Mathematical Olympiad; for several years he worked as a judge and a member of the Organizing Committee of the Lomonosov Tournament. Dorichenko served as a Teacher of Mathematics in Moscow schools 57 (1992–2011) and 179 (2002–present). Deputy Director of the mathematical section of Moscow school №179 (from 2007). Works at the Center of Pedagogical Mastery in Moscow. He also served as the Chair of the Central Jury and Problems Committee of the International Mathematics Tournament of Towns; and the Chair of the Jury of the Summer Conference of the Tournament of Towns. He is the Head of the Department of Mathematics a member of the Editorial Board of the magazine "Kvant" (2008–2016). In 2012 he founded the magazine «Kvantik» and has since served as its Editor-in-Chief. Dorichenko is presently the Secretary of the World Federation of National Mathematical

Competitions. He authored several books and multiple articles, published in magazines «Kvant», «Kvantik», «Mathematical Enlightenment», and «Mathematical Education».

Mary Falk de Losada (María), Colombia, is a Professor at the Universidad Antonio Nariño in Bogotá. She did her graduate work in mathematics and math education at Harvard University and the University of Illinois at Chicago. She is cofounder with Ricardo Losada and Rafael Mariño of the Colombian Mathematics Olympiads, the Iberoamericans Mathematics Olympiad, and with many others of the World Federation of Mathematics Competitions (WFNMC). She has received the David Hilbert Award from WFNMC and the José Celestino Mutis Prize from the Colombian Mathematics Society for her work in mathematics Olympiads in Colombia and Latin America. She was elected member at large of the ICMI Executive Committee and has worked in various capacities on the organization of ICMEs since 1996, as well as currently being a member of the International Mathematics Union Circle. In 2003, she served as the President of the International Jury of International Mathematical Olympiad in Colombia. María served as the President of WFNMC, and at this time is the Immediate Past President, serving on the Executive Committee and chairing the WFNMC awards committee that determines the recipients of the Paul Erdős Award. She has published several books in Colombia for future math teachers as well as many of the yearly collections of problems from the Colombian Math Olympiads. Retired as professor of the National University of Colombia, she continues to divide her time between the Colombian Math Olympiads and a doctoral program in mathematics education at the Universidad Antonio Nariño in Bogotá. Married for over fifty years to Ricardo Losada, they have five children—all in fields related to mathematics —and nine grandchildren.

Robert Geretschläger, Austria Robert (full name: OStR Prof. Mag.rer.nat. Dr.phil. Robert Rudolf Alfred Geretschläger; quite a mouthful) has been active as a mathematics teacher, author and competitions coordinator for over 35 years. Along with his full-time teaching duties at Bundesrealgymnasium Keplerstrasse in Graz, Austria, and his work at the Karl-Franzens University in Graz, he has also worked on curriculum development in Austria and co-authored a series of high school mathematics textbooks. He is the author of "Geometric Origami" and co-author of such books as "The Circle" along with numerous compilations of competition problems. His main interests lie in the preparation and organization of mathematics competitions. He has been responsible for the introduction and organization of several competitions in Austria, such as the Mathematical Kangaroo, the

International Tournament of the Towns, and the Mediterranean Mathematics Competition, and has been actively involved in the Austrian Mathematical Olympiad for over 30 years. Since the year 2000. He has represented Austria at the International Mathematical Olympiad, first as Deputy Leader of the Austrian team, and starting in 2006 as a member of the problem selection committee, and the Austrian Team Leader. He is also a member of the international boards of the Kangourou sans Frontières (KSF) and presently a Vice President of the World Federation of National Mathematics Competitions (WFNMC).

Matúš Harminc, Slovakia, a 1986 Ph.D. in Group Theory and Algebra, is an Associate Professor at the Institute of Mathematics, Faculty of Science, Pavol Jozef Šafárik University, Košice. After graduation, hc worked at the Mathematical Institute of the Slovak Academy of Sciences. From 1975 to 1986 he taught on several Slovak faculties. He gives courses on Number Theory, Mathematical Methods of Problem Solving, and the Introduction to Mathematics for pre-service teachers. His main research areas are teaching methods and the solving of math word problems.

Martina Jesenská, Slovakia, works at the Giles Academy, Boston, Lincolnshire, England. In 2015, she graduated from Pavol Jozef Šafárik University in Košice with the master's degree in Teaching Mathematics. She is presently on her second year of teaching Mathematics.

Romualdas Kašuba, Lithuania, is Ph.D. from the University of Greifswald, Germany, teaches Mathematics, Communications Skills, and Ethics at Vilnius University. He has a long history with various Mathematical Olympiads (MO): being a jury member of the Lithuanian MO since 1979, the Deputy Leader of the Lithuanian IMO team since 1996, and the leader of the Lithuanian team at the Baltic Way team-contest since 1995. In 1999, he initiated the Lithuanian Mathematical Olympiad for youngsters and has been involved in Lithuanian Kangaroo movement from the same year, as well as being the leader of the Lithuanian MEMO team since 2009. Besides being the author of several books in Lithuanian, he has composed several booklets in English ("What to do when you do not know what to do," Parts I and II, as well as "Once upon a time I saw the puzzle," Parts I, II, and III). Furthermore, he rewrote and expanded a book with a similar title in Russian, which was published in Moscow in 2012 (2nd ed. 2014). During 2008–2016 he represented Lithuania at ICMI, 2010–2014 was the Board member of MCG (International Group for Mathematical Creativity and Giftedness), and is the member of the Editorial Board of MCG Newsletter.

Nikolai Konstantinov, **Russia**, a 1963 Ph.D. in mathematics and physics, worked at the Physics Department at Moscow State University, the Institute of Theoretical and Experimental Physics, the Institute of Patent Information, Institute of Economics of the USSR Science Academy, Moscow Institute of Open Education, and at several Moscow schools: 7, 57, 91, and 179. He is currently a scientific supervisor at the Moscow school 179. In 1968 he coauthored a computer program that animated and modeled the movement of a cat (the program operated by solving differential equations). In 1979 founded the multidisciplinary Lomonosov Tournament for secondary students and has been its chair ever since. During 1967–1980, he was a member of the jury of the Soviet Union Mathematical Olympiad. In 1980 he founded the International Mathematics Tournament of Towns and has since been its president. In 1989 he organized the International Summer Conference of the Tournament of Towns. He founded the Independent Moscow University IMU) and is currently a trustee of IMU and of the Moscow Center of Continuous Mathematical Education. Konstantinov has been a member of the advisory board of "Kvant" magazine, and of the editorial boards of Russian magazines "Mathematical education" and "Mathematical Enlightenment", Enrichment series (Australian International Centre for Mathematics Enrichment). Konstantinov's publications include two articles on set-theoretic geometry, an article on computer science, over ten articles on mathematical economics, and a series of articles on mathematical education (including specific courses of calculus taught at mathematical schools). Konstantinov is a 1992 recipient of the Paul Erdős Award from the World Federation of National Mathematics Competitions (WFNMC) "for his significant contribution in developing the Tournament of the Towns' Contest in Russia"; a 2002 Award of the Moscow City Council "for services to the city." In WFNMC he is a representative of Europe.

José Heber Nieto Said, **Venezuela**, is a Professor at Universidad del Zulia. Born in Uruguay, he earned his M.Sc. degree from Universidad de Buenos Aires in Argentine, and Dr.H.C. from Universidad del Zulia, Venezuela. He has written several papers and books on mathematics, especially on combinatorics, problem solving and mathematical Olympiads. His research interests include algebraic and analytic combinatorics. He has been involved with mathematical Olympiads for the past twenty years: he has been a coach, Leader or Deputy Leader of the Venezuelan Team and participates actively in the organization of Olympiads in Venezuela and the Caribbean region. He is particularly fond of problem solving and regularly submits problems and solutions to math competitions and to the problems sections of several journals and on-line forums.

Rafael Sánchez Lamoneda, Colombia, a 1986 Ph.D. in Mathematics from Brandeis University, USA, is an Assistant Professor at the Universidad Antonio Nariño, Bogota, Colombia. He was born in Venezuela. He has written several papers and books on Algebra and Mathematical Olympiads problems. He was Chair of the Venezuelan Institute for Scientific Research's Mathematical Department. In 2010 he won the Paul Erdős Award in 2010, given by the World Federation for National Mathematical Competitions, and in 1993 Best Mathematical Paper Prize from the Venezuelan National Council for Scientific Research. He is member of the Venezuelan Mathematical Olympiads Commission since 1978, Coach, Leader and Deputy Leader of Venezuelan teams to international mathematical competitions, including the International Mathematical Olympiad (IMO). From 2012 to 2016 he was an elected member for the International Mathematical Olympiad Advisory Board, IMOAB, and since 2012 President of the IMO Ethics Committee. He is advisor of the Venezuelan Academy of Physical, Mathematical and Natural Sciences. He was President of the Venezuelan Mathematical Association and actual president and founder of the Venezuelan Association for Mathematical Competitions. His research interests include Characteristic-Free Representation Theory of Gl (n), Schur and Weyl Modules, Homological Algebra and Mathematical Olympiads problems.

Ingrid Semanišinová, Slovakia, is a 2007 Ph.D. in Theory of Mathematics Education, is an Assistant Professor on the Faculty of Science, Pavol Jozef Šafárik University, Košice, and teaches Didactics of Mathematics, History of Mathematics, Problem Solving in Mathematics Education and Mathematics for Geographers. Her research topics are aimed at pre-service teacher preparation, activities for students and developing creativity in the classroom. She has six years' experience with teaching sixth to ninth grade mathematics at a local middle school. She is a member of the regional committee of Mathematical Olympiad.

Vinayak M. Sholapurkar, India, is the Head, Center for Postgraduate Studies in Mathematics, S. P. College, Pune, India. He completed his Ph. D. in Operator Theory from Savitribai Phule Pune University. He has an extensive experience of teaching graduate level mathematics courses and has been involved in training Math Olympiad students at the regional and National levels, as well as in the International Mathematical Olympiad training camps. He was the National Coordinator for Mathematical Olympiads in India from 2013 to 2016. Dr Sholapurkar served as an Observer with Indian Team in IMO, Cape Town, 2014; and led the Indian Team in EGMO, Busteni, Romania, 2016. He has been working as the National

Coordinator for Madhava Mathematics Competition for Undergraduate students in India since its inception in 2009. He delivered talks at several National and International conferences and workshops and is a co-author of a popular book "An Excursion in Mathematics" meant for the preparation for Math Olympiads. His publications reflect his keen interest in the area of Operator Theory, his field of specialization as also, Functional Analysis. Some of these papers are: Rigidity theorems for spherical hyperexpansions (with Sameer Chavan), Complex Analysis and Operator Theory, Vol. 7(5), (2013), 1545-1568; Completely monotone functions of finite order and Agler's conditions (with Sameer Chavan), Studia Mathematica, Vol. 226(3), (2015), 229-258; Completely hyperexpansive tuples of finite order (with Sameer Chavan), J. Math. Ana. and Appl. Vol. 447(2017), 1009-1026.

Kar-Ping Shum, **China**, earned his Ph.D. in Mathematics from the University of Alberta in 1971 and was awarded Honorary Doctor of Mathematics by Gomel University, Belarus in 2002. He is the Founding Chairman of Hong Kong IMO committee (1986–present) and has been the leader and inspiration for the organization of the IMO in Hong Kong in 1994 and again in 2016. He has been for years affiliated with the Chinese University of Hong Kong, where he served as Professor of Mathematics (2000–2002), Research Chair Professor (2003–2007), Honorary Professor (2008–2011). He is Honorary Director of the Institute of Mathematics and Professor of Mathematics, Yunnan University at Kunming (2011–present). He was also Honorary Professor, National Research Council, The Chinese Academy of Sciences in Beijing, 1990–1997; President of Hong Kong Mathematics Society, 1985- 1989; President of the Southeast Asian Mathematical society (1990–1991 and 2001–2002); Chair of the Federation of the Hong Kong Institutions staffs associations (1996–present). He is Editor-in-Chief of Asian European Journal of Mathematics (2007–present); Theory of Semigroup and Its Applications (2008–present); Southeast Asian Bulletin of Mathematics (1989–present). Editor of a number of Journals in Taiwan, Mainland China, Vietnam, Thailand, Malaysia, Iran, Korea, Malaysia, USA, Hungary, Italy, India and Pakistan, etc.

Alexander Soifer, **USA** (the editor and an author), a 1973 Ph.D. from Moscow, is a Professor of Mathematics, Film Studies, and Art History at the University of Colorado (1979–present). In 1978 he left Russia as a refugee in search of freedom, which he found on the shores of the New World. In 1988, Soifer was charmed by Paul Erdős into switching from Abelian group theory to Euclidean Ramsey theory and 'Erdősian' discrete geometry. Soifer's works (some 400 papers and 12 books) include *The Mathematical Coloring Book: Mathematics of Coloring and the Colorful Life of Its Creators*; *The*

Scholar and the State: In Search of Van der Waerden; *Mathematics as Problem Solving*; *The Colorado Mathematical Olympiad: The First Two Decades and Further Explorations*; *The Colorado Mathematical Olympiad: The Third Decade and Further Explorations*; *Geometric Etudes in Combinatorial Mathematics*; *How Does One Cut a Triangle?*—all published by Springer. He co-wrote and edited the monograph *Ramsey Theory Yesterday, Today, and Tomorrow*, Birkhäuser, 2011. Soifer contributed chapters to *Chromatic Graph Theory*, Cambridge U Press, 2015; and *Open Problems in Mathematics*, edited by John F. Nash. Jr.; Springer, Switzerland, 2016. He is a recipient of the Paul Erdős Award (2006) and President of the World Federation of National Mathematics Competitions (2012–present), and has been on its Executive Committee since 1996, serving previously as Senior Vice President and Secretary. Soifer is the publisher and editor of the research quarterly *Geombinatorics*, 1991–present. He created *The Colorado Mathematical Olympiad* and has run it for over 33 years, 1984–present. Soifer was first to serve on the judges of both the Soviet Union (1970–1973) and the United States (1996–2005) Mathematical Olympiads.

Peter Taylor, **Australia**, a 1972 Ph.D. in Applied Mathematics from the University of Adelaide, was an academic at the University of Canberra from 1972 to 2012, including Executive Director of the Australian Mathematics Trust from 1994 to 2012. He is now an Emeritus Professor at the University of Canberra. His many publications include *Challenging Mathematics in and beyond the Classroom*, Springer 2009, being the Final Report of ICME Study 16 he co-chaired, book co-edited with Ed Barbeau. In 2015 he was made an Officer (AO) of the Order of Australia, in 1994 presented the Paul Erdős Award from the World Federation of National Mathematics Competitions (WFNMC), and in 1994 BH Neumann Award (Australian Mathematics Trust). He has been on the Executive of WFNMC since 1996 (President 2000 to 2004) and has been on the Executive of the International Group for Mathematical Creativity and Giftedness since 2012.

Iliana Tsvetkova, **Bulgaria**, is a teacher of Mathematics at Sofia High School of Mathematics. In this position she has gained a vast experience in working with mathematically talented students. An important part of Iliana's work is the preparation of students for National and International mathematics competitions. Her students won many prizes including gold, silver and bronze medals at 2004 and 2012 IMO. Iliana herself has been a Team Leader of Bulgariathein many international mathematics competitions: Po Leung Kuk Primary Mathematics World Contest (PMWC), Elementary Mathematics International Competition (EMIC), International World Youth Mathematics Competition (IWYMIC), World Youth *Mathematics* Intercity

Competition (WYMIC), World Mathematics Team Championship WMTC, Tuymaada Olympiad, Jautikov Olympiad. She has shared her experience working with mathematically outstanding students in a number of articles some of them written for “Mathematics Competitions”, the journal of the WFNMC, and the National Council of Teachers of Mathematics (NCTM). Iliana Tsvetkova was awarded as “Teacher of the Year” in 2005 and 2013; this is the most prestigious prize for teachers of mathematics in Bulgaria.

Author Index

A. Soifer (ed.), *Competitions for Young Mathematicians*,
ICME-13 Monographs, DOI 10.1007/978-3-319-56585-9

Subject Index

A. Soifer (ed.), *Competitions for Young Mathematicians*,
ICME-13 Monographs, DOI 10.1007/978-3-319-56585-9

MIX
Papier aus verantwortungsvollen Quellen
Paper from responsible sources
FSC® C105338

If you have any concerns about our products,
you can contact us on
ProductSafety@springernature.com

In case Publisher is established outside the EU,
the EU authorized representative is:
Springer Nature Customer Service Center GmbH
Europaplatz 3, 69115 Heidelberg, Germany

Printed by Libri Plureos GmbH
in Hamburg, Germany